# Large-Scale Forest Restoration

Landscapes are being degraded and simplified across the globe. This book explores how forest restoration might be carried out to increase landscape heterogeneity, improve ecological functioning and restore ecosystem services in such landscapes. It focuses on large, landscape-scale reforestation because that is the scale at which restoration is needed if many of the problems that have now developed are to be addressed. It also shows how large-scale forest restoration might improve human livelihoods as well as improve conservation outcomes.

A number of governments have undertaken national reforestation programs in recent years; some have been more successful than others. The author reviews these to explore what type of reforestation should be used, where this should be carried out and how much should be done. For example, are the traditional industrial forms of reforestation necessarily the best to use in all situations? How can forest restoration be reconciled with the need for food security? And, are there spatial thresholds that must be exceeded to generate economic and environmental benefits?

The book also examines the policy and institutional settings needed to encourage large-scale reforestation. This includes a discussion of the place for incentives to encourage landholders to undertake particular types of reforestation and to reforest particular locations. It also considers forms of governance that are likely to lead to an equitable sharing of the costs and benefits of forest restoration.

**David Lamb** taught forest ecology at the University of Queensland, Australia, for 30 years and has since undertaken consultancy work on forest restoration for a range of international organisations and aid agencies, including the World Bank, the Food and Agriculture Organization of the United Nations (FAO), the Australian Agency for International Development (AusAID) and the German Society for International Cooperation (GIZ). He is a member of the Commission on Ecosystem Management of the International Union for the Conservation of Nature (IUCN).

## The Earthscan Forest Library

This series brings together a wide collection of volumes addressing diverse aspects of forests and forestry and drawing on a range of disciplinary perspectives. Titles cover the full range of forest science: biology, ecology, biodiversity, restoration, management (including silviculture and timber production), geography and environment (including climate change), socio-economics, anthropology, policy, law and governance. The series aims to demonstrate the important role of forests in nature, in peoples' livelihoods and in contributing to broader sustainable development goals. It is aimed at undergraduate and postgraduate students, researchers, professionals, policy-makers and concerned members of civil society.



**Rainforest Tourism, Conservation and Management**
Challenges for Sustainable Development
*Edited by Bruce Prideaux*

**Large-Scale Forest Restoration**
*David Lamb*

**Forests and Globalization**
Challenges and Opportunities for Sustainable Development
*Edited by William Nikolakis and John Innes*

**Smallholders, Forest Management and Rural Development in the Amazon**
*Benno Pokorny*

**Managing Forests as Complex Adaptive Systems**
Building Resilience to the Challenge of Global Change
*Edited by Christian Messier, Klaus J. Puettmann and K. David Coates*

**Evidence-based Conservation**
Lessons from the Lower Mekong
*Edited by Terry C.H. Sunderland, Jeffrey Sayer, Minh-Ha Hoang*

**Global Environmental Forest Policies**
An International Comparison
*Constance McDermott, Benjamin Cashore and Peter Kanowski*

Additional information on these and further titles can be found at
http://www.routledge.com/books/series/ECTEFL

**"Covering plantation forestry, natural forest regrowth, and ecological restoration of degraded forests, Dr. Lamb explains how to achieve effective, large-scale reforestation or restoration in tropical and extra-tropical regions alike. This book comes at just the right time, and is based on a lifetime of experience and reflection. Don't miss it."** – *James Aronson, Restoration Ecologist, CEFE / CNRS, France, and Missouri Botanical Garden.*

**"Large-scale forest restoration is needed globally to improve human livelihoods and ecological functioning. David Lamb uses his wealth of experience to explore the socioeconomic, legal, and historic context of the challenge. Authoritative case studies illustrate the complexity of restoration in practice."** – *John A. Stanturf, Senior Research Ecologist, US Forest Service, USA.*

**"David Lamb has thought deeply about large-scale forest restoration, and provides us with carefully researched insights on where and how to go about it. His analysis is informed by lessons learned from around the world over the past 100 years. This book is immensely informative and timely – a valuable resource as we mobilize our collective efforts to restore millions of hectares of deforested and degraded land to benefit people and the planet."**
– *Robert Winterbottom, Senior Fellow, Restoration Initiative, World Resources Institute, USA.*

# Large-Scale Forest Restoration

**David Lamb**

Routledge
Taylor & Francis Group
LONDON AND NEW YORK

earthscan
from Routledge

First published 2014
by Routledge
2 Park Square, Milton Park, Abingdon, Oxon OX14 4RN

And by Routledge
711 Third Avenue, New York, NY 10017

*Routledge is an imprint of the Taylor & Francis Group, an informa business*



*British Library Cataloguing-in-Publication Data*
A catalogue record for this book is available from the British Library

*Library of Congress Cataloging-in-Publication Data*
Lamb, David, 1941-
Large-scale forest restoration / David Lamb.
pages cm. – (The Earthscan forest library)
Includes bibliographical references and index.
(pbk : alk. paper) 1. Forest restoration. 2. Landscape ecology. I. Title.
SD409.L325 2014
634.9'56–dc23
2014014365

ISBN: 978-0-415-66318-2 (hbk)
ISBN: 978-0-203-07164-9 (ebk)

Typeset in Baskerville
by Saxon Graphics Ltd, Derby

Printed and bound in the United States of America by Publishers Graphics, LLC on sustainably sourced paper.

# Contents


| | | |
|---|---|---|
| | *List of figures* | viii |
| | *List of tables* | x |
| | *List of boxes* | xii |
| | *Preface* | xiv |
| | *Acknowledgements* | xvii |
| 1 | The need for large-scale forest restoration | 1 |
| 2 | Lessons from the last hundred years | 33 |
| 3 | Natural forest regrowth | 68 |
| 4 | Types of planted forests | 101 |
| 5 | Where in the landscape should forest restoration take place? | 133 |
| 6 | How much forest restoration is needed? | 161 |
| 7 | Creating multi-functional landscapes: choices and trade-offs | 185 |
| 8 | Making it happen: policies and institutions | 207 |
| 9 | Final discussion | 238 |
| | *References* | 253 |
| | *Index* | 291 |

# Figures


| | | |
|---|---|---|
| 1.1 | There are many kinds of planted forests: (A) a *Cunninghamia lanceolata* plantation in China; (B) a 60-year-old *Flindersia australis* plantation in Australia | 4 |
| 1.2 | Change in forest area 1990–2010 in relation to GDP per person in various regions of the world | 25 |
| 1.3 | Expected changes in rural and urban populations between 1950 and 2050 in various regions of the world | 28 |
| 1.4 | Forest regrowth on former agricultural fields in southern Sweden abandoned perhaps 100 years ago | 29 |
| 2.1 | Wood produced by planted forests in 2005 and in 2030 for the three scenarios projected by Carle and Holmgren | 62 |
| 2.2 | Forest transitions can follow different pathways | 65 |
| 3.1 | Influences on the composition of regrowth forest patches | 71 |
| 3.2 | Hypothetical regrowth pathways | 72 |
| 3.3 | Natural forest regrowth on former farmland in the middle hills of Nepal | 79 |
| 3.4 | Trees regenerating from old stumps and roots in a farmer's field in Sub-Saharan Africa | 81 |
| 3.5 | Naturally regenerated tree seedlings on a hillside in Hong Kong | 92 |
| 3.6 | Small forest patches can be exposed to the risk of frequent wildfires, especially in areas with seasonal climates | 95 |
| 3.7 | Fuelwood being collected from regrowth forests | 97 |
| 4.1 | A conceptual outline of alternative ways of restoring forest cover at a deforested site | 104 |
| 4.2 | Enrichment of an older Swedish conifer plantation with seedlings of broad-leaved species | 109 |
| 4.3 | Understorey of native species developing beneath the canopy of a *Pinus massoniana* plantation now incorporated within Tam Dao National Park, Vietnam | 110 |
| 4.4 | Different types of multi-species plantings established for production purposes | 114 |

4.5 Successional development diverted after a wildfire stimulated germination of *Acacia* seed stored in the topsoil 124
4.6 A row seeder used to sow a mixture of seed of tree and understorey species on former farmland in Western Australia 128
4.7 A large area of newly restored *Eucalyptus* woodland containing a variety of species established on former farmland in the southern region of Western Australia 128
5.1 The net benefit of planting strips of trees amongst crops or pastures in dryland salinity areas 140
5.2 Some steep slopes such as these in the Loess Plateau of China may be difficult to deliberately reforest and may have to be left to regenerate naturally 144
5.3 Relationship between topography and priority areas for reforestation in the Loess Plateau region of China 145
5.4 Better and worse reforestation options for improving biodiversity conservation 150
5.5 Species are lost when forests are fragmented 151
6.1 The hypothetical effect of increasing the area reforested on (A) water yield, (B) sedimentation, (C) biodiversity and (D) carbon stores 165
6.2 The overall number of species present increases as an agricultural landscape is reforested because more of the original species requiring a forest habitat are able to recolonise the area 176
6.3 More carbon is initially sequestered in biomass using plantations grown on successive short rotations than in a single long rotation occupying the same time period 180
8.1 A privately owned tree nursery in Vietnam selling seedlings to farmers 215

# Tables

1.1 Global estimates of the amount of land potentially available for reforestation 11
1.2 Recent activities of international bodies favouring increased reforestation 14
1.3 Examples of some recent large-scale national reforestation programmes 15
1.4 The context in which future reforestation will take place 18
1.5 Percentage of respondents in different countries surveyed by the Pew Global Attitudes Project in 2007 nominating environmental problems as a top environmental threat 25
2.1 Area of planted forest and expansion between 1990 and 2010 35
2.2 Changes in area of plantations (protective and productive) and natural forest between 1990 and 2010 35
2.3 Area of planted production forests in selected countries 39
2.4 Changes over time in the areas (000 ha) of planted production forests owned by different landholders 59
2.5 Comparison of projected future global production forest areas made by FAO and Indufor 61
2.6 Plantation areas (000 ha) allocated to production and protection functions 64
3.1 Examples of situations where natural forest regrowth has occurred over large landscape areas 74
4.1 The advantages and disadvantages of using exotic or native species in planting programmes 102
4.2 Attributes of the three main types of plantings used to reforest former agricultural or 'degraded' lands 106
4.3 Common causes of plantation failures 107
4.4 Potential advantages of multi-species plantations 112
4.5 Different types of multi-species plantings for production purposes 113
4.6 Some comparative advantages and disadvantages of direct sowing and planted seedlings 126

5.1 Elements of the landscape matrix of importance for forest restoration planning 136
5.2 Prioritising the choice of different types of natural forest remnants for protection using reforested buffer strips 152
5.3 Preferred locations for reforestation within the agricultural matrix depending on the primary objective 158
6.1 Consequences of deforestation and reforestation on hydrological flows 164
6.2 The consequence of large-scale forest recovery in Western Europe for species of vertebrate fauna 176
7.1 Some questions for those undertaking large-scale forest restoration 186
7.2 Effect of future climate changes on intact and modified ecosystems 199
8.1 Common misconceptions that can hinder reforestation policy development 208
8.2 Policy areas needed to enable reforestation by private landholders 211
8.3 Financial and non-financial incentives to encourage reforestation where it might not otherwise occur 216
8.4 Potential difficulties in making payment schemes to owners of new plantations supplying ecosystem services act as an incentive for reforestation 220
8.5 Desirable attributes for institutions involved in large-scale reforestation 227
9.1 Building resilience in large-scale forest restoration programmes 247
9.2 Forms of forest cover likely to be present in 300 years' time 251

# Boxes


1.1 Defining forests, reforestation and restoration 3
1.2 Definitions of land degradation 8
1.3 Ecosystem services 13
1.4 The land grabbers 20
2.1 The second rotation problem in South Australian pine plantations 56
3.1 Why natural regeneration can sometimes fail to restore forest cover 89
4.1 The catalytic effect of timber plantations 110
4.2 Spruce-beech mixtures in Germany 117
4.3 Restoring species-rich forests in the Atlantic region of Brazil 122
4.4 Diverted successional development 124
5.1 What is a landscape? 135
5.2 Dryland salinity in Western Australia 141
5.3 The Grain for Green Program, China 145
5.4 The great Indian hedge 153
5.5 Example of a very large-scale corridor 156
6.1 Rewilding Europe 178
7.1 Multi-species plantings on a large scale in Hunan, China 193
7.2 Differences in perceptions of the value of biodiversity in the agricultural landscapes of Ghana 201
8.1 Local modifications to national land allocation programmes in Vietnam 214
8.2 Reverse auctions 224
8.3 Typology of participation showing the ways governments might involve people in the management of rural development projects such as reforestation 229

*The trees of the Niû Mountain were once beautiful. But can the mountains be regarded any longer as beautiful since, being in the borders of a big state, the trees have been hewn down with axes and hatchets? Still with the rest given them by the days and nights and the nourishment given them by the rains and dew, they were not without buds and sprouts springing forth. But then the cattle and sheep pastured upon them once and again. That is why the mountain looks so bald, and when people now see it, they think it was never finely wooded. But is this the nature of the mountain?*

Mencius
Chinese philosopher in the Warring States period
(403–221 BC) describing the impact of deforestation

# Preface

Most people have an instinctive liking of trees even though they might not necessarily wish to live in a forest. And most people probably experience a feeling of unease and regret when they see a forest being cleared. Yet, despite these feelings, very large areas of forest have been lost, especially over the last 100 years. Some of this land has been used to produce food but, in hindsight, much forest clearing has been unnecessary. It now seems that some kind of a psychological threshold may have been crossed and that there is increasing interest in restoring some of these lost forests. More people are coming to feel that the cost of past clearing, in terms of biodiversity losses, soil erosion and stream sedimentation, has simply been too great.

Various kinds of reforestation have been practiced for many years in most parts of the world. Much of this has been done at a relatively small scale on individual farms or rural estates. But this pattern is now changing. About 100 years ago governments and then timber companies, mostly in temperate regions, began creating large industrial timber plantations. The species used and the types of plantation created were different than those found on most farm woodlots. Products from these industrial plantations first supplemented timber coming from natural forests. Subsequently, they have begun to replace these supplies. But at the same time as this began happening, reforestation objectives also started to change. Many of the more recent reforestation projects now aim to generate mostly conservation benefits and not just supply timber. Some are being implemented on steep areas within landscapes to protect watersheds while others are being established in belts or corridors to assist the movement of wildlife or stabilise soils. These are being implemented at landscape scales with some even being planned at continental scales. In addition there are plans in Europe and North America to encourage forests as part of a 'rewilding' movement to recreate large areas of Pleistocene habitats in order to conserve threatened species from that era. The type of plantings used in these projects differs from previous industrial plantings and includes species mixtures and more indigenous species rather than the monocultures of fast-growing exotic timber species more commonly used in the past. The scale of these new plantings is also more ambitious than most of the plantings carried out previously.

The purpose of this book is to explore these newer forms of reforestation and the 'scaling-up' process that is underway especially on degraded lands. Large-scale changes of any kind are often fraught with problems. One group of problems concerns the technicalities of implementing reforestation over extensive areas. Some might point to large industrial pulpwood plantations as showing that this is possible. These simple systems have certainly been effective in producing considerable quantities of timber but are rather less effective at providing a wider range of ecosystem services. Moreover, there are concerns that these types of plantation are ecologically brittle and have low levels of resilience. Extensive but uniform plantings of just one species expose them to high degrees of ecological and economic risk. But what other options are available? And where in the landscape should these plantings be done to maximise their ecological and economic effectiveness?

A second set of problems concerns the social impacts of large-scale changes to land use practices. Those in favour of reforestation often assume the benefits are self-evident but people owning the land may be less certain of these benefits. What are the opportunity costs of reforestation and who will pay these? Forest restoration on a large scale may have considerable environmental benefits but it raises the issue of how this will be done in an equitable way and in a manner that ensures some degree of social justice is achieved.

A third group of problems concerns the ways in which forest restoration is implemented and managed. Is tree growing simply another form of agriculture and are the policies developed to manage agricultural land use practices sufficient? Or might an entirely new set of policies and institutions need to be developed given the capacity of new forests to supply a wide variety of ecosystem services? If new institutions are needed then how will these be integrated within existing administrative structures and how will current bureaucracies view them?

These changes in reforestation practices are taking place at a time when some serious global issues are coming to a head. One of these is food security and concerns over whether there will be enough land to grow food to meet the needs of a rapidly growing human population. Will these concerns limit our ability to restore forest cover? A second issue involves the changes in global climates that are now developing. There is a scientific consensus that ways must be found to limit further growth in atmospheric carbon to limit these changes. New forests can help achieve this by sequestering and storing atmospheric carbon but substantial areas of land will be needed to do so. Will there be enough land for food production and carbon sequestration? Or is the dilemma more apparent than real and might the extent of degraded land that is mostly marginal for agriculture be enough to allow considerable reforestation without limiting future food production?

The book begins by reviewing current assessment of land degradation and how much of this may be available for reforestation. It also explores in more detail the context in which any reforestation will take place. The second chapter examines the history of reforestation over the last 100 years to determine what lessons have been learned and whether this provides a sufficient basis for an enhanced future reforestation programme. The third chapter discusses a form of reforestation which is often overlooked by those wishing to increase forest areas, namely natural regrowth. It is widespread and low-cost but is it reliable enough and capable of restoring the original forests? The fourth chapter considers other forms of reforestation that rely on planting seedlings or sowing seed. It discusses silvicultural systems primarily devised to produce forest products as well as others designed to restore ecosystem services. Chapters 5 and 6 discuss the importance of the spatial location and spatial extent of these various forms of reforestation on ecological functioning and the delivery of ecosystem services while Chapter 7 considers some of the trade-offs that might have to be made. Chapter 8 then considers the policy settings and institutions needed to encourage landowners to undertake reforestation. These must take account of the need to make more use of native species and to establish these on degraded landscapes. The book concludes with a review of what might be the next steps forward and a brief review of what might be possible within the future environmental context.

Large-scale forest restoration is unlikely to be neat and methodical. Rather, it will be a messy process involving a variety of silvicultural approaches including some that are yet to be developed. It will involve state forestry agencies as well as large and small landholders. In some cases it will be the result of a carefully planned programme but in others it will be the result of benign neglect, allowing natural regeneration to occur, or a series of unconnected small decisions by different landholders scattered across the landscape mosaic. But, perhaps for the first time in history, there appears to be growing political support around the globe for efforts to reverse years of deforestation. This support is fragile and may disappear as other problems move to the forefront of global political and economic debates. The task of ecologists and silviculturalists is to make the most of this opportunity and, in collaboration with economists and other social scientists, to develop techniques that maximise the effectiveness of any reforestation and increase the likelihood that these new forests become accepted as valuable and necessary components of the rural landscape.

# Acknowledgements

Over the years I have been fortunate to have met and worked with a wonderful group of colleagues. They are scattered across the world but share a common view of the need to reforest degraded lands and to do it in ways that generate conservation and livelihood benefits. The interactions occurred in dusty fields, forests, seminar rooms and conference halls while I was at the University of Queensland and a member of the Commission on Ecosystem Management of the IUCN.

The following have kindly provided insights on reforestation or have read parts of the draft manuscript and offered comments. Some have also provided photographs and I have acknowledged these in the text (the other unacknowledged photographs being my own). I am very grateful to all of them for their assistance and hope they will think I have done justice to their views and feedback. Needless to say, the responsibility for any errors remaining is mine and not theirs. They include Sasha Alexander, Dominic Blay, Sampurno Bruijnzeel, Margaret Chapman, Stephen Elliott, Vera Lex Engel, Josef Ernstberger, Jennifer Firn, Don Gilmour, Manuel Guariguata, Andrew Ingles, Justin Jonson, Rod Keenan, Stephen Kelleher, Thomas Knoke, Cora van Oosten, Tony Rinaudo, Tim Rollison, Jeff Sayer, Ulli Schmidt, Gill Shepherd, John Stanturf, Tint Thuang, Tony Whitten and Bob Winterbottom.

Finally, I am also most grateful to Marg, Andrew and Kate who have, as usual, assisted in more ways than they can know.

# 1 The need for large-scale forest restoration

## Introduction

One of the dominant environmental features of the twentieth century has been the dramatic decline in forest cover. This has been going on for many years but the rate has accelerated dramatically in the last 50 years. Huge areas have been deforested in both temperate and tropical regions. Significant amounts of forests still remain but, in many places, these persist in landscape mosaics made up of residual forest patches embedded in a matrix of other land uses. Further, much of this remaining forest is regrowth rather than undisturbed old-growth forest meaning that its value for purposes such as wildlife conservation has been reduced. This deforestation process and the variety of direct and indirect causes responsible for it have been reviewed by Kaimowitz and Angelson (1998), Lambin *et al.* (2003) and Chomitz (2007).

Many of these former forest lands have since been used for agriculture and this has bought great wealth and food security to large numbers of people. Some of these agricultural lands have now been settled for many years and have achieved ecological stability. But not all land clearing has been successful and some cleared lands are now only marginally suitable for agriculture. Some have been largely abandoned. There are a variety of reasons for this. In certain cases it was because the soils or environmental conditions were always unsuitable for any kind of agriculture meaning that failure was inevitable. In other cases inappropriate cropping methods or over-grazing caused damage. Sometimes economic or social constraints have led to the decline of what were initially successful agricultural enterprises.

The process of deforestation and forest fragmentation has had many ecological consequences. One of these has been the loss of considerable biodiversity, especially amongst those species requiring specialised forest habitats. But other changes have arisen from efforts to maximise agricultural production. These changes have included soil losses, soil compaction, acidification and overall fertility declines. Rivers draining these lands have sometimes been affected by sedimentation, chemical pollution (from agricultural chemicals) or salinity (UNEP 2007). At the same time, many cleared lands have become dominated by exotic weed or pest species. Some of these landscapes have undoubtedly crossed an ecological threshold and moved to a new state condition from which recovery will be difficult.

These changes have also had significant socio-economic consequences. Despite the degradation, large numbers of people continue to live in these landscapes. One estimate suggests one-third of the world's population now suffer from the effects of land degradation with many of the most affected people being found in the world's drier regions (UNEP 2007). A large number of these people traditionally depended on the natural capital contained in forests for food, building materials, medicines and other resources. Forest loss, together with land degradation, means these people are increasingly vulnerable to natural hazards and future socio-economic changes. Often such people are caught in a 'downward spiral' and are forced to make increased demands on the limited resources still available as their populations grow. Further resource degradation occurs as these demands increase. Scherr (2000) argues that under such circumstances, the best strategy is to develop ways of increasing people's access to natural resources, enhance the productivity of these resources and involve communities in resolving management dilemmas.

Many farmers have grown trees on their land but larger-scale reforestation only became more widespread in the early years of the twentieth century when some governments began creating industrial timber plantations to make up for the loss of natural forests (see Box 1.1 for a definition of forests, reforestation and restoration). Over time private growers have taken over many of these plantations and created new ones. But, in more recent years, the emphasis has begun to change as the extent of natural forest losses have increased and the adverse ecological consequences have become more obvious. There is now increasing interest in undertaking reforestation to provide ecosystem services such as watershed protection and biodiversity conservation rather than to simply produce timber; indeed the production of goods such as timber is becoming a secondary objective in many locations. Some of the reforestation schemes now being proposed by governments and the international community are very large.

This book explores how this forest restoration might be done to generate better functional outcomes as well as improve human livelihoods. In particular, it considers the ways in which various forms of large-scale reforestation of cleared land might be implemented. This first chapter begins this exploration of how large-scale forest restoration might be undertaken by initially clarifying what is meant by 'large-scale' and then by exploring how much degraded land there is and the extent to which it might be available for reforestation. It also describes the context in which future forest restoration will be carried out.

**Box 1.1 Defining forests, reforestation and restoration**

There is considerable debate over the terms used to describe various types of forest and different forms of reforestation (Carle and Holmgren 2003). Some dispute whether plantations are really forests while others point to the role of planted forests in many temperate regions contributing to conservation objectives (Meijaard and Sheil 2011). In fact there is a considerable variation in the types of planted forests and even the most simple of these can sometimes change and evolve over time to the point where they may seem like natural forests to a layman (Figure 1.1). Evans (2009) discusses some of the difficulties in defining the types of forests found along the continuum between simple, even-aged plantation forests managed on short rotations for production purposes and undisturbed natural forests. Between these extremes can be found a variety of forests differing in composition, the number or proportion of species planted, the number of age classes present (i.e. whether the forests are self-sustaining), the tree density and the spatial arrangement of the trees. The definitional difficulties include accounting for differences in the starting point, structural complexity and degree of naturalness of the forest or the purposes for which they have been established.

A particular complication involves planted forests that, over time, are colonised by additional tree species. Are these plantations or forests? Over 5 million ha of planted forests in the western USA are excluded from national statistics because of this colonisation process (Fernandez *et al.* 2002). And what of naturally regenerating forests that have been enriched by planting to increase the proportion of certain favoured species? Such developments have led to some referring to certain forests as being 'semi-natural planted forests' (Evans 2009). Unfortunately different countries use these terms in different ways. Thus Finland classifies most of its forests as semi-natural although they are planted using native species while the USA also uses mostly native species but refers to the new forests as plantations.

In the present context the following terminology shall be used (based largely on Evans 2009).

***Types of forest***

(i) *Plantation*: forest stands artificially established (by planting or direct seeding) and containing one or more species. These may be established for production or protection purposes.

(ii) *Semi-natural planted forest*: forests containing a mixture of planted and naturally regenerated species. Examples are: (a) even-aged plantations that, over time, have been converted to uneven-aged forests because of natural regeneration and/or external colonists reaching the site and establishing under the tree canopy; and (b) natural regeneration that has been enriched using planted seedlings.

(iii) *Regrowth forests*: forests that have regenerated largely through natural means following a natural or human-induced disturbance. The composition, tree density and structure depend on the time since the disturbance and may change significantly over time.

*Figure 1.1* There are many kinds of planted forests: (A) a *Cunninghamia lanceolata* plantation in China that was established and maintained as a monoculture; (B) a 60-year-old *Flindersia australis* plantation in Australia planted as a monoculture but now an uneven-aged multi-species forest because of enrichment by colonists from adjoining rainforest (Photos: J. Firn).

It is difficult to prescribe a certain threshold size that must be reached before planted trees become a plantation or part of a semi-natural forest. It seems sensible to exclude single rows of trees planted along roadsides or fence lines even though these may contribute significantly to rural plantings.

***Reforestation***

Reforestation shall be taken to include the re-establishment of forests or woodlands at sites where woody plants were the former natural vegetative cover by deliberately planting or sowing seeds as well as by natural regeneration (excluding situations where production forests are being reforested as part of the normal silvicultural cycle). That is, it is the opposite of deforestation. Evans (2009) recognised two basic methods of re-establishing trees at cleared sites:

(i) *Afforestation*: the act of creating a forest at a site where trees have been absent for more than 50 years.
(ii) *Reforestation*: the act of re-establishing trees at sites deforested within the previous 50 years irrespective of whether or not the same species are replanted.

The problem with this distinction is that site histories are often unknown. Largely for convenience, the term reforestation shall be used here to cover both approaches.

***Restoration***

Reforestation can be done in a variety of ways. Restoration is sometimes used to describe forms of reforestation where the emphasis is on multi-species plantings or on some degree of 'naturalness' but the distinction is not always clear and two terms (reforestation and restoration) will mostly be used here interchangeably. On the other hand, the term 'Ecological Restoration' will be restricted to the process of attempting to restore a forest ecosystem previously present at a now degraded site. This particular type of restoration is discussed further in Chapter 4.

## The meaning and significance of 'large-scale' forest restoration

Some explanation of the term 'large-scale' is needed. This is clearly a subjective term and can mean different things to different people depending on their viewpoint and circumstances. Thus, a 1,000 ha plantation area might be regarded as being relatively small-scale by some industrial growers but be seen as relatively large-scale by a community living on a 2,000 ha island. For the purposes of this book large-scale will be taken to mean:

(i) a large contiguous planted area exceeding, say, 10,000 ha;
(ii) a number of small, separate forest plantings that collectively add up to a large area even though they are scattered across a wider landscape;

(iii) a forest area large enough to sustain a breeding population of a top-order predator;
(iv) a forest area large enough to influence hydrological processes within a watershed (e.g. to lower water tables in order to reduce salinity).

Large-scale reforestation is important because this is the scale at which interventions are needed to influence and restore many of the ecological processes and functions that were changed when the original forests were removed. Small isolated patches of restored forest may be locally useful but may have little effect on restoring regional hydrological processes, combating large-scale erosion or increasing the populations of endangered species. And while a single small patch of forest might be able to supply forest products for localised use, it is unlikely to be commercially attractive unless there are other forests nearby to provide sufficient resources for a sustainable industry.

Large-scale projects have had a chequered history. Many, such as dams, land settlement and irrigation schemes and, indeed, some reforestation programmes, have failed in the past leading to further environmental degradation, massive losses of funds and much damage to public confidence (Scott 1998). Some have left local communities worse off than before. But this book will argue that the legacy of many years of land mismanagement and degradation is so great that only a systematic and large-scale programme will be able to address the problem. Large-scale reforestation will need the collective efforts of government agencies, industrial corporations, non-government organisations (NGOs) and individual landholders. It will also require a variety of silvicultural methods far more diverse than those commonly used in industrial timber plantations. Finally, it will necessitate the development of supportive and flexible policies, institutions and systems of governance able to facilitate the process and adapt it where circumstances require this.

## How much degraded land is there?

The obvious place in which to undertake reforestation would be degraded lands. But just what are 'degraded lands'?

### *Defining degradation*

It is surprisingly difficult to define degradation. Part of the reason for the difficulty is because, like the term 'large-scale', degradation is also a perceptual term that means different things to different people. A wildlife conservationist might see a recently cleared piece of once-forested land as being degraded while a farmer owning the same piece of land would see a productive cropland; indeed such a farmer might say the land had only acquired value by being cleared. On the other hand, that same farmer might agree the land was degraded once over-grazing had occurred, sheet erosion became widespread or land slumps were common. But farmers also differ amongst themselves in their perceptions of degradation. Potter (1987) describes how one group of farmers in Indonesia practicing shifting cultivation saw increasing amounts of the aggressive grass species, *Imperata cylindrica*, as being indicative of increased degradation because it meant the forest regrowth phase of shifting cultivation on which their farming system depended was threatened. By contrast, other farmers in the same place who owned cattle saw the spread of the grass across their farms as being quite beneficial and to be encouraged. Degradation was clearly a matter of perception and depended on what each group hoped to obtain from the land.

Such differences in values and perception mean it is hard to get agreement over whether degradation has occurred, and, perhaps more importantly, whether anything should be done to overcome it. A variety of definitions of land degradation have been offered by various authors (Box 1.2). These all reflect the fact that degradation means changes have occurred in the availability of products, goods and services that people might obtain from land or forests. A broad review of overall forest degradation is given by the Food and Agriculture Organization of the United Nations (FAO 2011).

Irrespective of these differences in perspective, degradation is not an either-or matter; rather, there are differences in the degree of degradation that can occur. Some degraded land may have become marginal for agriculture but may still be productive, at least to some degree, for other purposes. Other more heavily degraded land may be incapable of supporting much vegetative growth of any kind. This gradient in productivity or utility is likely to be matched by the number of people who perceive that degradation has indeed occurred. Wildlife conservationists may be the only people seeing new agricultural lands as being degraded but when the land loses it topsoil, suffers gulley erosion and sediments begin to pollute rivers they will be joined by farmers and graziers, irrigators, hydro-electric generators and water supply managers. Perceptions about the severity of degradation may also change over time depending on rainfall. If, for example, a series of above-average rainfall years increases vegetative cover then there may be a decline in the number of those regarding an area as being degraded.

### *Estimates of the amount of degraded land*

While the definitions of degradation in Box 1.2 convey a sense of what land degradation means they are not especially useful in the task of mapping areas of degraded land or quantifying its extent. This is because none of them specify the threshold condition at which degradation begins. Early attempts to map degradation used ground surveys and relatively subjective assessments of landscape condition.

**Box 1.2 Definitions of land degradation**

Most definitions refer to a reduction in the productive capacity of land caused by changes in soil fertility, erosion, weeds or recurrent fires due to inappropriate human actions although some authors also link degradation to changes caused by natural events. Thus:

> Land degradation is a loss of capability to meet the demands made upon it.
> (Blaikie and Brookfield 1987)

> Degradation is alterations to all aspects of the natural (or biophysical) environment by human actions, to the detriment of vegetation, soils, landforms and water (surface and subsurface, terrestrial and marine) and ecosystems.
> (Conacher and Conacher 1995)

> Land degradation is any change or disturbance to the land perceived to be deleterious or undesirable.
> (Johnson *et al.* 1997)

> Degraded land is a long term decline in ecosystem functioning and productivity caused by disturbances from which the land cannot recover.
> (Bai *et al.* 2008)

> Land degradation is the reduction in the capacity of the land to provide ecosystem goods and services and assure its functions over a period of time for its beneficiaries.
> (Biancalani *et al.* 2011)

> Land degradation leads to a significant reduction of the productive capacity of land. Human activities contributing to land degradation include unsustainable agricultural land use, poor soil and water management practices, deforestation, removal of natural vegetation, frequent use of heavy machinery, over-grazing, improper crop rotation and poor irrigation practices. Natural disasters, including drought, floods and landslides also contribute.
> UNEP (Accessed 16 May 2014. Available at: www.unep.org/dgef/LandDegradation/tabid/1702/Default.aspx)

> Land degradation and desertification constitutes a permanent decline in the provision of all services that land would otherwise provide. They adversely affect food security, water security, biodiversity and many ecosystem services as well as recreational, heritage and cultural values.
> (UNCCD 2012)

None of these defines a threshold condition at which point degradation can be said to have occurred (or beyond which, recovery by natural processes will be difficult). Nor can any be readily used for mapping the extent of degradation.

One of the first global assessments was that by Oldeman *et al.* (1990). Their widely quoted study was named the Global Assessment of Soil Degradation (GLASOD) and was based on 'expert' but, nonetheless, subjective assessments. They estimated there were 1.96 billion hectares of land across the world damaged by water or wind erosion, nutrient depletion, salinity, physical impediments or contamination of some kind. Of these, water and wind erosion were by far the most serious problems. Such assessments are largely maps of perceptions and Sonneveld and Dent (2009) concluded the study was only moderately consistent and not very reproducible.

An obvious way to improve assessments would be to try to quantify degradation in some way. Remote sensing techniques are now being used in some surveys to assess degradation and track changes over time (Bai *et al.* 2008, Simula 2009). But these techniques, too, have attracted criticism and it is clear they need to be complemented by on-ground surveys (Dent *et al.* 2009, Wessels 2009). In a recent review Lambin and Meyfroidt (2011) argued that there are no credible measures of the total area of degraded land across the globe and that many previous assessments of annual increases in the extent of degraded land are likely to be over-estimates. They suggested that around 1–2.9 million ha of additional land is probably being degraded each year implying that an additional 30–87 million ha of land will be degraded between 2000 and 2030 in addition to that which already exists.

But degradation is not just a bio-physical problem because it involves socio-economic changes as well. That is, degradation may not be amenable to a 'simple' techno-fix but will require rather more complex changes in policies, institutions and systems of governance if it is to be addressed (Andersson *et al.* 2011). Assessment of the extent of degraded land and, more importantly, just how degradation might be overcome will require a detailed understanding of the historical drivers of land use change and the circumstances of the present day land users.

## How much land might be available for reforestation?

Irrespective of how much degraded land has accumulated it is clear that much of it is not necessarily available for reforestation. On the other hand, some other land that might be seen as being marginal for agriculture may be available. Dauber *et al.* (2012) identified a variety of 'surplus' lands that might possibly be available for reforestation. These included fallow land normally part of a production cycle, areas referred to in Europe as 'set-aside' land on which production has been suspended for political reasons, abandoned land, former mine sites and economically or ecologically marginal agricultural land. It is difficult to determine the overall area covered by such lands, especially given the imperfect definition of some of these categories and because opportunity costs of reforesting marginal lands might vary from year to year depending on market prices of agricultural crops and forest products. Nonetheless, a number of attempts have been made.

### *Global assessments of land potentially available for reforestation*

Estimates of the amount of abandoned or under-utilised land across the globe that might be available for reforestation are shown in Table 1.1. Some of the earlier estimates were relatively subjective but, over time, more use has been made of remote sensing tools and computer models of vegetation cover. Thus Grainger (1988) estimated there were 2,077 million ha of degraded tropical lands of which 758 million ha could be suitable and available for reforestation. Shortly afterwards, Houghton (1990) used assessments of biomass carbon stocks and concluded there were 500 million ha of grazing lands, savannahs and grassland plus a further 365 million ha of fallow lands that might be available in the tropics. He assumed this land might be available for reforestation because it was thought to be relatively unused at the time. Both of these studies were restricted to tropical lands. Nilsson and Schopfhauser (1995) undertook a broader global study and concluded there were only 345 million ha available for reforestation. This was based on aggregated regional estimates that potentially provide more realistic accounts of the land actually available for reforestation. Campbell *et al.* (2008) also attempted a global analysis and estimated lands available for reforestation using historical land use data, satellite-derived land cover data and global ecosystem modelling. They estimated 269 million ha of croplands and 479 million ha of pastures have been permanently abandoned across the globe over the last 300 years. Allowing for forest regrowth and urbanisation, they estimated there are now 385–472 million ha of abandoned agricultural land across the globe that could be suitable for reforestation.

Each of these estimates is large but several more recent studies point to even larger amounts of land potentially available for reforestation. Benitez *et al.* (2007) used several global land cover data sets to estimate the amount of land potentially available. The excluded land deemed suitable for agriculture, where the population density was presently more than 200 persons per square kilometre and land with an elevation higher than 3,500 m. This indicated there could be between 2,600 and 3,500 million ha of land suitable for reforestation. Laestadius *et al.* (2011) tried to base their estimate on the different types of reforestation that might have to be used in different situations. They firstly modelled the potential global forest and woodland area that would be present if soils and climate were the only limiting factors and humans were absent. They then mapped the actual distribution of forests now present. Their final step identified restoration opportunities within the deforested but suitable landscapes. This involved taking account of population densities and land use practices and estimating how much land might be available for 'wide-scale restoration' (i.e. larger plantings) where there were less than ten people per square kilometre and 'mosaic restoration' (i.e. patches of reforestation within a mostly agricultural landscape) where there were between ten and one hundred people per square kilometre. Based on this analysis they estimated there might be two billion ha of land potentially available for reforestation or restoration. Finally, based on a review of a number of studies Dauber *et al.* (2012) estimated there might be an area between 250 million ha and 1,580 million ha depending on the reforestation objectives.

There is considerable variation in the size of the estimates in Table 1.1. This is not surprising given the differences in methods used and in the assumptions made. But the over-riding message is clear – the areas potentially available for reforestation are huge. Indeed those of Benitez *et al.* (2007) and Laestadius *et al.* (2011) exceed the areas of the world under crops (estimated by Ramankutty *et al.* [2008] to be around 1,500 million ha). For comparative purposes it is also useful to note that the area of tree plantations in 2010 was 264 million ha (FAO 2011).

It is important to note that the amount of land actually available for reforestation may fluctuate over time. For example, a study carried out in the European Union estimated a further 12.6–16.8 million ha of farmland across the region were likely to be abandoned by 2030 and could then be available for reforestation (Keenleyside and Tucker 2010). Large areas of former farmland have also been abandoned recently in Ukraine and parts of eastern Europe (Kuemmerle *et al.* 2011, Alcantara *et al.* 2012). But much of this land has been abandoned because of political and economic events and could possibly revert to agriculture if circumstances changed again.

The obvious conclusion arising from all of these assessments is that there are sizeable opportunities across the globe for further reforestation. But perhaps this should be put more forcibly – the magnitude of these estimates of deforested and under-utilised lands suggest there is a clear need for reforestation on a much larger scale than has occurred hitherto.

*Table 1.1* Global estimates of the amount of land potentially available for reforestation (areas marked with asterisk are tropics only)

| *Reference* | *Method* | *Area (million ha)* |
|---|---|---|
| Grainger (1988) | Aggregated regional estimates | 758* |
| Houghton (1990) | Modelling | 865* |
| Nilsson and Schopfhauser (1995) | Aggregated regional estimates | 345 |
| Campbell *et al.* (2008) | Historical data, remote sensing, modelling | 385–472 |
| Benitez *et al.* (2007) | Remote sensing, modelling | 2,600–3,500 |
| Laestadius *et al.* (2011) | Remote sensing, modelling | 2,000 |
| Dauber *et al.* (2012) | Synthesis of literature | 250–1,580 |

### *Local assessments*

Large-scale assessments based on global databases and remote sensing are useful for global policy discussions. But local assessments are needed when planning where and how reforestation might actually be done. There are various reasons why marginal agricultural land or apparently 'idle' or 'degraded' land may not be available for reforestation. One important constraint concerns land tenure. If it is ambiguous or contested then long-term land uses such as reforestation will be risky. But even where tenure is more settled, not all land users will be necessarily interested in reforestation even though this might be preferred from a national perspective. For example, in comparison to agricultural crops a land owner might think the initial costs of reforestation are too high and the time before any financial return is too long. Another landowner might simply tolerate the presence of some degradation on their land provided the area affected is a relatively small proportion of the total farm area. In both cases the opportunity costs of reforestation are simply too high. Tree growing might also be unattractive for cultural reasons such as it not being a traditional land use activity in the community: 'it ruins good farmland' (Schirmer and Bull 2014). Indeed, earlier in their careers, some farmers may have been involved in clearing forests from these same lands to create their farms and the idea of now planting trees runs counter to all they have done in their working lives.

However, reforestation might be quite an attractive land use in other situations. For example, a rising market for forest products might make tree growing commercially attractive to large corporations as well as smaller landholders. This is likely to be especially true at sites close to roads where goods can easily reach markets. And some farmers with land unsuitable for cropping might be willing to reforest it if the initial costs were modest and they knew how to do so. Reforestation could also be attractive to governments where degraded watersheds needed protection or where the habitats of particular wildlife needed to be restored to ensure their survival. Ideally, such reforestation might be funded by the goods supplied, or from the ecosystem services provided by reforestation.

## Changing attitudes to reforestation

Until recently the main driver of reforestation was the perceived need to establish a timber resource to replace that lost when natural forests were cleared. These timber plantations mostly used exotic species grown in monocultures. Planting was usually carried out by a government forestry department or by private companies encouraged to do so by favourable policies and, in many cases, various forms of incentives. But a change in attitude began to develop during the second half of the twentieth century. Reforestation then began to be undertaken to restore ecosystem functioning and supply various ecosystem services rather than to simply produce timber (see Box 1.3). This appears to have been prompted by a variety of factors. Amongst these were a growing awareness of the extent of soil erosion, declines in water quality and questions over the sustainability of agricultural production. In addition there has been rising unease over the growing loss of global biodiversity caused by deforestation and forest fragmentation. And, most recently, there has been concern over the contribution of deforestation to global climate change.

Various NGOs were amongst the first to draw attention to these trends but their arguments are now being accepted by governments and by many key international institutions. These include the UN Framework Convention on Climate Change (UNFCCC), the UN Convention on Biological Diversity (CBD), the UN Convention to Combat Desertification (UNCCD), the UN Environmental Program (UNEP) and the UN Forum on Forests (UNFF) (Table 1.2). Though differing in their focus, each of these bodies is now calling for actions to reverse forest and land degradation and to increase forest cover. All of them emphasise using forms of reforestation to improve livelihoods, the supply of ecosystem services such as carbon sequestration and water supplies for urban areas, rather than just timber production.

A number of multi-party agreements are also being developed. Two recent initiatives deserve special mention. One is the Bonn Challenge which was an initiative of the German government and the International Union for the Conservation of Nature. This seeks to promote the restoration of 150 million ha of land by 2020. At the 2012 Rio+20 meeting held in Brazil, the USA and Rwanda committed themselves to helping meet this target. At subsequent international meetings India, Costa Rica, El Salvador, the Brazilian Mata Atlantica Restoration Pact and the Mesoamerican Alliance of Peoples and Forests also expressed support meaning there are now commitments or expressions of intent covering 50 million ha (IUCN

**Box 1.3 Ecosystem services**

Ecosystem services are the benefits that people obtain from ecosystems (Millennium Ecosystem Assessment 2003). They include:

(i) *Provisioning services*: food, timber, fresh water and fibre. Food and timber in particular are sometimes referred to as 'goods' to distinguish them from other ecosystem services.
(ii) *Regulating services*: the regulation of local climate, soil erosion, water yields, floods and biological controls of pests and diseases.
(iii) *Supporting services*: habitats for species, soil formation, photosynthesis and nutrient cycling.
(iv) *Cultural services*: recreation, tourism, aesthetic enjoyment and spiritual fulfilment.

*Table 1.2* Recent activities of international bodies favouring increased reforestation

| *Body* | *Activity* |
|---|---|
| Convention on Biological Diversity (CBD) | At the Conference of the Parties 10 in Nagoya, Japan, 2010 the meeting agreed to the Aichi strategic plan for biodiversity conservation between 2011 and 2020. The plan calls for:<br>• Target 14: restoration of ecosystems that provide vital ecosystem services for the poor and vulnerable, particularly water;<br>• Target 15: restoration of 15 per cent of all degraded ecosystems by 2020 in order to promote resilience, enable adaption to climate change and combat land degradation and desertification (Convention on Biological Diversity 2010).<br>This was reaffirmed at the subsequent Conference of the Parties 11 at Hyderabad, India, 2012 (Convention on Biological Diversity 2012). |
| United Nations Convention to Combat Desertification (UNCCD) | Seeks to achieve zero net land and forest degradation by 2030; to be achieved by arresting further degradation and by restoring and rehabilitating existing degraded lands (UNCCD 2012). |
| United Nations Environmental Program (UNEP) | Recommends restoring degraded ecosystems to secure ecosystem services, addressing challenges in disaster risk reduction, water supplies, carbon sequestration and food security (Nellerman and Corcoran 2010). |
| UN Forum on Forests (UNFF) | Expressed concern about degradation and called on member states to implement reforestation (UN Forum on Forests 2006). |
| UN Framework Convention on Climate Change (UNFCCC) | Has adopted a global goal to slow, halt and then reverse the loss of forest cover (UN Forum on Forests 2006). The Reducing Emissions from Deforestation and Forest Degradation (REDD+) mechanism offers a possible means to fund forest restoration (Alexander *et al.* 2011). |
| United Nations Conference on Sustainable Development, Rio+20 | 'We call for enhanced efforts to achieve the sustainable management of forests, reforestation, restoration and afforestation, and we support all efforts that effectively slow, halt and reverse deforestation and forest degradation.'<br>The final outcome document specifies the need to facilitate ecosystem restoration to enable sustainable development (UN General Assembly 2012). |
| Asia Pacific Economic Cooperation (APEC) | The Leaders Declaration at the Sydney meeting in 2007 agreed on the need to increase forest cover in the Asia-Pacific region by 20 million ha before 2020. This was reaffirmed at the first meeting of APEC ministers responsible for forestry at Beijing 2011. (Accessed 16 May 2014. Available at http://www.apec.org/Meeting-Papers/Ministerial-Statements/Forestry/2011_forestry.aspx). |
| The Bonn Challenge | A ministerial roundtable hosted by Germany and the IUCN in 2011. It aims to restore 150 million ha by 2020 (IUCN 2012). |
| Gabarone Declaration | Ten African nations pledge ahead of Rio+20 to ensure the benefits of natural capital are quantified and integrated into development plans through, *inter alia* ecosystem restoration (Conservation International 2012). |
| Great Green Wall for the Sahara and Sahel | A scheme involving 20 African countries to combat desertification and improve rural development. Originally conceived as a belt of trees but has evolved to a more complex programme of improved land management (Accessed 16 May 2014. Available at www.fao.org/partnerships/great-green-wall). |

2012). A similar set of aspirations are included in the Gabarone Declaration of 2012 signed by the presidents or ministerial representatives of ten African countries. Amongst other objectives this calls for 'Ecosystem restoration measures, as well as actions that mitigate stresses on natural capital' (Conservation International 2012).

A number of national governments have also begun to develop national forest restoration plans. Some of these are a continuation of earlier programmes aimed at increasing timber resources but, increasingly, these plans are beginning to include reforestation specifically aimed at generating ecosystem services. In some cases this is the primary purpose. Examples of some of the programmes are given in Table 1.3. The Australian, Indian and Philippines initiatives are examples where the intent is to create a new supply of forest products while also generating ecosystem services with the balance between these two objectives depending on the location of

*Table 1.3* Examples of some recent large-scale national reforestation programmes

| *Country* | *Scale (mill ha)* | *Dates* | *Notes* |
|---|---|---|---|
| Australia | 4–5 | 1997–2020 | Mainly forest products but some services (Anon. 2001, Stanton 2001) |
| Brazil | 15 | By 2050 | Biodiversity restoration by the Atlantic Forest Restoration Pact (Calmon *et al.* 2011) |
| China | 82 | 2000–2050 | Various reforestation for timber production and environmental protection (Chokkalingam *et al.* 2006, Cao *et al.* 2011) |
| India | 10 | 2010–2020 | Greening India Mission to improve ecosystem services (Ravindranath and Murthy 2010) |
| Philippines | 1.5 | 2011–2016 | National Greening Program to generate income from sale of forest products as well as protect watersheds (Accessed 16 May 2014. Available at http://ngp.denr.gov.ph/) |
| Rwanda | 2 | (2035?) | Part of a plan to achieve country-wide reversal of land, water and forest degradation by 2035 (Kamanzi 2011) |
| South Korea | 4.25 | 1960s–present | Initially for forest products but has become more concerned with generating ecosystem services (Lee and Suh 2005, Tak *et al.* 2007, Lee 2012) |
| Vietnam | 5 | 1998+ | The Five Million Hectare Reforestation Program which includes 3 million ha of production forests and 2 million ha of protection forests (De Jong *et al.* 2006) |

particular plantings. The South Korean scheme is one where the original objective of supplying timber was gradually superseded by one where the emphasis is on supplying ecosystem services. But perhaps the most interesting programmes are those where, from the start, the emphasis has been almost entirely concerned with the supply of one or more ecosystem services. The Brazilian scheme is one such programme. This is being undertaken in the Atlantic Forest region and is specifically designed to restore biodiversity. The law in this province specifically requires that landholders restore some of their cleared farmland and that they use more than 80 native species to do so (Calmon *et al.* 2011). Some have argued that such requirements are unnecessarily prescriptive (Aronson *et al.* 2011) but the intent of this very large scheme to conserve regional biodiversity is clear.

However, there is no doubt that the biggest new forest restoration schemes are those being implemented in China. These follow a number of earlier large-scale timber plantation initiatives (Chokkalingam *et al.* 2006). Interestingly, the primary purpose since 2000 has become environmental protection (69 million ha) rather than timber production (13 million ha). The former include the Grain for Green Project (or Sloping Land Conversion Program) which aims to reforest marginal farmland, including steep land with slopes greater than 25 degrees (32 million ha), the Three Norths Shelterbelt Program to control desertification around northern regions (27.5 million ha), the Natural Forests Conservation Program (4.4 million ha) and the Sand Control Program (5.2 million ha). Details of these schemes are given in Chokkalingam *et al.* (2006) and Cao *et al.* (2011).

While there can be debate over the effectiveness of these programmes and whether they will achieve their targets, they do illustrate that governments have begun to respond to the mounting environmental costs of land degradation and that a significant change in attitudes towards reforestation is now underway. Tree planting is no longer seen simply as a means of increasing timber resources for industrial purposes but is becoming recognised as a way of generating ecosystem services. This has major implications for the way reforestation is carried out. It is unlikely that the forms of reforestation developed to suit the needs of large industrial growers will be sufficient for these new objectives and that a rather more diverse range of silvicultural approaches will be needed. And some of the more degraded sites being reforested may need special techniques (e.g. ripping, mounding, liming, fertiliser applications) to allow reforestation to take place. These issues will be discussed in subsequent chapters.

## Restoring ecosystem functioning and ecosystem services

It is generally agreed amongst ecologists that the supply of ecosystem services depends on effective ecosystem functioning and that this, in turn, depends on a system's biodiversity (Balvanera *et al.* 2006, Quijas *et al.* 2010, Isbell *et al.* 2011). However, there is considerable uncertainty about the nature and strength of these relationships and especially about the impact that any changes in biodiversity may have. A review by Cardinale *et al.* (2012) attempted a summary of our present state-of-knowledge. They argued that the present consensus concerning biodiversity and ecosystem functioning is that:

(i) Biodiversity losses reduce the efficiency by which ecological communities capture biologically essential resources (e.g. nutrients, water, light), produce biomass, decompose and recycle biologically essential nutrients.
(ii) There is mounting evidence that biodiversity increases the stability of ecosystem functions through time. In other words, more biodiversity is likely to enhance ecosystem resilience.
(iii) The impact of changes in biodiversity on any single ecosystem process is nonlinear and change accelerates as biodiversity loss increases.
(iv) Diverse communities are more productive because they contain key species with certain functional traits. That is, the presence of certain key species, as well as the overall diversity of species, are both important drivers of ecosystem functioning.
(v) Loss of diversity across trophic levels has the potential to influence ecosystem functions even more strongly than the loss of diversity within a particular trophic level. In some situations the loss of a top-order predator can cascade through a food web and affect the amount of standing plant biomass.

The consequences of these relationships for the provision of ecosystem services are less clear (Balvanera *et al.* 2006, Benayas *et al.* 2009, Cardinale *et al.* 2012). In the short term at least, agricultural experience shows that some provisioning services such as the production of certain foods or fibre can be enhanced by using a single preferred species although this will reduce other ecosystem services. But the reverse is also true and the production of some forest goods can be increased by reforestation using larger numbers of tree species (Hooper *et al.* 2005, Pretzsch 2005, Paquette and Messier 2010, Clough *et al.* 2011). In both cases the most important factor is not biodiversity *per se* but the identity and traits of the particular species involved in the monoculture or multi-species assemblages.

However, scale is also important. So, for example, Pasari *et al.* (2013) found the capacity to sustain multiple ecological functions across a larger area requires maintaining both local diversity as well as regional diversity (the variety of distinct communities and the overall number of species across the landscape as a whole). Likewise Gamfeldt *et al.* (2013) concluded that increased levels of biodiversity was positively correlated across a wide range of scales with services such as tree biomass, soil carbon storage, berry production and game production potential. Importantly, they noted that no single tree species was able to promote all services. This means that plantation monocultures are likely to have important limitations when multiple services are needed. Even a patchwork of adjacent monocultures will be unable to optimise the provision of ecosystem services at a landscape scale. They suggest a better solution would be to use a mosaic of different species mixtures. Isbell *et al.* (2011) also pointed to the importance of biodiversity through its role in maintaining functionality despite variations in environmental conditions.

In summary, traditional industrial plantations may be very efficient producers of timber but will be much less effective in restoring a wide range of ecosystem functions or services that are becoming of more interest to land managers, especially at larger spatial scales. Some of the alternative silvicultural systems that might be used to generate these functions and services will be discussed in subsequent chapters. In the meantime it is useful to consider the context in which any reforestation is likely to take place.

## Factors potentially limiting future reforestation

A number of changes are underway that will affect the ways in which any future reforestation takes place. Some of these are likely to make reforestation more difficult to achieve while others will make it easier. Table 1.4 shows some of the main factors. Of the factors most likely to place limitations on the extent of reforestation, the most prominent is undoubtedly the need to grow more food to support the rising human population.

### *The need for more agricultural land in the future*

Human populations are increasing and projected to rise to 9.3 billion by 2050 and possibly 10 billion by 2100 (United Nations 2011). This will

*Table 1.4* The context in which future reforestation will take place

| *Conditions making reforestation/restoration more difficult* | *Conditions favouring reforestation/restoration* |
|---|---|
| Growing populations | Increased concern amongst the general public about environmental issues |
| Rising need for food | Development of markets for ecosystem services as well as goods |
| Increasing demand for agricultural land | Urbanisation and out-migration |
| Continuing rural poverty | Legal obligations to resolve environmental damage |
| Uncertain land tenure | The need to combat global climate change |

substantially increase the need for greater food production. This rising demand is being amplified by the fact that human longevities are gradually rising and people's diets are changing as their economic circumstances improve. A larger number are now choosing to eat more meat. The result is that even with recent productivity gains, roughly one in seven people still lack sufficient food or are chronically malnourished (Foley *et al.* 2011). Much of this is because existing food supplies are not being distributed to those who need it and the problems are political rather than technical (De Schutter 2011a, FAO 2012). Nonetheless, some commentators suggest food production will have to increase by 70–100 per cent by 2050 (Cribb 2010, Foley *et al.* 2011, Hertel 2011). One consequence of these events is an increase in competition for land within many countries meaning it may be more difficult to find land available for reforestation (Smith *et al.* 2010).

There have already been some consequences. In 2007–2008 the price of rice and other commodities increased sharply leading some producer countries to cease exporting in order to maintain supplies for their own populations (*Economist* 2008). This meant importing countries were suddenly confronted by the possibility of food shortages. In some places this led to civil disturbances and Lagi *et al.* (2011) linked variations in the price of food over the period from 2007 to 2011 with the timing of civil disturbances in North Africa and the Middle East over that same period. In addition, some countries with limited areas of agricultural land have also begun to seek future food security by actually purchasing land in other countries. Perhaps inevitably, the rising demand for agricultural land has led to some large financial investors 're-discovering' the agriculture sector, as well as the potential importance of biofuels and the carbon market, and entering the market for agricultural lands (Hertel 2011). The global extent of these foreign land acquisitions is not known although the World Bank estimates 57 million ha were purchased between 2008 and 2009 when the food price index began to rise. The NGO Oxfam estimates the overall total could be even as high as 500 million ha (*Economist* 2011, Pearce 2012). This scramble for agricultural land has been described as a 'land grab' (Box 1.4) and has obvious implications for those interested in forest restoration and especially for programmes seeking to undertake reforestation on a large scale (including that they might be seen as land grabbers themselves).

The amount of land actually needed for future food production is now a matter of some debate. The debate concerns not only: (i) the magnitude of the projected demand; but also (ii) the capacity of agricultural systems to meet that demand. Forecasts of the magnitude of the future demand depend on assumptions that must be made about future population growth rates, consumption rates (including of meat), the proportion of agricultural land actually used for food production (rather than for other purposes such as biofuels) and the productivity of this land when used for various foods. Smith *et al.* (2010) used eight different global models to assess how

**Box 1.4 The land grabbers**

Large-scale buying of agricultural land by foreign entities has been underway for some time but the rate appears to have sharply increased after the rise in global food prices in 2007–2008. It has occurred in many countries but seems to have been especially prevalent in Sub-Saharan Africa. The buyers have been state-owned enterprises from countries such as Saudi Arabia and China as well as multi-national agribusinesses and others who have simply sensed an investment opportunity (Pearce 2012). In most cases host governments have apparently welcomed the investments because of benefits such as export income, economic growth and infrastructure that they assume will flow from such projects. Local governments and, sometimes, local leaders have often been supportive. Both the investors and host governments have commonly assumed customary-owned land is being under-utilised.

Many of these large land-acquisition projects have run into problems or failed because of disputes over land ownership. Either customary ownership was not recognised by the State when the land was sold to the foreign investor or it was assumed it was not being used because there were no houses or permanent crops present. Sometimes land was sold by a supposed 'leader' on behalf of the community without the community's knowledge about what was being proposed or without their agreement. German *et al.* (2011) describe the variety of problems that have occurred in parts of East Africa. Most of these are currently ongoing.

Deininger (2011) has reviewed this wave of agricultural land buying and concluded that many of the buyers were domestic investors and not just foreigners. He also found that it has mainly occurred in countries with limited land rights protection and weak frameworks for consultation. This, together with unclear or duplicative institutional responsibilities, neglect of social, economic and environmental issues during the planning phase of projects as well as failure to monitor and enforce agreements has often led to 'very negative impacts on the ground'. He notes that areas not suited to large-scale agricultural expansion need to be protected from encroachment. While most land grabs have been undertaken to carry out large-scale agricultural developments some have also been done for the sake of large timber plantation projects (Gerber 2011, German *et al.* 2011, Kroger and Nylund 2012).

much land would be needed for crops and pasture between 2010 and 2050. Most models suggested an expansion of cropland and grassland would be needed to meet demand and that this would be at the expense of 'unmanaged' natural forest. However, there was considerable uncertainty about the projections and also about regional differences. Nor was it clear how much marginal or degraded land might be used in preference to clearing intact natural forest. Much depended on the models used and the assumptions made, especially about the demand for land to produce biofuels.

A different conclusion about the magnitude of the future demand was reached by Ausuble *et al.* (2013). Rather than the demand increasing, they concluded that the global demand for land for food production has now peaked and that 145 million ha of agricultural land is likely to be relinquished and become available for other uses over the next 50 years. Savings could also be made in the way food is transported and stored. If true, the overall level of future competition for land – especially marginal and degraded lands – may be modest and at least some of this relinquished land would almost certainly be available for reforestation. However they, too, cautioned that their forecast is based on forecasts of food production and consumption and is also vulnerable to unexpected demands for land for uses such as biofuels.

A second way of exploring the demand for agricultural land is to consider the capacity of agricultural systems to meet any demand. Just how might future food supplies be enhanced by improvements in agricultural productivity? Part of the debate concerns the role of agribusinesses in securing food security and whether small farms will – or should – also have a significant role to play. Collier (2008) points to the success of large-scale agribusinesses in Brazil and argues these types of agricultural systems will be essential in meeting future food requirements (although many are also interested in growing energy crops for biofuels rather than food). The advantages of such large-scale industrial enterprises are that they can make the best use of new technologies and combine these with the efficient use of capital and market chains. The opposing view is that the output per unit area from small farms is often greater than from larger farms (Cornia 1985). Further, opponents of large agribusinesses note there is considerable scope for many smallholders to increase productivity if only they had access to appropriate technologies to improve the efficiency of fertiliser and water use (Foley *et al.* 2011, Hertel 2011). A number of researchers have concluded that in many cases productivity gains in the order of 25 per cent might be possible (Pretty *et al.* 2006, Deininger 2011, Tilman *et al.* 2011).

A significant additional advantage of assisting smallholders to improve their agricultural productivity is that the benefits are more likely to be more equitably shared than is the case with food produced by larger-scale farming systems (De Schutter 2011b). A reliance on large-scale agribusiness also tends to mean an over-reliance on export crops and a greater risk of rural communities being deprived of their customary lands. The choice between large- and small-scale approaches is, of course, an artificial one and it is highly likely that both approaches will be used in future. Indeed, there is no reason why these two approaches could not be complementary (Wegner and Zwart 2011).

What is the significance of this debate for reforestation? The importance of this lies in the different types of landscape generated by large agribusinesses or small-scale agriculture systems. If large-scale agribusinesses become more widespread the resulting landscapes will tend to be more homogenous and standardised than where there is a mosaic of smaller farms. Trees are more difficult to fit into these industrial farming systems than they are into small-scale farms where a wider variety of crops are often used. On the other hand, most landscapes used by industrial farming groups have areas of land less suited to cropping and some reforestation may be possible on steeper areas or along riverine strips within the cropping areas. Whether this occurs or not will depend on whether farm managers seek to optimise short-term productivity or achieve longer-term sustainability.

Reforestation is more likely to be possible in landscape dominated by small-scale farming systems. Many of these farmers will be interested in growing more than one crop and perhaps practicing some form of agroforestry. Provided they have tenure, some of these farmers may also be interested in reforestation on part of their land to diversify their incomes and reduce risk (Kremen and Miles 2012). These trees may include timber trees as well as trees able to produce a variety of goods such as fruit, resins, medicines, etc. This tree planting might take the form of boundary strips or small forest patches. Much will depend on the size of individual landholdings and on the level of household revenues (see below).

Overall the key conclusion is that, in both types of agricultural systems, there is scope for balancing the need for food production with forms of reforestation that enable other economic and environmental gains to be made (McNeely and Scherr 2003, Vandermeer and Perfecto 2006, Fischer *et al.* 2008, Meyfroidt and Lambin 2011). Some trade-offs may be needed but these will depend on the type, extent and spatial distribution of any reforestation. The nature of these trade-offs together with the policies needed to encourage reforestation will be discussed further in subsequent chapters.

### *Poverty*

Poverty is another factor potentially limiting the extent of reforestation. Many degraded areas are occupied by poor people but they are rarely able to afford to take the longer-term view that reforestation requires. There are a variety of definitions of poverty but the common elements are that poor people have limited access to resources, including financial resources, and are highly vulnerable to unexpected changes in economic or environmental conditions (World Bank 2001, Fisher *et al.* 2008, Roe 2008). Many poor people are necessarily risk averse. Such people cannot afford to use land available to them for anything other than food production and cannot afford either the initial costs of tree planting or the wait until benefits are generated.

But there are degrees of poverty and the empirical evidence shows that some farmers with relatively modest incomes do undertake some forms of tree planting under certain conditions (Holmgren *et al.* 1994). As noted above, these plantings might include various types of agroforestry or farm boundary plantings and help diversify the range of products produced on farms and reduce risks. They can also be attractive because they take little effort to maintain once they have been established. But there are constraints and farmers with very small farms who are entirely dependent on the produce or income being generated on their land are less likely to engage in reforestation.

### *Uncertain systems of land tenure and property rights*

Tenure is commonly defined as being the relationship, whether legally or customarily defined, amongst people or communities with respect to land (FAO 2002). The forms of tenure are crucial in defining who can access and use land for various purposes.

Tenure can vary in its duration (short-term or indefinite) and assurance (i.e. the extent to which it is recognised by the State and/or the community). The importance of tenure is obvious: most people intending to plant trees will not do so unless they can be reasonably confident that they will benefit – financially or otherwise – from doing so.

Since trees take so long to grow compared with most other crops, then tenure must be measured in decades rather than simply a year or two if it is to be effective. There is considerable empirical evidence from a variety of locations and socio-economic situations illustrating the importance of tenure for reforestation (Holmgren *et al.* 1994, Oviedo 2005, Fenske 2011, Kassa *et al.* 2011).

Tenure and property rights are often taken for granted in most developed countries where land managers either own their land outright or have long-term leases over it. It is less widely recognised in many developing countries where governments have frequently failed to recognise customary forms of land ownership and resource management. Even when landholders have tenure, the title may be uncertain and there may be no central land register. Both factors can lead to conflict. There is some evidence that this is changing (White and Martin 2002, Wily 2005) but until tenure and property rights are common and uncontested, reforestation will remain unattractive for many land users.

Of course the provision of land tenure does not necessarily mean that land will be reforested. Sikor and Muller (2009) have reviewed some of the recent history of land reform and argue that problems often arise when land allocation is a largely top-down process that ignores the aspirations of community members for whom tree planting is often of second-order importance. Other factors such as alternative economic crops, cultural preferences and technical capacities will also be important in determining choices. In short, the absence of tenure will limit the likelihood that reforestation will occur but tenure alone will be insufficient to ensure that reforestation actually does take place.

The lack of tenure is usually not an impediment for large industrial reforestation projects. This is because these are undertaken by governments or by companies who have secured some form of legal control over land, often, in the case of large timber companies, with government assistance. But, even here, problems can still occur. One such problem is when the relationship between a company and a government breaks down or when the ownership of the land they have acquired is contested. Such can be the case when customary owners dispute a government's right to award control of what they consider is their land to a third party. In these circumstances sabotage and the deliberate lighting of wildfires to burn plantations can occur even though the company has 'legal' tenure.

Finally, mention must be made of the seemingly paradoxical fact that tree planting can sometimes be prompted by uncertain tenure. This can occur when trees are planted as a way of asserting land ownership claims. Thus farmers might plant trees along their farm boundaries to define the extent of their land claim. While the practice is common it is essentially a localised response and is unlikely to be significant for large-scale reforestation.

## Factors potentially favouring future reforestation

In contrast to these disincentives there are also a number of trends likely to make widespread reforestation more attractive.

### *Changing public perceptions about the need for reforestation*

There is increased interest amongst the public about environmental issues and about the need to combat forest and land degradation. This is likely to make it easier to undertake reforestation on a larger scale. Evidence of this increased interest comes from several sources. One is the dramatic rise in the number of conservation organisations across the globe in the second half of the twentieth century. For example, the IUCN was formed in the late 1940s to bring government conservation agencies and non-government conservation bodies from across the globe together. After a modest start it now has over 1,000 government agencies and conservation groups as its members. Its growth has been matched by the growth of a large number of other, more specialised, conservation groups working at international or national levels. A more explicit piece of evidence comes from a recent global survey of attitudes towards pollution and environmental issues (Table 1.5). This found concern was widespread in both developed and developing countries. Interestingly, the level of concern was lower in some African countries where it appeared people were more worried about inequality and other issues than environmental matters. The survey covers attitudes to issues other than just deforestation but is indicative of widespread environmental concerns.

It can be difficult to translate attitudes into action. A recent analysis of changes in national forest cover in relation to national wealth suggests actual reforestation is more likely in more prosperous societies (Figure 1.2).

*Table 1.5* Percentage of respondents in different countries surveyed by the Pew Global Attitudes Project in 2007 nominating environmental problems as a top environmental threat (Pew Research 2007)

| *<20%* | *21–30%* | *31–40%* | *41–50%* | *51–60%* | *>60%* |
|---|---|---|---|---|---|
| Lebanon 13 | Turkey 27 | USA 37 | Bolivia 42 | Canada 54 | Sweden 66 |
| Pakistan 18 | Jordan 30 | Poland 33 | Brazil 49 | Argentina 53 | China 70 |
| Ethiopia 7 | Kuwait 22 | Egypt 40 | Chile 44 | Peru 55 | Japan 70 |
| Ivory Coast 14 | Palestine 28 | Morocco 31 | Mexico 45 | France 52 | South Korea 77 |
| Kenya 17 | Israel 26 | Indonesia 32 | Venezuela 45 | Italy 51 | |
| Mali 19 | Bangladesh 30 | Malaysia 37 | Britain 46 | Ukraine 57 | |
| Nigeria 17 | Ghana 22 | | Germany 45 | | |
| Senegal 13 | South Africa 22 | | Spain 46 | | |
| | Tanzania 24 | | Bulgaria 45 | | |
| | Uganda 22 | | Czech Republic 49 | | |
| | | | Russia 43 | | |
| | | | India 49 | | |
| | | | Slovakia 50 | | |

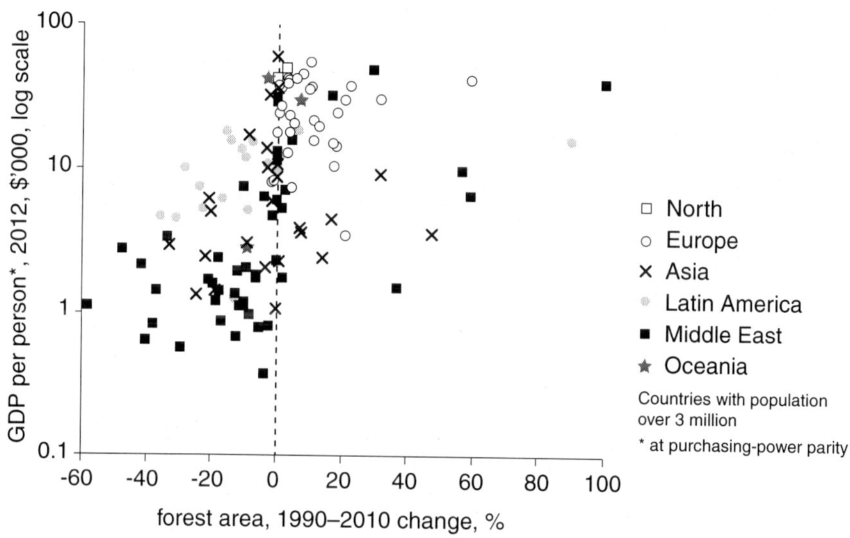


*Figure 1.2* Change in forest area 1990–2010 in relation to GDP per person in various regions of the world (redrawn from *Economist* 2013).

While this may reflect a demand by citizens of these richer countries to restore and protect forests it almost certainly also reflects the fact that richer countries do this partly at the expense of poorer countries. That is, richer countries increase pressures to harvest timber from poorer countries. In some cases this can contribute to increased deforestation in these countries.

But it would be wrong to interpret environmental concern as being an issue concerning only wealthier countries. Many poorer countries are now strengthening their environmental protection laws and increasing the extent of protected areas within their national boundaries. Steinberg (2009) points to the substantial recent growth of environmentally focussed NGOs in many developing countries and their demands for improved environmental management. Some of the countries reporting the highest percentages of citizens actively involved in environmental activism were in developing countries. A number of these are supported financially and intellectually by companion groups in developed countries and have gone on to act as community monitors of environmental quality. Many have become lobbyists promoting higher standards and demanding policy responses. Other indications of changing public perceptions are the success of programmes such as the UNEP campaign to plant a billion trees which quickly achieved a total of 12 billion trees worldwide (Coll 2011). One consequence of these changing public perceptions has been the national responses outlined in Table 1.3. Interestingly, this increased public interest in reforestation is not aimed at creating industrial-style timber plantations but in developing forms of reforestation to achieve broader goals such as biodiversity conservation and watershed protection.

A particularly interesting recent survey indicating how public attitudes are changing was one carried out in China (Cao *et al.* 2009). A total of 5,000 people were asked about their views on environmental degradation and the need for conservation. Around 91 per cent believed environmental quality had degraded over the preceding decade (compared with 44 per cent holding this view ten years earlier). A large majority (78 per cent) supported the government's Grain for Green reforestation programme and 73 per cent said they were personally willing to donate money for conservation purposes. There was a strong link between environmental attitudes, education and income level and those living in urban areas were more concerned than those living in rural areas although rural people in parts of western China where environmental degradation has been severe were also particularly concerned. The authors concluded there had been a major change in public attitudes in recent decades and widespread agreement of the need for greater effort in environmental conservation.

### *Improving markets for forest goods and ecosystem services*

A second trend favouring increased reforestation has been an improving market for forest products and ecosystem services provided by forests. The trend towards growing trees in plantations to supplement the timber produced from natural forests has been underway for many years but there is now increased interest in reforestation to supply markets for ecosystem services. Again, reforestation to stabilise hill slopes and protect watersheds is also something that governments have promoted for some years but markets are now being developed to pay forest owners who hitherto have provided these services without financial compensation. Some of the earliest markets involving payments for ecosystem services developed in

Costa Rica (Wunscher *et al.* 2008, Daniels *et al.* 2010) but the concept has become more widely accepted and is now being explored in a variety of countries. The services included watershed protection, biodiversity, pollination and, potentially the largest of all, carbon sequestration (Wunder 2005, Wunder and Wertz-Kanounnikoff 2009). In the past there has been a tendency to assume that any type of reforestation will be able to generate these services but there is increasing recognition that this is not the case and that more specialised forms of reforestation will be needed for particular purposes.

The underlying basis of these markets and some of their limitations have been outlined by Wunder (2005). Amongst the limitations are the potentially high transaction costs involved in dealing with large numbers of smallholders and ensuring an equitable distribution of payments. There is also concern about whether the payments will be sufficient to overcome the opportunity costs of reforestation. This means this new market may be sufficient to encourage reforestation in some situations where agriculture is marginal but not in others where agriculture is highly profitable.

### *Increasing urbanisation and out-migration*

A third factor potentially favouring increased reforestation is the global trend towards urbanisation. Large numbers of people are leaving the countryside and moving to live in towns and cities. A number are also involved in out-migration to work in an overseas country (Kato 1994, Rudel *et al.* 2002, Aide and Grau 2004, United Nations 2011). In some cases such as Nepal it is largely the males who are leaving, leading to a feminisation of the residual workforce (Don Gilmour; pers. comm.). In some locations this means rural population densities are beginning to stabilise while in others they mean rural population densities are actually dropping. Figure 1.3 shows some regional trends. The consequences of these population movements depend on both ecological and socio-economic circumstances. In some locations natural forest regrowth is occurring where farms have been abandoned (Figure 1.4). The rate at which this is occurring and the composition of these secondary forests depend on how long the area has been farmed and how easy it is for plants to recolonise the site. In some places where recolonisation is able to occur, forests resembling the original forest are able to develop. In others where recolonisation is more difficult, rather more depauperate forests develop which lack some of the more slowly dispersed species.

But urbanisation does not necessarily lead to land being abandoned. In many cases a departing farmer may sell or lease their land to a neighbour. These farmers may continue to use the land for cropping but some may find their enlarged landholdings enable them to concentrate cropping on their better land and contemplate reforesting steeper or poorer quality land. The type of reforestation they carry out depends on whether there is

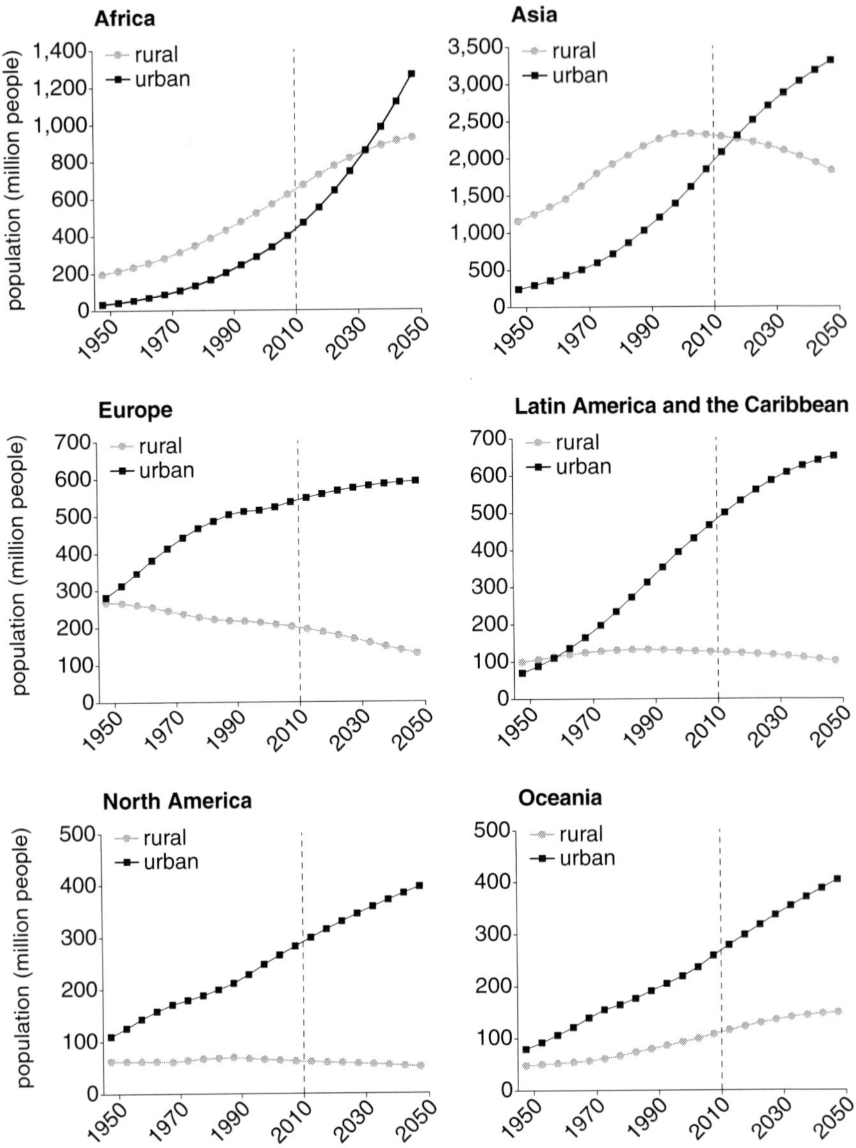


*Figure 1.3* Expected changes in rural and urban populations between 1950 and 2050 in various regions of the world (United Nations 2011).

*Figure 1.4* Forest regrowth on former agricultural fields in southern Sweden abandoned perhaps 100 years ago. The stone wall was used to enclose the field.

a strong market for goods such as timber or whether the landholder is undertaking reforestation for other purposes. In the former case they may use industrial-style timber plantings but in the second case a rather more varied range of reforestation methods might be used.

***Legal obligations***

A fourth factor likely to lead to greater reforestation is the legal requirement for owners of certain types of land to do so. There are two situations where this can be the case. One is where governments have declared that all land in a certain location must be reforested. For example, the Chinese government's Grain for Green programme aims to reforest land on slopes greater than 25 degrees and has a funding mechanism in place to do so (Xu *et al.* 2004, Bennett 2008). Another situation where reforestation may be obligatory is on former mine sites. In this case the mine owner is obliged to rehabilitate the site once mining is complete. There is often considerable uncertainty over just what type of reforestation is acceptable under these situations and the circumstances under which legal responsibilities can be relinquished. Since a common objective is to create long-lived and self-sustaining forests, simple plantation monocultures of the type used in industrial forestry operations are usually not appropriate in these cases. The topic is discussed further in Chapter 4.

### *The need to combat global climate change*

The final reason for thinking that reforestation might be more attractive in future is because of the widely acknowledged role of forests in sequestering and storing carbon.

Agricultural clearing has released large amounts of carbon once stored in forest biomass (Houghton 2013). Reforestation, especially of degraded lands, is a way in which some of this might be taken up and stored once more in plant biomass. The areas to be used include present degraded areas but others may become available in future if climate change shifts the present distribution of the world's food production areas (Schmidhuber and Tubiello 2007).

Estimates of the potential of new forests to store carbon in biomass range from 1 to 19 t C $ha^{-1}$ $yr^{-1}$ depending on the type of reforestation and where it is done (Richards and Stokes 2004). However, there is a difference between what is technically possible and what is economically feasible. The potential also depends on the carbon market price as well as on reforestation establishment and transaction costs (Benitez *et al.* 2007, Cacho *et al.* 2008, Golub *et al.* 2009, Birch *et al.* 2010). One study that highlighted how even modest carbon prices could generate economically viable reforestation over very large areas was that carried out in temperate Australia by Paterson and Bryan (2012). This showed it could be more profitable to establish trees than continue with agricultural production (cereal cropping and sheep grazing) over one-third of the area (two million ha) once the carbon price exceeded the modest level of US$ 27 per t $CO_2$-e and was more profitable over the whole area once the price exceeded US$ 58 per t $CO_2$-e. The results were sensitive to variations in the establishment costs and discount rates and the authors were careful to point out that policy decisions on any carbon sequestration–food production trade-off will necessarily involve consideration of a variety of other factors such as landholder preferences, changes to regional water resources and impacts on salinisation and biodiversity. However, the study demonstrates just how attractive reforestation could become if an appropriate carbon market is established.

A second area where climate change may increase the attractiveness of reforestation concerns the need to maintain the current protected area network. Many protected areas represent islands in a 'sea' of agricultural crops. A changing climate may mean that some of these are no longer suitable habitats for certain species. But without altitudinal or latitudinal corridors these species will be unable to migrate and will become locally extinct. In such cases appropriate forms of reforestation may enable these species to respond to the changing environment and persist in the landscape.

This highlights the fact that perhaps the main role of trees and forests in helping to combat climate change will probably take place at the landscape level. Large new blocks of forest will store considerable amounts of carbon but the maintenance of biodiversity and food security will depend on the establishment of diverse and multi-functional landscapes in which patches of restored forests will play a significant role (Scherr *et al.* 2012). The types of forest needed, their locations within these landscapes and the extent of reforestation needed in these various areas will be discussed in subsequent chapters.

It must be said that, as well as providing opportunities for reforestation, climate change also presents a significant risk as well. At the time of writing a collective global response to reduce the human activities driving climate change has not yet developed. This means there is increasing likelihood that major changes to global ecosystems and biodiversity will take place (Bellard *et al.* 2012). Agricultural systems will be amongst the ecosystems affected (Schmidhuber and Tubiello 2007). If global agreements can be reached then the changes to these systems induced by climate change may be tolerable. But, if not, then large-scale human migration, threats to food security and other changes induced by climate change could severely limit any opportunities for large-scale reforestation.

## Risks associated with large-scale reforestation

Tree planting is usually seen as a 'good thing' and the benefits seem obvious to foresters and ecologists. Nevertheless, there are some potential risks and it is necessary to identify these at a relatively early stage in a book such as this. These risks include:

- Reforestation will displace large numbers of customary land owners whose tenurial claims are not recognised by the government or who have not agreed to having their lands used for reforestation.
- Natural forests (including undisturbed as well as regrowth or secondary forests) may be cleared and replaced by planted forests. Such replacement may have a major impact on regional biodiversity.
- Large-scale reforestation will convert a heterogeneous agricultural landscape mosaic into a simplified and homogenous landscape dominated by recently planted trees. The effects of this will be even more negative if, as is usually the case in industrial plantations, the planted forests contain trees of a small number of species growing in structurally simple stands. This homogeneity will limit the capacity of many native biota to move across and persist in the landscape.
- Large-scale reforestation will increase evapo-transpiration and adversely affect regional water resources.
- Reforestation will over-emphasise the use of exotic species at the expense of native species and that some of these planted species may become invasive.

In short, the social and ecological consequences of reforestation depend on what type of reforestation is undertaken, its spatial context and how it is planned and managed. Ways of avoiding or minimising these risks and some of the trade-offs to be made, especially in multi-functional plantations, will be discussed in subsequent chapters.

## Conclusions

Large areas across the globe have been deforested and degraded. This has had adverse consequences for human livelihoods, future food security, water resources and biodiversity conservation. There is increasing interest in reforestation in order to restore ecosystem functioning and produce ecosystem services rather than just produce timber.

There is a need for reforestation to be carried out on a large scale because of the extent of land degradation. Many desired outcomes are scale-dependent. These include the restoration of hydrological cycles as well as the maintenance of the populations of many wildlife species.

It has proved to be difficult to quantify and map the world's degraded areas even though these may still be increasing at the rate of 2.9 million ha each year. It is also difficult to estimate how much of such land might be available for reforestation although recent estimates put the area at around two billion ha.

Any future reforestation will take place in a rapidly evolving world. It may have to accommodate a rising demand for agricultural land needed to satisfy a substantial increase in the need for food. If so, this will make competition for land for reforestation more intense. At the same time public concern about mounting environmental problems, markets for ecosystem goods as well as forest products and improving systems of tenure and property rights are likely to make reforestation easier to carry out than in the past.

This suggests it should be possible to design new multi-functional landscape mosaics that provide food as well as forest-based goods and services. These new landscapes could include production forests as well as protection forests and might be established by government agencies and large industrial growers as well as smaller landholders.

A variety of types of reforestation will be needed to meet these various objectives and the relatively simple forms developed to satisfy the needs of industry will not necessarily be sufficient for purposes as diverse as watershed protection, carbon sequestration or biodiversity conservation.

There are risks involved in any large-scale enterprise. But much has been learned in the past 100 years – both positive and negative – about how to undertake reforestation over big areas. The next chapter will review some of these lessons.

# 2 Lessons from the last hundred years

## Introduction

Although the twentieth century has been marked by an unprecedented degree of global deforestation it is also true that a large amount of reforestation has occurred over the last 100 years. It might be expected that reforestation only becomes attractive once a country has lost a certain proportion of its original forest area. But this is not always the case; some reforestation is now being carried out in countries that still have significant areas of natural forest as well as in those where deforestation has been severe. This chapter explores why reforestation has been undertaken in the past and what lessons can be learned from these experiences that might be used in future reforestation programmes.

## Establishing tree plantations

Tree planting for commercial purposes is a relatively unattractive land use activity compared with many alternatives available to a landholder because of the high initial costs and the long delay before the receipt of any financial returns. Given these unattractive features, why has it been done? The most obvious reason is because natural forest areas have shrunk and those areas remaining are unable to now meet the rising demand for forest products. Consequently, there is an improving market price for forest products. Plantations have a number of particular commercial advantages. One is that they are usually more productive than most natural forests. This is because sites can be prepared by ploughing or mounding, the soils can be fertilised and competition within the plantation can be managed by weeding and thinning. Eucalypt plantations are a good example. Many eucalypt plantations have mean annual increments (MAIs) $>20\ m^3\ ha^{-1}\ y^{-1}$ while the MAI of a typical natural eucalypt forest may be $<6\ m^3\ ha^{-1}\ y^{-1}$. Similar results can be found for most commonly used plantation species.

A second advantage offered by plantations is that the rising demand for forest products is often concentrated on certain specific products rather than on trees in general. Examples include species suitable for high-quality timber production or those suitable for wood pulp production having specified physical properties. It makes commercial sense to grow just these preferred species in monocultural plantations rather than having to select these preferred species from the wider variety present in most natural forests.

A third advantage of plantations is that these can be established in areas close to roads or markets. This makes transport costs much lower than from most natural forests, many of which are now found in relatively isolated locations or in steep mountain areas. All these factors mean the profitability of tree plantations can be high.

Plantation establishment has a long history. There are early accounts of various species being grown in plantations in the Middle East in ancient times (Evans 2009) and China has a rich silvicultural history involving tree plantings dating back several millennia. Some silvicultural texts on tree planting were written during the Han dynasty around AD 200 (Menzies 1996). In Europe efforts were made during the seventeenth century to encourage tree planting to meet the demand for naval timbers that could not be met from existing forests (Schama 1995, Evans 2009). Extensive reforestation was also practiced in Japan from the eighteenth century during the Tokugawa period (Totman 1989). This followed a period of extensive forest loss when it became clear that the remaining natural forest would be unable to meet the demands being placed on them for goods as well as ecosystem services. Finally, the highly valued and popular teak (*Tectona grandis*) has been grown in plantations in various parts of Asia for more than 500 years (Evans 2009).

Despite this long history and the merits of plantations, it was probably not until the twentieth century that industrial plantation establishment became widespread. This was probably because countries either still had sufficient natural forests or were able to import timber from elsewhere. The international timber trade boomed in the twentieth century but, at the same time, a number of countries also began thinking of ways of becoming more self-sufficient in forest products. Statistics on the extent of planting in the early years of the century are surprisingly difficult to find despite the fact that plantation areas were expanding. It was not until 1967 when FAO organised a symposium on Man-Made Forests and their Industrial Importance that some global data began to emerge. Based on documents provided at the meeting, it was estimated there might have been about 91 million ha of plantations established around the world by that time (Logan 1967). Less than 20 years later Spears (1983) estimated these had grown to cover 100 million ha (of which 30 million were in China).

By 2010 the area of plantations had more than doubled to 264 million ha with the biggest plantations being found in the Asia-Pacific region (Table 2.1). This represents about 5.4 per cent of agricultural land (i.e. arable land, permanent crops and permanent pastures). The annual increase in plantation areas was around 2.1 per cent over the period 2000 to 2010. At the same time natural forest cover continued to decline at an annual rate of about 0.13 per cent (Table 2.2). This means that over the 20 year period

*Table 2.1* Areas of planted forest and expansion between 1990 and 2010 (FAO 2011)

| *Region* | *Area in 2010 (000 ha)* | *Annual change (%) 1990–2000* | *Annual change (%) 2000–2010* |
|---|---|---|---|
| Africa | 15,409 | 1.06 | 1.75 |
| Asia and Pacific | 119,884 | 2.02 | 2.85 |
| Europe* | 69,318 | 1.01 | 0.60 |
| Latin America and Caribbean | 14,952 | 1.79 | 3.23 |
| Near East | 15,082 | 1.27 | 1.49 |
| North America | 37,529 | 4.13 | 2.46 |
| World | 264,084 | 1.88 | 2.09 |

* including the Russia federation

*Table 2.2* Changes in area of plantations (protective and productive) and natural forest between 1990 and 2010 (FAO 2011)

| | *Area (mill. ha)* | | | *Annual change (%)* | |
|---|---|---|---|---|---|
| | *1990* | *2000* | *2010* | *1990–2000* | *2000–2010* |
| Plantations | 178.3 | 214.8 | 264.1 | 1.88 | 2.09 |
| Natural forest | 4,168.4 | 4,085.1 | 4,032.9 | –0.20 | –0.13 |
| Plantations as percentage of natural forest | 4.2 | 5.2 | 6.5 | | |

between 1990 and 2010 natural forest shrunk by 135.5 million ha while plantations increased by 85.8 million ha. By 2010 plantations had grown to represent 6.5 per cent of the area covered by natural forests. Carle *et al.* (2009) estimate these planted forests were then probably generating two-thirds of the world's industrial roundwood. The highest rates of increase in plantation area were in the Asia-Pacific, Latin American and Caribbean regions.

The silvicultural methods used to establish timber plantations have gradually evolved over time and now share some common features. Thus nursery methods, seedling standards and planting densities are much the same in many places. Likewise, a remarkably small number of species are now used in most plantations. But the history of reforestation has been different in different countries. Likewise, the policies and institutions needed to create these plantations have varied enormously. Some of these approaches are outlined in the following case studies.

## Examples of reforestation practice in the twentieth century

The examples below describe some of the situations in which reforestation took place in the twentieth century. The first three case studies (United Kingdom, South Korea, Algeria) involve reforestation in countries where natural forest cover had been limited or largely absent for some years. The second three (Australia, Brazil, Chile) involve reforestation in countries where considerable natural forest still remained at the time that reforestation commenced.

### Case study 1: Reforestation after centuries of deforestation in the United Kingdom

The United Kingdom provides an example of where substantial reforestation only began many years after most of the original forests had been removed. Britain's once extensive forests were heavily cleared during the early Iron Age and Rackham (1986) suggests half of the original forests of England had been lost by 500 BC. Regrowth forests and some original forest persisted after that time although the cover of England and Wales had probably declined to 7 per cent (and to even less in Scotland) by 1600 (Williams 2006). Forest cover continued to decline until it reached perhaps 5 per cent by the beginning of the twentieth century. The remaining patches of forest mostly persisted on the country estates of wealthy landholders. Periodic attempts were made to encourage reforestation to provide naval timbers, especially at times of military crisis, but these had little long-term impact. Mather (1993) refers to the nineteenth century in particular as a long period of (reforestation) darkness.

However, then the situation suddenly changed. Starting in the early 1920s, a reforestation programme led to a doubling of Britain's forest cover (to more than 10 per cent) by the end of the century (Mather 1993). In Scotland the forest cover rose from 4 per cent to 17 per cent (Mason 2007b). This took a long period of sustained activity and involved incremental increases via a series of sequential funding opportunities rather than from a single extended reforestation scheme. It also occurred over a period when there were marked changes in British society which led to substantial changes in the objectives and nature of the reforestation programme.

The trigger for reforestation was the fact that the First World War devastated the nation's remaining forests. Immediately prior to the war Britain had been importing 92 per cent of its timber, an amount that had increased five-fold over the previous 70 years (Mather 1993). This had led to much discussion about the need for reforestation. After the war the government decided it had to rebuild a timber resource and try to achieve a degree of self-sufficiency (Grundy 2005). An additional objective was to provide rural employment. It was thought this planting could be done without compromising food supplies by targeting marginal agricultural areas and hill grazing lands. A Forestry Commission was established in 1919 and the new Commission set about buying land for plantations on the open market. Over time this meant that most planting was carried out in some of the extensive upland areas of England, Wales and Scotland. Subsequently, this largely strategic objective was replaced by a requirement that reforestation should also provide a satisfactory return on the funds invested. This change replaced the earlier objective of fostering rural employment although employment continued to be important, especially in some of the more remote rural areas.

Planting began in the 1920s and, apart from a period during the Second World War, continued for the next 80 years. The peak occurred in the early 1970s when more than 20,000 ha were being planted each year. Most of the plantings were done using conifers although the natural forests of Britain were originally dominated by broad-leaved hardwoods in the south and conifers in the north. The species used in replanting represented a mix of the native Scots pine (*Pinus sylvestris*) as well as exotics including lodgepole pine (*Pinus contorta*), sitka spruce (*Picea sitchensis*), Norway spruce (*Picea abies*), Douglas fir (*Pseudotsuga menziesii*) and larch (*Larix* spp.). Sitka spruce became the most widely planted species across Britain (Mason 2007a, b).

Despite the apparent success of the programme it began to attract adverse public comment. The criticisms took several forms. One was that the large blocks of even-aged trees were unattractive and lacked any public amenity. A second was that they were almost entirely coniferous and lacked native broad-leaved species. A third was that, on top of these undesirable outcomes, the apparent overall return on government funds was modest (although this was difficult to assess because of the role of the Forestry Commission as both a regulator as well as a plantation manager).

These critiques led to a major change in silvicultural objectives. Rather than maximising timber production, the focus of managers changed to one of managing the new forests for a much more diverse range of outcomes including improved biodiversity conservation and recreational opportunities. This was done by increased planting of native broad-leaved species, especially in southern parts of the country, and by modifying some of the existing plantations. Even-aged monocultures were gradually converted to uneven-aged stands and some conifer plantations were enriched with broad-leaved species (Malcolm 1997, Mason 2007a, b). Techniques to establish new native woodlands entirely for conservation purposes were also developed (Rodwell and Patterson 1994). In addition, policies were changed to enable reforestation to provide more employment in more remote areas, help farmers diversify their incomes and to foster forest-based industries (Grundy 2005).

Over time Forestry Commission plantings gradually declined and the role of the private sector increased. In the early 1950s the private sector was responsible for 26 per cent of total plantings but this rose to 42 per cent in the 1960s and 67 per cent in the 1970s. After the 1980s the Forestry Commission gradually withdrew from further new plantings and began to sell some plantation areas to private owners (Grundy 2005). This involvement by the private sector was assisted by a variety of incentive schemes involving grants and taxation deduction which attracted landholders as well as city investors. The nature of these changed over time (Aldhous 1997). At one stage the tax concessions were especially attractive because the higher the forest owner's marginal rate of tax, the greater the subsidy. Since the highest rate of tax reached 83 per cent the taxpayer only had to supply 17 per cent of planting costs on an after-tax basis. This was a powerful incentive to undertake reforestation. In fact it was so attractive that the concession was later modified because it was criticised as excessively favouring high income tax payers. Subsequent assistance to private growers has taken the form of various grants depending on the species being planted and the size of the plantings (Aldhous 1997).

This national reforestation programme did not make Britain self-sufficient in timber – imports of forest products in 2000 were still about 80 per cent of consumption (Grundy 2005). On the other hand, it did achieve the objective of increasing the country's forest resource and by 2005 plantations had grown to 1.9 million ha of which 0.9 million ha were established by the State and the remainder by private owners (Table 2.3). This represented the largest change in land ownership and usage in Britain in modern times.

### *Conclusion*

Reforestation was initiated by the State after a very long period during which forest cover was low and timber was largely obtained from imports. The private sector became involved after reforestation techniques had been developed and financial incentives made it attractive to do so. But public attitudes changed over the period from requiring goods, such as timber, to wanting the new forests to provide ecosystem services, such as recreational opportunities and biodiversity.

*Table 2.3* Area of planted production forests in selected countries (Evans 2009)

| | *Planted production forest area (000 ha) in 2005* | *Increase 1990–2005 (000 ha)* | *Increase over 1990 total (%)* | *Percentage forest cover (planted and natural) in 2010** |
|---|---|---|---|---|
| Algeria | 12 | 6 | 100 | 1 |
| Australia | 1,800 | 743 | 72 | 19 |
| Brazil | 5,400 | 314 | 6 | 62 |
| Chile | 2,700 | 920 | 53 | 22 |
| South Korea | 1,400 | 616 | 82 | 63 |
| United Kingdom | 1,900 | 40 | 2 | 12 |
| Global | 110,600 | 31,395 | 40 | 31 |

*FAO (2011)

### Case study 2: Transforming a degraded landscape in South Korea

South Korea was heavily deforested during the early years of the twentieth century. Many areas were completely deforested, especially mountain areas, and erosion was widespread and severe. After the end of the Korean War in 1953 some small-scale reforestation was undertaken involving mostly fast-growing exotic species. But in 1973 the government commenced what was to be the first of several large national forest development plans. The aim of these was to reforest the 2.6 million ha of degraded land then estimated to be present (Lee *et al.* 2004, Lee and Suh 2005, Tak *et al.* 2007, Bae *et al.* 2012). The first plan (1973–1978) established 1 million ha of new forest. It was carried out by local forestry associations and county forestry cooperatives which were organised in every village. These received advice and training from the central government. At that time privately owned (i.e. non-state) forests represented 73 per cent of the total forest area. The first plan formed part of a national economic development plan referred to as the 'Sae-Ma-Ul' or 'new community' movement. This was seen as a way of promoting rural development through self-help and involved improving rural infrastructure as well as reforestation. Villagers were strongly encouraged to participate although financial incentives were provided (Bae *et al*. 2012). An initial goal of this programme was to improve rural community livelihoods by growing trees for fuel and other species such as chestnuts for food. But the primary objectives were to create a new forest resource and stabilise hill slopes to prevent erosion. A relatively small number of fast-growing species were used including poplars, *Robinia pseudoacacia* and *Pinus rigida*, as well as chestnuts (*Castanea creanata*).

The second plan (1979–1987) changed the focus from forests able to produce fuelwood to forests able to supply building timbers. During this plan, a further 1.14 million ha of new forests were established. A rather broader range of species were used including poplars, larch (*Larix kaempferi*) and various native pines (*Pinus koraiensis* and *Pinus thunbergii*) as well as the exotic *Pinus rigida*. The third plan (1988–1997) once again changed the emphasis, this time to make greater use of the more valuable commercial timber species such as larch (*Larix kaempferi*) and Korean pine (*Pinus koraiensis*). The emphasis also changed from just production to giving greater weight to creating 'multi-functional forests'. By now Korea was well on the way from being a relatively poor country to becoming one of the world's major industrial economies. It was also finding it cheaper to import more of its timber requirements (Tak *et al*. 2007). At the same time a growing use was being made of the new forests for recreational purposes.

The fourth plan (1998–2007) made sustainable management the focus of attention and substantially increased the diversity of species planted, lifting the number to 78 species. These included 23 timber tree species, several nut tree species and a variety of special purpose species, including shade tolerant and fire resistant species. More emphasis was given to using native species rather than exotics. By this time it was clear than natural regeneration was appearing in many of the nominal monoculture plantations including a variety of *Quercus* species and *Pinus densiflora* (Lee *et al*. 2004). These even-aged stands began to acquire multiple age classes. There is also evidence of certain wildlife species beginning to colonise the new planted forests (Tak *et al.* 2007).

By the early years of the new century South Korean forests had increased to cover 6.44 million ha or 63 per cent of the country. Of this 1.4 million ha were in production forests (Table 2.3). The overall forest biomass had grown from 20.6 Tg of carbon in 1954 to 239.1 Tg of carbon in 2007 (Li *et al.* 2010). These new forests included 1.42 million ha of state-owned plantations, 0.49 million ha of plantations owned by local governments and 4.53 million ha owned by private landholders. Much of the latter is owned by 2.1 million small farmers and households. Most of these privately owned forest plots are very small. Tak *et al.* (2007) describe 63 per cent of private landowners as having forests of less than one ha and 27 per cent having forest plots of between one ha and five ha. Few of these small forest areas are likely to be used for future timber production.

### *Conclusion*

The South Korean landscape has undergone a remarkable transformation within a relatively short time. This has led to forests now covering a very high proportion of the country. The initial reason for undertaking reforestation was to supply timber and other forest products. But South Korean society has changed over this same period and the future value of these forests is now likely to be due to their capacity to provide ecosystem services rather than just forest products.

## Case study 3: Building a 'green dam' to hold back the desert in Algeria

The Algerian experience differs from most other national reforestation programmes of the twentieth century. Unlike most countries, the primary objective of Algeria's early reforestation programme was to limit desertification rather than increase national timber resources. Algeria is a dry country and has always had limited areas of forest. There were probably around five million ha in 1830 at the beginning of the colonial era but these had shrunk to two million ha by 1930. A further one million ha were lost in the years leading to independence in 1962 and the natural forest cover is now about 1 per cent of the land area. Most of these natural forest remnants are found restricted to the mountain areas of northern Algeria. The dominant species in these forests include a variety of *Quercus* and *Juniperus* species as well as *Pinus halepensis* in the warmer semi-arid zone (Stewart 1993). The sad history of forest degradation and loss in Algeria is described in some detail by Zaimeche (1994).

Some reforestation trials were commenced in the late nineteenth century and these continued into the twentieth century even while this deforestation was still underway. Studies suggested that natural regeneration was not sufficient to allow forest recovery and, because of this, planting trials explored the use of cork oak (*Q. suber*), Aleppo pine (*P. halepensis*) and a variety of *Eucalyptus*, including *E. camaldulensis*, *E. globulus* and *E. gomphocephala*. After the Second World War, land rehabilitation entered a new phase between 1954 and 1962 when attempts began to integrate reforestation into the rural landscapes. These programmes were interrupted by the War of Independence with France which ended in 1962.

In the period of post-independence euphoria, reforestation was seen as a popular nation-building activity that could overcome years of degradation and forest loss (Zaimeche 1994). But the momentum of the previous decade had been lost. The technical capacity of the Forest Service had been greatly diminished by many staff leaving and the new government became increasingly authoritarian; rather than involving the community in reforestation, it asserted that it would undertake it and control what was done. A national reforestation programme of 60,000 ha per year was implemented using *P. halepensis* because it is tolerant of a variety of site conditions and is not grazed by goats. However, people came to see the State's reforestation policies as being disadvantageous to their interests and most of the plantings carried out during this period were burned by wildfires (Stewart 1993).

An ambitious and attention-grabbing new reforestation plan was proposed in 1974 when an NGO announced a scheme to create 'a green dam to hold back the Sahara' (Stewart 1993). The proposal was to establish a belt of trees in the steppe area on the high plains of the southern face of the Saharan Atlas, Hodna and Aures mountains (in the 200–300 mm rainfall zone). This belt of trees would stretch between the Moroccan and Tunisian borders, be 20–30 km wide and cover a distance of about a thousand kilometres; the total area would be around three million ha. An NGO initiated the idea but it was subsequently adopted by the Algerian government as a national project. Because of the scale of the project and its national status, planting was not in the hands of the Forest Service but was undertaken by relatively untrained army recruits. Some large areas were planted (mostly with *P. halepensis*) but seedling survival rates were often low. The programme was not at all popular with local people, especially shepherds, because the trees destroyed the natural pastures used for grazing (Vallejo 2005). The government at this time was increasingly seen as authoritarian and unconcerned with local communities (Zaimeche 1994). Perhaps unsurprisingly, wildfires and unregulated felling meant many reforested areas failed or were badly damaged and the scheme was eventually abandoned in 1990.

A revised and rather more modest programme subsequently began but under the supervision of the National Institute of Forest Research. This time care was taken to involve the local community in the scheme. Affected communities were involved in planting and rural development criteria were introduced so that tree planting could be linked with other activities. The diversity of species used was also increased to include a greater variety of native and exotic fodder plants and trees chosen to suit specific site conditions. The target area was revised down from three million ha to 300,000 ha and the original idea of a 'green dam' was changed to become instead a discontinuous strip of irrigated crops as well as fodder, fruit and timber trees. This meant specifically targeting areas with deep soils suitable for tree planting and water resources available for irrigation of agricultural crops (Vallejo 2005, OSS 2008).

### *Conclusion*

Reforestation has been difficult to achieve in Algeria despite a long history of forest loss. A hugely ambitious reforestation scheme to construct a 'green dam' was commenced in a difficult environment with little field testing of methods or species and without the involvement of local communities. Inevitably these people came to see the scheme as a threat to their livelihoods. Without their support the programme failed. The scheme has been replaced by a rather more modest initiative that seeks to generate benefits for those living in the area rather than attempting to simply build a 'green dam' (Stewart 1993).

## Case study 4: Pine plantations in the home of the eucalypt, Australia

Australia set about establishing plantations when it still had large areas of natural forests. These natural forests now cover 149 million ha or 19 per cent of the country although only 9.4 million ha are in public forests where timber harvesting is permitted. Many forests and the woodlands found in drier regions have very low annual growth increments (www.daff.gov.au/forestry). The dominant tree species in most of these forests belong to just two genera namely *Eucalyptus* and *Corymbia* (previously a sub-genus of *Eucalyptus*) although species-rich rainforests forming closed canopies are found in some areas. All of these species produce hardwood timbers. There are some native conifers but these are mostly restricted to smaller areas along the east coast and in the southern state of Tasmania.

Native forests growing on good soils began to be cleared following European settlement with much of this only starting in the second half of the nineteenth century. Forests are the responsibility of the various Australian states and all of them began trials testing the growth in plantations of native and exotic timber species at a relatively early stage. The earliest of these were commenced in the late nineteenth century (Carron 1990). Perhaps surprisingly, many of the species tested were pines, notwithstanding the fact that eucalypts have been widely used in plantations outside Australia. There were several reasons for this. One was that imports of coniferous timbers had begun early in Australia's history complementing the readily available local hardwood timbers and the states were keen to replace these and become more self-sufficient in softwoods. A second was that there was some uncertainty about using the regrowth eucalypt forests arising from the second cutting cycle in native forests. Logs from these regrowth forests tended to have severe growth stresses making sawing difficult as well as end splitting and excessive shrinkage during drying. Hillis (1978) described them as 'new woods' needing more research before they could be used for purposes other than pulpwood. Pine plantations were seen as a way of creating a new complementary timber resource and eventually several species of pine proved to be very suitable for planting in the soils available for plantations (Florence *et al*. 1990). These included *Pinus radiata* in temperate areas and *P. elliottii* and *P. caribeae* in tropical regions. The productivity of these plantations was often >15 $m^3\ ha^{-1}\ yr^{-1}$ (on rotations of 30–40 years) in comparison to <6 $m^3\ ha^{-1}\ yr^{-1}$ for most native eucalypt forests.

The majority of the early plantations were established on state-owned land, most of which was thought to be too marginal for agriculture. One consequence was that some areas of 'less productive' natural forest were cleared and replaced by monocultural plantations of exotic species (McDonald 1993). It was clear that a preferable option would have been to use agricultural land that had already been cleared but, at the time, the financial and political cost of purchasing such land was deemed to be too great. In more recent years, most plantations, especially those established by private industrial enterprises, have been established on former agricultural land.

Small areas of plantations were established in most states before the Second World War but the rate of planting increased after the war and especially after the 1960s. An implicit goal of self-sufficiency in wood products was adopted at a meeting of the states and federal government in 1967 and the target of having an annual national conifer planting rate of 33,000 ha was decided upon (Carron 1990). In 1967 the collective national plantation area was 330,000 ha but by 1987 this had increased to 905,000 ha (Dargavel and Semple 1990).

As noted above, most of the early plantations were established by government forestry departments but, over time, reforestation by private enterprises rather than public bodies became more widespread. At the same time many state governments began selling their state-owned plantations to private enterprises or passing them on to government-owned corporate entities. There had always been some investments in plantations by private timber companies and in 1987 they owned 30 per cent of the total plantation estate. But in the late 1980s and the 1990s changes in the taxation rules led to substantially increased plantation establishment by private investment companies. These were mostly financial companies rather than timber companies and they offered investors the opportunity to invest in short rotation eucalypt pulpwood plantations through 'managed investment schemes'. This proved highly attractive to many city-based investors although a subsequent change in the taxation rules in 2001 led to a dramatic decline in this particular form of plantation establishment (Schirmer and Kanowski 2005). The government continued to promote private investment in plantations when, in 1997, it launched a new policy which aimed to treble the national plantation area by 2020 (Mercer and Underwood 2002, Schirmer and Kanowski 2005, Stewart *et al*. 2011). The purpose was to increase the timber resource in order to create a globally competitive industry as well as improve employment and regional development (Stanton 2001). It is probably no coincidence that this policy initiative took place at a time when many native forests were being taken out of production and placed into the protected area network. Five years after it was launched the national plantation estate had reached 1.6 million ha with much of the increase (640,000 ha) being in eucalypt plantings rather than in pines (http://www.plantations2020.com.au). Most of these are being produced for fibre rather than sawlogs. Subsequent plantings to 2005 have increased the total area to 1.8 million ha (Table 2.3). The supply of timber from plantations now accounts for about two-thirds of Australia's log supply.

These increases in plantation area have not always been welcomed. Conservation bodies campaigned against the conversion of natural forests to plantations but many rural communities were also upset by the establishment of tree plantations on former farmland, even when the land was clearly marginal for agriculture. This was because of concerns that plantations would change the rural landscape and lead to a decline in rural communities and community services such as schools. This attitude remains common in many areas where large-scale reforestation has occurred (Gerrand *et al*. 2003, Stewart *et al*. 2011). This poses a dilemma for governments and means that ways will have to be found to make reforestation a socially acceptable land use alternative otherwise more forest products will have to be imported to meet demands.

For most of the twentieth century the primary motive for most plantation development in Australia was to create a new timber resource. More recently, other environmental factors have become important. Perhaps the most significant of these has been salinity and some significant reforestation has taken place in order to combat dryland salinity, especially in south-western West Australia (Vertessy *et al*. 2003, Booth 2012, George *et al*. 2012). This will be discussed further in later chapters. An increasing number of private landholders and NGOs have also embarked on developing forms of reforestation to generate various environmental benefits (Erskine *et al*. 2005, Erskine *et al*. 2007).

### *Conclusion*

Reforestation trials began in Australia when there was still a large natural forest estate and most of the initial plantations involved softwoods rather than local hardwoods. Government agencies undertook most of the initial reforestation but many government-owned plantations have been taken over by corporate groups in recent years. Commercial plantations have not always been welcomed by NGOs or rural communities although some landholders are becoming engaged in environmental plantings.

### Case study 5: Exotic monocultures in a global biodiversity hotspot, Brazil

Brazil is a heavily forested country, despite extensive agricultural clearing, and might be expected to not need plantations. But, as natural forests were cleared, the remaining forests became more remote and it became far more attractive to grow future timber supplies on former agricultural areas near the coast and closer to markets than using the natural forests of Amazonia (Fearnside 1998, Bacha 2006).

One of the first large-scale tree plantations was undertaken in 1928 in order to produce rubber using monocultures of the native Brazilian rubber tree (*Hevea brasiliensis*). The project was known as Fordlandia and the concession covered one million ha but it ultimately collapsed for a variety of ecological and social reasons. Another early large-scale project was the Jari pulpwood project in eastern Amazonia (Palmer 1991). In this case the intent was to establish a large self-contained pulpwood project by clearing the natural forest and replacing it by a plantation of exotic species. This project also ran into difficulties and a combination of silvicultural problems, mismanagement and political issues meant the original owner eventually withdrew. Despite these early problems, the project has since been revived and continues to this day although in a much modified form.

Meanwhile, more modest and traditional approaches to plantation establishment were undertaken by several government agencies. These aimed to generate sawlogs and timber for charcoal for the steel industry as well as for wood pulp. By 1960 there was around 2.1 million ha of such plantations most of which were established using *Pinus* or *Eucalyptus* (Carrere and Lohman 1996). However, the situation changed dramatically in 1965 when the federal government offered taxation incentives to attract private investors to the forest industry (Bacha 2003, 2006). These programmes were immediately successful and plantation establishment rates increased sharply reaching 400,000 ha per year by 1978. Since 1967 most of these plantations have been established on previously cleared agricultural land because of changes in Brazil's environmental laws (Brockerhoff *et al*. 2012). The incentive programme was withdrawn in the late 1980s although some states and industry groups continued to provide assistance packages to support small growers. In the late 1990s there was then a gradual decline in the annual planting rate to around 150,000 ha as international pulp market prices slumped (Fearnside 1998). Those participating in these plantings included large pulp companies, companies interested in producing charcoal and smallholders. The pulpwood producers made up perhaps 60–70 per cent of plantings (Bacha 2003). By 2005 the plantation estate covered 5.4 million ha (Table 2.3) rising to 7.4 million ha in 2010 (FAO 2010) and Brazil had become a major player in the international pulp markets, supplying more than 50 per cent of the world's eucalypt pulp.

Two types of land were used for plantation development. Most appears to have been previously cleared and low-cost agricultural land although some areas of natural forest were converted as well (Kröger 2012). Large pulp mills require large plantation areas usually within a radius of <100 km to support them and this has sometimes been difficult to acquire. There seems little doubt that in some cases common land has been taken over and small farmers displaced to achieve this although this seems to have been more common on low-cost land (Kröger and Nylund 2012).

The species used in most plantations were largely exotics despite Brazil being one of the most biodiverse countries on Earth. The Jari project initially favoured the exotic species *Gmelina arborea* but in recent years extensive testing followed by tree breeding has led to the use of *E. grandis*, *E. urophylla* and various eucalypt hybrids as well as pines. Other widely grown species include teak, rubber (*Hevea brasiliensis*) and various *Acacia*. Those used for pulpwood are usually grown on rotations of less than 12 years and sometimes as short as 5–6 years (Fearnside 1998, Cossalter and Pye-Smith 2003). Research has led to astonishingly high growth rates in excess of 80 $m^3 ha^{-1} yr^{-1}$ have being recorded in some plantations although routine productivity levels of less than half this are probably more common. Even so, productivity rates like this are still very high by international standards.

Meanwhile, reforestation is also beginning to be carried out on agricultural lands by other landholders including farmers and NGOs. Many of these are making more use of native species. Although some of these plantings have been established to generate income, most have been established to restore ecosystem services and to fulfil a legal requirement that all landholders have a proportion of their landholdings occupied by forest (Wuethrich 2007, Rodrigues *et al*. 2009, Rodrigues *et al*. 2011).

### *Conclusion*

Brazil has developed some very large and highly profitable plantation forests while still having extensive areas of natural forest in which clearing continues. Most of the plantations have been developed by private corporations with the help of various government incentives. The majority are located in the east of the country where exporting is easier. Most use exotic species and, by international standards, have very high levels of productivity. Government policies are now prompting individual landholders to also undertake reforestation to supply ecosystem services.

## Case study 6: Creating a large new commercial forest resource in Chile

In recent years Chile has made a major investment in plantation forests despite still having significant areas of natural forests. Its experience demonstrates both the advantages and disadvantages of large-scale reforestation. The natural forests of Chile presently cover 22 per cent of the country but were once more extensive. Many areas were degraded by clearing and logging with more than 2.5 million ha being cleared between 1920 and 1930 alone. Clearing continued up until the 1980s and most remaining natural forests are now in the temperate southern parts of the country. Native tree species were originally thought to be so slow-growing that regeneration after logging would be impossible, at least from a commercial perspective (Clapp 1995). Under these circumstances plantations were an obvious alternative. The first plantings were carried out in the nineteenth century but substantial plantings only really began in the 1940s when a state-owned development agency began promoting plantation development by private owners. The motive was to supplement the supply of timber coming from the natural forests and to create a new industry based on a plantation resource. Annual plantings were around 6,000 ha in the 1940s rising to 20,000 ha per year in the early 1950s (Lara and Veblen 1993). A variety of incentives were given to growers by the State including tax exemptions, loans and partnerships. Planting rates subsequently slumped as growers lost interest in tree planting because of declining domestic demand (Clapp 1995).

After the 1970s, the State became more directly involved in plantation establishment and state-owned plantations increased in area from only 7,000 ha in 1970 to 366,000 ha in 1985 (Morales 2005). In 1973, following a coup, a military government replaced the civilian government and stressed the role of the private sector once more. Large financial subsidies were given to companies to cover establishment costs. These often exceeded 70 per cent of costs and at one stage even reached 90 per cent (Morales 2005). Favourable taxation treatment was also provided to private companies. Not surprisingly the planting rates by the private sector soared (92,000 ha being planted in 1981 alone) while that by the public sector virtually ceased (Lara and Veblen 1993, Morales 2005).

The outcome has been a massive increase in the plantation area that has tripled the commercial forest resource within a remarkably short time. In 1990 the plantation area had reached 1.45 million ha and it continued growing, eventually reaching 2.66 million ha in 2000. This was made up of 365,700 ha of state-owned plantations and 2.3 million ha of private plantations. Subsequently, most of the government-owned plantations were privatised and the role of the government has become one of overseer and regulator (Morales 2005). Since then the resource has become the basis for a thriving timber industry that includes pulp and paper mills as well as sawmills and veneer mills. Most of the output is exported and by 2000 was contributing 13 per cent of Chile's export income. By 2005 the plantation area covered 2.7 million ha (Table 2.3).

Throughout this period the majority of plantations were created using two exotic species, namely *Pinus radiata* and *Eucalyptus globulus*. In 2001 *P. radiata* made up 74 per cent of the plantation area with the eucalypt contributing only 18 per cent (Morales 2005). The pines are grown on rotations of 20–25 years and the *E. globulus* plantations are grown on rotations of 15–25 years. Native species have not been a significant part of the plantation programme.

The tumultuous political and economic history of Chile during this period has been described in detail by Clapp (1995). But he concludes:

> In a generation, Chile has created one of the world's most competitive forest resources, at a relatively low environmental cost. It is not a natural forest capable of supporting multiple use, but as an economic forest it is almost a masterpiece.
>
> (Clapp 1995)

Notwithstanding Clapp's suggestion that the environmental costs have been low, many have argued that, on the contrary, these have been considerable. Indeed Echeverria *et al.* (2006) argue that most of the new plantations have developed at the expense of native forest. The expansion of the plantation estate has also caused social problems (Gerber 2011). Much of the land used for plantation establishment was once occupied by the indigenous Mapuche people who were the customary owners. Some have still retained their land but a large number were pressured to sell to timber companies or become tenants of major landholders (Clapp 1998). It is often assumed that plantation programmes are able to provide significant amounts of rural employment but only modest employment was created, at least in the early stages of the Chilean programme. Thus employment reached one job in silviculture and utilisation per 39 ha which is much less than the more common figure of one position per 12–20 ha that international experience suggests it might have reached (Lara and Veblen 1993). Most of these jobs were casual and poorly paid as trade unions were strongly discouraged during the period of the military government (Clapp 1995). WRM (2011) argue that there is extensive rural poverty in the areas of the country in which the plantation programme was most concentrated and reforestation has failed to overcome it. In more recent years, as the forests have begun to be harvested, the social benefits may have begun to improve as trees have matured and harvesting has increased (Morales 2005). More smallholders have also begun to grow trees on their land extending plantation ownership beyond the handful of large timber companies (Clapp 1995). Morales (2005) notes there were 5,841 plantation owners with less than 100 ha and 2,880 of these had plantations of less than 10 ha.

It is often argued that the creation of plantations takes pressure off natural forests. This has not occurred so far in Chile although it may do in future. Throughout this period the native forests continued to be used to produce fuelwood and, subsequently, woodchips (Lara and Veblen 1993, Morales 2005). Natural regeneration after such logging is unpredictable because of irregular seeding by many species. Nonetheless, native species are being harvested and sold to Japan as woodchips. Chips from the Chilean eucalypt plantations may take over this market when these reach maturity but it is not clear that logging in natural forests will then cease. Indeed Clapp (2001) suggests new markets in North America may become available, meaning that the plantations have simply supplemented logging in natural forests. On the other hand Gregersen *et al.* (2011) suggest recent moves to protect and rehabilitate natural forests will lead to a cessation of logging in most of these residual natural forests.

### *Conclusion*

Despite having a significant natural forest resource, Chile decided it was more commercially profitable to replace part of this with an industrial plantation forest estate. A major reforestation programme has been undertaken over a long time period. This was mostly done by the private sector although with substantial financial assistance from the government. The initiative is now providing significant export income for Chile. The scheme has generated employment but has also had some significant negative social impacts, especially on indigenous people. It has also caused a substantial loss of natural forest and biodiversity.

## Issues emerging from these case studies

A large number of countries have undertaken some form of large-scale reforestation during the last 100 years but these six case studies are probably representative of the general experience. The main issues emerging are as follows:

- *Objective of reforestation:* the primary motive behind most reforestation schemes has been to rebuild a timber resource in order to compensate for the loss of natural forests (e.g. United Kingdom and South Korea) or, where the country still had extensive forests, to create a particular type of timber resource not supplied by the native forests (e.g. Australia, Brazil and Chile). Self-sufficiency was sometimes a declared additional aim although this appears to have become less important over time. The protective role of these new forests was usually also valued but relatively few countries undertook reforestation on a large scale simply to generate environmental benefits. Of the six case studies above, Algeria was the only one that did so, although this was also an important secondary objective in South Korea. It was only towards the end of the twentieth century that reforestation for largely environmental reasons became more widespread.
- *Land used:* in most cases the land used had already been deforested although in some cases degraded residual forest was also cleared and replanted (e.g. Australia, Brazil and Chile). In most cases the land used was thought to be marginal for agriculture although sometimes higher

quality agricultural land was used in order to assemble a financially viable minimum plantation area (e.g. where plantations had to be within a certain distance from a pulp mill). In some cases land has been purchased on the open market (e.g. United Kingdom) but in other cases there is evidence that it has been acquired forcibly and under controversial circumstances. There were examples of this amongst the case studies (e.g. Chile) but Gerber (2011) has found many other examples from around the world of conflict arising from large-scale plantation establishment.

- *Species used:* most of the species used in countries reviewed in these case studies were exotics and these came from a remarkably small number of genera (e.g. *Pinus* and *Eucalyptus*) although most countries have tested a broader range of species. The most favoured species have been those with the highest growth rates and productivities although there has also been a tendency to use species able to tolerate a variety of site conditions rather than more site-demanding species. A global survey of the species used in plantations supports the conclusion of most of the world's plantations being based on a relatively narrow genetic pool (Evans 2009). This genetic diversity is further limited by the fact that many of the most favoured species have undergone a breeding programme and been hybridised or subjected to genetic modification in order to increase productivity and enhance pulping and fibre characteristics. Many industrial plantations are now beginning to use clonal material.
- *Types of plantations established:* the majority of plantations were simple monocultures. Those grown to produce pulpwood were commonly grown on short rotations, often of less than 10 years, while those grown to produce sawlogs had rotations of 20 years or more depending on whether they were being grown in temperate or tropical environments. At the end of the rotation most of these plantations are completely felled and replanted. Over time a common set of management techniques have evolved to improve productivity and enhance the homogeneity of log sizes. These include the use of tubed seedlings (although some open rooted seedlings have also been used); the use of planting densities of around 1,000–1,500 trees per ha and, in the case of sawlog plantations, prescriptions covering the ways the trees should be pruned and the stands thinned. Weed control and fertilisers have also been widely employed to boost productivity.
- *Plantation owners:* state forestry agencies were responsible for initiating reforestation activities in most countries. By undertaking species trials and developing silvicultural techniques appropriate to the sites being planted they were able to reduce risks and demonstrate the potential profitability of tree growing to timber corporations and others. This helped legitimise reforestation as an acceptable land use activity. In many countries private growers subsequently became involved in reforestation either by establishing plantations themselves or by

purchasing plantations established by state forestry agencies (e.g. United Kingdom. Australia and Chile). South Korea appears an important exception since private landowners were heavily involved in initial reforestation efforts. Most private growers were corporate entities but some smallholders have become involved as well (see further below).

- *Policies used to encourage plantations:* governments have usually used a variety of incentives and subsidies to make reforestation attractive to private growers and overcome the perceived disadvantages of forestry as a land use. These include the high initial costs of reforestation investments, the long delay before any financial returns are received and the need for a critical area of plantations to be established to support the subsequent establishment of industrial processing facilities. This assistance has included cheap land, cash grants and taxation incentives. Help has often been provided in the form of improved regional infrastructures such as new road or rail systems. In most cases there appear to have been relatively few conditions attached to the provision of this support. South Korea is, again, an important exception; reforestation was made part of the national economic development programme and citizens were strongly encouraged to participate, at least in the earlier initiatives.
- *Distribution of benefits from reforestation:* the benefits of reforestation have not always been equitably distributed even though a declared objective has often been to create employment opportunities in rural areas. This has occurred in some cases but in others the employment benefits have often gone to temporary employees from outside the area. Most benefits from industrial plantations appear to flow to growers and processors. The planting stage does provide employment but the need for labour then declines until harvesting and processing begins. Not all rural communities are necessarily in favour of large-scale tree plantings, especially when industrial plantations replace other agricultural enterprises (Australia, UK, Chile and Brazil). In some cases hostility to plantation establishment has led to these plantings being damaged (e.g. Algeria). In the US the public have often been opposed to plantation monocultures and to clear-cutting these at the end of rotations (Fernandez *et al.* 2002).
- *Changes over time:* the objectives of many of the reforestation programmes sponsored by governments have often changed over time in ways that could not be forecast by the original proponents. In many cases this has occurred before the planted trees have even been harvested. The major change has been towards seeking greater environmental benefits (e.g. United Kingdom and South Korea). The change in objectives has sometimes led to monocultures being enriched to enhance biodiversity and monocultures being replaced by mixed-species plantings. In some cases plantations established to provide timber have become more valued as recreational forests and will never be harvested.

## Other lessons from the last 100 years

In addition to those emerging from these particular case studies, a number of other lessons emerge from experiences elsewhere around the world over the last 100 years.

### *Sustainability of industrial plantations*

Not all plantation efforts have been successful. Failure can occur because species are planted at inappropriate sites, because plantations are not protected or because they are not properly managed. This issue is discussed further in Chapter 4. But problems can also develop over the longer term in plantations that seemed initially productive.

#### *Declining productivity*

One that has been widely discussed amongst foresters is the 'second rotation problem'. In this case the productivity of a plantation declines after the first harvest takes place and the replanted trees do not grow as well as those grown in the first rotation. There are several possible explanations that might account for the decline. One is that a gradual change occurs in the mycorrhizal populations or in certain microbial processes that limit growth as the plantation ages. Such changes appear to be involved in the so-called replant problems that occur in some orchard crops, such as peach and apple, and could conceivably affect timber trees (e.g. Isutsa and Merwin 2000). A second possible cause is that the availability of soil water declines; the capacity of plantations to lower water tables is well known (Vertessy *et al.* 2003, Bruijnzeel *et al.* 2005, Van Dijk and Keenan 2007) and could mean that the resources available to second rotation trees are less than those available to trees of the first rotation. A third possible cause is that nutrient availability changes over time. Nutritional problems might arise from the export of nutrient in logs harvested at the end of the first rotation (especially if these are young trees with a high proportion of nutrient-rich sapwood as is the case with pulpwood). Alternatively, it might be due to the trees of the first rotation accessing and fully utilising soil nutrient resources that are then no longer available to trees in subsequent rotations. One example of nutrient limitations inducing a second rotation productivity decline is outlined in Box 2.1. While second rotation problems have undoubtedly occurred Evans (2009) suggests the problem is not widespread and that most timber plantations have maintained or even increased productivity over successive rotations.

**Box 2.1 The second rotation problem in South Australian pine plantations**

Extensive plantations of *Pinus radiata* were established in the sandy soils of South Australia. Over time evidence mounted that the productivity of the second rotation was substantially less than the first rotation (Keeves 1966). There were no obvious signs of stress and the trees did not respond to fertiliser applications once the decline had been recognised (after about age 10 years). Subsequent research showed that, nonetheless, the problem was related to a nutritional deficiency. At the time it was normal practice to clear fell trees at about 40 years, burn the debris and replant. This had the effect of volatilising a considerable amount of nitrogen stored in litter and logging debris. The new trees then suffered a nitrogen limitation at a critical stage of their establishment because of the poor quality soils being used. This limitation and dependence on the supply of nitrogen from the soils diminished after canopy closure (around age 5 years) when the trees were increasingly supplied by recycled nitrogen. This was the reason for the lack of response to fertiliser applied at year 10; it was simply too late. Research by Woods (1990) showed that avoiding burning logging debris and applying nitrogen fertiliser at the time of replanting was sufficient to sustain productivity in second and subsequent rotations.

*Pests and diseases*

A second potential limitation on sustainability is that arising from the fact that diversity has been reduced, meaning that the arrival of a particular insect pest or plant disease could rapidly spread and destroy the whole plantation. The risk is accentuated by the fact that trees in most industrial plantations are closely spaced and grown in monocultures with genetic diversity being further reduced by tree breeding programmes that currently target only productivity. In addition, many plantation areas include sites that are less suitable for the preferred species but which are still planted because they fall within the plantation boundaries. Trees growing on these less suitable sites are likely to be stressed in some way which can make them more vulnerable to pests and diseases than those growing in natural forests. Evidence is accumulating to show that plantations with a single species are more prone to damage by insect (Jactel and Brockerhoff 2007, Nair 2007). Some have suggested that exotic plantation species are more prone to insect pests than native species but, in a survey of the evidence, Nair (2007) found otherwise.

In some cases pests or diseases can cause plantations to completely fail. A well-known tropical example is the damage caused by the tip borer *Hypsipyla* sp. to members of the Meliaceae (Speight and Wiley 2001). But in most cases damage is limited or can be reduced by careful management. The tropics are where the simplification exemplified by plantation monocultures is greatest and yet the tropics also provide an example of a successful long-term monocultural tree plantation. This is rubber which has been grown in large-scale plantations for over 100 years without serious or extensive failures (Lee 1999). It is notable that this success has been accompanied by a considerable monitoring and research effort that is likely to be greater than that undertaken in most timber tree plantations.

### *Role of plantations in reducing pressure on natural forests*

The high productivity of plantations in comparison with the productivity of natural forests has been used to suggest that plantation may be a means of reducing the pressure to clear and replace natural forests (Sedjo and Botkin 1997, Evans 2009).

The historical evidence is that 85.8 million ha of forests have been established since 1990 but that 135.5 million ha of natural forests have been cleared over the same time period. Simultaneously, the protected area network also increased, reducing the forest area available for production. So, the question is whether plantations were at all influential? Would more natural forest have been cleared or would less forest have been added to the protected area estate?

Much depends on the location of the natural forests with respect to roads and markets as well as the topography and soils they occupy. In many parts of the world a natural forest on gentle topography, having good soils and located near a road or market is likely to be at risk of being cleared and used for agriculture. This may occur even when the State owns the land. In such cases the profitability of agriculture is likely to be far greater than that of timber from a sustainably managed natural forest (or, indeed, from a timber plantation). The most probable outcome, then, is that these natural forests would be cleared for agriculture and the possibility of obtaining timber from plantations at another location would be unlikely to influence the land use decision. Angelsen and Kaimowitz (2001) concluded that increased production arising from an improved agricultural technology may make deforestation even more profitable and be the trigger for further deforestation rather than being an incentive for forest conservation.

Would the situation be different where conditions were more marginal for agriculture? Again, much depends on the location, markets and land ownership. If there were already profitable timber plantations established nearby on former agricultural lands it could be financially attractive to expand these into residual natural forest areas, provided these areas were extensive, had a gentle topography and good road access. That is, under these circumstances existing plantations would not necessarily protect the natural forests but, on the contrary, might facilitate the continued expansion of the plantation estate. This appears to have happened in a number of locations, including when: (i) the State owned the land and wished to encourage the private timber industry as well as; where (ii) a state forestry agency was itself undertaking reforestation but could not afford to buy privately owned marginal agricultural land. The pressure to clear natural forest would be highest if there was a good export market for forest products or a large industrial centre nearby (e.g. a ply or pulp mill). Any decision would be strongly influenced by transport costs and it is unlikely that natural forests in steeper or more remote areas would be converted to plantations because these would be too expensive to manage. Cossalter and Pye-Smyth (2003) note FAO estimates that in 2001 around 6–7 per cent of plantations were being established at the expense of tropical forests and specific examples of this from Australia and Chile are given above.

The trade-off becomes more nuanced when public pressure begins to place a greater value on the ecosystem services supplied by natural forests. In this case recognition of the benefits of forests for watershed protection, biodiversity conservation or carbon sequestration make the opportunity (and political) costs of clearing much higher. Under these circumstances plantations may reduce the pressure on natural forests because they provide an alternative source of timber and allow for a win-win outcome.

In the particular case of Chile, Clapp (2001) reviewed evidence that the pressure on state-owned natural forests might decline as the plantations began being harvested. He found the plantations did not protect the natural forests. Rather, when the new plantations began taking over the markets once supplied by the native forests, new markets were found which would allow logging in the natural forests to continue. Where clearing of native forests for plantations has diminished, such as in Australia or Brazil, the reason seems to be largely because of changes in government policies caused by public pressure to value ecosystem services and not because plantations had replaced the timber producing function of natural forests.

In short, the evidence that reforestation helps take pressure off native forests is not particularly strong although as Mather noted (1993: 207) it is difficult to imagine global forests being conserved without some tree planting being done.

### *The increasing role of smallholders*

Most of the reforestation undertaken in the last 100 years was initially undertaken by state forestry agencies or sponsored by governments even though corporate groups subsequently became more active. But the situation began to change in the later decades of the twentieth century when it became clear that smallholders (households and family farms) were also becoming active in reforestation. Of course many had always been engaged in various forms of agroforestry but this began to expand to include farm plantations. The first indication of this activity, at least at a global level, was the report by Del Lungo *et al.* (2006) based on country responses to a questionnaire. This showed that smallholders were collectively responsible for establishing about 50 million ha of plantation by 2005 (Table 2.4).

*Table 2.4* Changes over time in the areas (000 ha) of planted production forests owned by different landholders (Source: Del Lungo *et al.* 2006)

| *Owner* | *1990* | *2000* | *2005* |
|---|---|---|---|
| State | 85,972 | 76,223 | 77,352 |
| Corporate | 21,033 | 25,998 | 27,176 |
| Smallholder | 15,178 | 37,509 | 49,980 |
| Total | 122,183 | 139,730 | 154,508 |

The survey showed that the area reforested by smallholders was greater than that established by corporate landholders and represented 32 per cent of global plantations. Since 1990 the area of plantations owned by the State has declined by 10 per cent, that owned by corporation has increased by 29 per cent, while that owned by smallholders has increased by nearly 330 per cent. There is reason to believe these estimates may understate the real significance of smallholders. First, some governments (e.g. Indonesia and Myanmar) report that all plantations are owned by the State which seems unlikely to be true. Second, many smallholder plantations are, by definition, 'small' making them difficult to locate and include in these national assessments. But, these limitations notwithstanding, the clear implication of the data is that smallholders have a large and growing role to play in reforestation. Some of these growers are planting trees because they need the forest products for local consumption (e.g. fuel and fodder) while others are doing so because they believe there is a market opportunity and have been encouraged by government forestry agencies to do so.

The quality and productivity of plantations owned by smallholders may not be as high as that in state or industry-owned plantations. But many smallholders have developed commercial partnerships with government bodies or industrial plantation companies. This can be an attractive partnership for both parties: the corporate partner does not have to solve the problem of acquiring land while the smallholder has access to technical advice and a guaranteed market for their trees. Mayers and Vermuelen (2002) and Nawir *et al.* (2003) have outlined some of the potential pitfalls in out-grower schemes and the ways these might be overcome.

### *The changing industrial market*

Many of the early national reforestation programmes were designed to produce sawlogs and it was expected that many plantations would have to be grown on rotations of 30–40 years in order to allow trees to reach a sufficient size. But this has changed as processing technologies have evolved allowing processors to use much smaller-sized logs. This, together with the development of technologies to produce reconstituted timber panels from small trees means the need for long rotations is declining.

Much of the material produced from plantations composed of the more commonly used species could be described as commodity timbers. Such timbers are perfectly adequate for most purposes but these are not necessarily suitable for some of the more specialised end-uses such as high-quality furniture. Natural forests continue to be the main source of these 'cabinet' timbers and, with some exceptions (e.g. teak), few of them are currently grown in plantations. The capacity of natural forests to continue to supply these specialised timbers is in doubt because the area of natural forests continues to decline and the productivity of second cutting cycles in those that have persisted will be much less than in the initial harvest. Further, composition of the logged forests may be different. This means there could be an important niche market for plantations able to supply such material.

Technical developments in pulp and paper mills are also leading to changes in the area of plantations needed to sustain them. Mills must be large to be viable and plantation areas of more than 50,000 ha are now becoming more common. Small isolated plantations have always been problematic from a financial viewpoint, even for sawlogs, since log transport costs limit their profitability. This limitation is becoming even greater as fuel costs rise. Companies can find it difficult to acquire large contiguous areas of land near to mills or markets. In such cases joint ventures and out-grower partnerships with local landowners can be mutually attractive – the company obtains land for plantations while the small landowners benefit from a stable and long-term market.

But changes can also occur in areas with mature plantation industries established to support pulp and paper mills. This is the case in the USA where much of the present timber industry is based on plantations that have developed over a long period of time in the southern states. Some privately owned forests and plantations in this region are now being cleared because of growing urbanisation and the effect this is having on land prices. Falling pulpwood prices are also reducing the ability of some larger industrial growers to compete with imports (Fernandez *et al.* 2002). Nonetheless, projections suggest that the privately owned plantation area across the region will increase over the next few decades even though the overall forest area may decline (Zhang and Polyakov 2010). But the ownership of plantations and the nature of the market for forest products will both change. More plantations will be owned by non-industrial forest growers (encouraged by government incentives) than the large industrial growers of the past and future production is likely to be directed towards biofuels such as cellulosic ethanol and wood pellets. Most of these newer plantations are expected to be in areas with low populations and slow rates of population growth (Zhang and Polyakov 2010).

***Future production trends***

The area of production forests and the timbers arising from these are likely to increase over the next few decades. A major analysis by Carle and Holmgren (2008) projected areas and production rates based on three alternative scenarios (Table 2.5). Scenario 1 was the 'pessimistic' scenario in which there were no productivity increases and the rate of plantation establishment slowed. Scenario 2 was the 'business as usual' scenario where plantation areas continued to increase at historical rates but there was no productivity increase. Scenario 3 was the 'enhanced productivity' scenario in which plantation areas increased at the same rate as in Scenario 2 but, in addition, productivity increased as a result of genetic, management and technological improvements. The results of modelling suggested that by 2030 the area covered by plantations would range from 303 million ha (Scenario 1) to 345 million ha (Scenarios 2 and 3). Timber volumes were forecast to rise from 1,400 million $m^3$ in 2005 to reach 1,589 million $m^3$ (Scenario 1), 1,689 million $m^3$ (Scenario 2) or 2,145 million $m^3$ (Scenario 3). These trends are shown in Figure 2.1.

*Table 2.5* Comparison of projected future global production forest areas made by FAO (Carle and Holmgren 2008) and Indufor (2012)

| | *FAO* | *Indufor* |
|---|---|---|
| Current plantation (million ha) | 260[1] | 54 |
| Projected plantation area (million ha) | | |
| 2022 | | 67 |
| 2030 | 302–345 | |
| 2050 | | 91 |
| Timber produced (bill $m^3$ y) | | |
| 2005 | 1.40 | |
| 2012 | | 0.52 |
| 2022 | | 0.71 |
| 2030 | 1.59–2.14 | |
| 2050 | | 1.0–2.0 |
| Timber demand ($m^3$ y) | | |
| 2012 | | 1.5 |
| 2030 | | 1.7–5.0 |
| 2050 | | 2.0–8.0 |
| Proportion of demand met by plantations | 'much of world's future demand' | 25–35% by 2050 |

[1] cf 264 million ha in 2010 (FAO 2011)

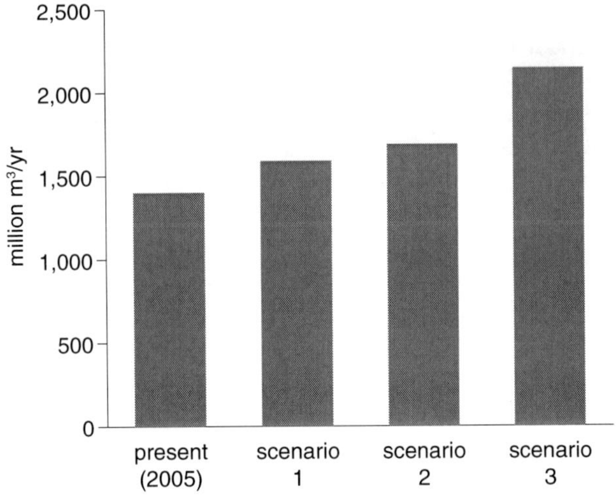


*Figure 2.1* Wood produced by planted forests in 2005 and in 2030 for the three scenarios (1: Pessimistic; 2: Business as usual; 3: Enhanced productivity) projected by Carle and Holmgren (2008).

What does this mean in terms of the likelihood that plantations might one day meet all of the world's timber needs? There will always be special products that can only be supplied from natural forests. In addition, decisions about local resource usage are usually made on the basis of local information rather than global averages. But, based on these assumed levels of production, Evans (2009) concluded that planted forests may meet much of the world's future demand for timber. He even concluded they might possibly exceed it and make inroads into markets for non-renewable construction materials such as cement, steel and aluminium. Victor (2006) also forecast that planted forests might come to meet much of the global demand for timber in the next few decades provided that modest improvements in plantation productivity and timber processing continue to be made.

But a quite contrary conclusion was reached by Indufor 2012 in a report prepared for the Forest Stewardship Council who suggest there are only 54 million ha of fast-growing, high-yielding and intensively managed timber plantations (cf the estimate by Carle and Holmgren [2008] of 260 million ha of all types of forest plantations in Table 2.1). They estimated that the area of these most productive plantations will almost double to 91 million ha by 2050 (cf up to 345 million ha possible by 2030 in the Carle and Holmgren 2008 study). But they believe that the demands placed on these plantations will also increase. They developed three alternative production scenarios (ranging from 'optimistic' to 'pessimistic') and three alternative demand scenarios. Depending on the scenario chosen the annual supply of roundwood by 2050 was estimated to vary between 1.0 billion $m^3$ and 2.0

billion $m^3$ (Table 2.5). Likewise, the annual demand was projected to lie between 2.0 billion $m^3$ and 8.0 billion $m^3$ by 2050. The capacity of plantations to meet future demands depends on which combination of these alternative supply and demand scenarios is chosen. They concluded the most likely outcome would be that plantations will supply only 25–35 per cent of the 2050 market for industrial roundwood with the remainder being supplied by natural and semi-natural forests in the boreal and temperate forest zone.

Any projections of supply and demand over long time periods like these are notoriously difficult and depend on the assumptions made. However, the Indufor analysis clearly regards the majority of the world's plantations as being unlikely to make a significant contribution to meeting future demands for industrial timbers whatever value they may have in supplying local timber needs. Given the areas involved, even a small improvement in productivity could yield large amounts of additional timber and change the supply–demand balance.

### *Planting forests to provide ecosystem services*

In recent years there has been increased interest in improving plantation management standards. This has been assisted by the development of codes of practice to regulate the ways plantations are established and managed (e.g. ITTO 1993) and the formation of the Forest Stewardship Council (Eden 2009) which aimed to certify that standards were being achieved. But perhaps the most interesting change that began occurring towards the end of the twentieth century was the increasing interest in reforestation to supply ecosystem services instead of just timber. These include such things as watershed protection, habitats for wildlife, carbon sequestration and pollination (Bauhus *et al.* 2010). Many of these new plantings are described as 'protection forests' while Mather (1993) refers to them as 'post-industrial' forests.

The areas occupied by these newer forms of plantings are becoming quite large. By 2005 they were estimated to cover 30 million ha and were about 21 per cent of the total planted forest area (Table 2.6). Over the previous 15 years they had increased by 23 per cent. There are large regional differences in the relative proportion of plantations explicitly designated as protection forests with the highest proportion being found in Asia where they make up 32 per cent of the total. In Europe the proportion is 22 per cent, in Africa it is 18 per cent, in North and Central America it is 6 per cent, while in Oceania and South America it is less than 1 per cent. These data are almost certainly imperfect measures of the extent of reforestation undertaken for protective purposes since much is often carried out by organisations other than the forestry agencies who responded to these surveys. Likewise, many forests are established to provide timber as well as have a protective function. Nonetheless, the data clearly show that not all reforestation is undertaken for production purposes.

*Table 2.6* Plantation areas (000 ha) allocated to production and protection functions (Evans 2009)

| | *1990* | *2000* | *2005* | *% increase 1990–2005* |
|---|---|---|---|---|
| Productive | 79,165 | 99,356 | 110,560 | 40 |
| Protective | 24,562 | 28,628 | 30,259 | 23 |
| Protective as % of total | 24 | 22 | 21 | |

It is not clear how functionally effective these plantations actually are since many have been established using the species and silvicultural methods developed for industrial timber plantations. This issue will become increasingly important as markets for ecosystem services develop and begin to drive reforestation programmes. The question of which silvicultural methods are needed to generate particular ecosystem services is discussed further in later chapters.

## The forest transition

When countries go from having net deforestation to a net increase in forest cover they are said to have undergone a forest transition (Mather 2007, Rudel *et al.* 2010). Well-known examples include China, Vietnam and South Korea. There are a variety of ways in which the transition can occur. Most only begin once natural forest cover has been significantly reduced although it can sometimes begin when large areas of forest still remain. Likewise, it may take place once forest cover reaches a perceived critical threshold or it may occur a long time after this threshold has been reached (Figure 2.2). In discussions such as these much depends on how a 'forest' is defined. Most analyses rely on FAO data which define forest as an area exceeding 0.5 ha with trees taller than 5 m and having a cover of more than 10 per cent. But no distinction is made in this definition between undisturbed old-growth forest, regrowth forest and planted forests. Further, statistics can be confused by some state forestry agencies declaring land presently devoid of trees to be state 'forest' simply because it once had trees in the past and the agency hopes it will do so again in the future. This means a country may be still losing natural forest cover but have undergone the forest transition because of the scale of its reforestation programme.

The forest transition can be thought of as an example of an environmental Kuznet's curve in which deforestation is carried out to fund development but reforestation is eventually carried out once economic circumstances permit. Kauppi *et al.* (2006) have argued that forest cover only increases once the national GDP of a country exceeds a certain threshold (US$4,600) but the evidence shows the GDP of countries undergoing the forest transition is quite variable suggesting other factors are involved. A number of possible

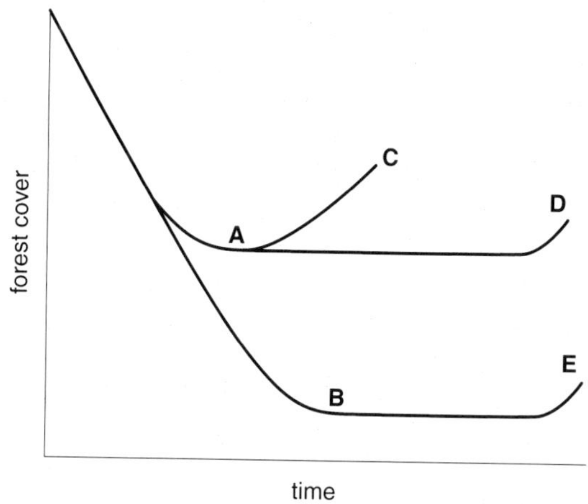


*Figure 2.2* Forest transitions can follow different pathways. Reforestation may begin while significant amounts of natural forest still remain (A) or only after most natural forest is cleared (B). Recovery can begin quickly once forest cover has reached a certain threshold (C) or only after a long delay (D or E).

'pathways' have been suggested (Rudel *et al.* 2005, Lambin and Meyfroidt 2010). These include:

- *State forest policy pathway:* reforestation is driven by changes in state forestry policies. These may be policies established to increase timber supplies, replace imports or create rural employment. Many government reforestation policies initially involved the creation of state-owned plantations but, over time, private investment in reforestation has been encouraged by most governments. This has included incentives for corporations as well as smallholders.
- *Forest scarcity pathway:* reforestation is largely driven by an improving market for forest products caused by a decline in natural forest cover and a decrease in the availability of forest resources. This change makes reforestation commercially attractive, especially in places where timber imports are limited. When this pathway was originally proposed it was seen as dealing largely with forest products but Lambin and Meyfroidt (2010) argue it equally applies to reforestation triggered by a decline in the availability of ecosystem services such as watershed protection.
- *Smallholder reforestation pathway:* this is a variation of the forest scarcity pathway. Many smallholders have traditionally practiced various forms of agroforestry including the development of agroforests (Clarke and Thaman 1993, Michon 2005). More recently, with a rise in timber markets caused by a decline in forest cover, many smallholders have also begun to grow timber trees as a means of diversifying incomes and using land marginal for agriculture.

- *Economic development pathway:* urbanisation and the movement of farmers to city areas can lead to farmland being abandoned. Over time natural regrowth can develop on this former farmland. In this case the transformation is not caused by a shortage of forest resources but rather by social changes.
- *Globalisation pathway:* this can occur when a national economy is closely integrated into global markets for capital, labour, commodities and ideas; in this case reforestation is induced by changes taking place or initiated some distance from the site. For example, it may be driven by taxation regimes, exchange rates or land prices that attract overseas capital. Alternatively, it may be promoted by the international conservation movement. Globalisation also assists the displacement of deforestation from countries undergoing the forest transition to other forested countries.

The notion of a single pathway is probably misleading since several (and perhaps all) of these pathways probably operate at some time or other in most countries that undertake reforestation. But government policy initiatives are critical and many plantations have only been established once suitable government policy settings have been put in place. Reforestation is a long-term activity and can be a risky investment unless the correct species have been identified and appropriate silvicultural methods developed. These tasks have usually been carried out – at least initially – by government forestry research agencies. Likewise, governments have usually had a critical role in providing financial incentives in the form of grants, subsidies or taxation concessions to encourage private enterprise to consider investing in reforestation rather than in some other agricultural or industrial activity. This has allowed timber corporations as well as smallholders to take advantage of market opportunities.

The notion of pathways may be useful as a conceptual tool but they are not particularly useful in any predictive sense. For example, they cannot be used to identify the timing of the points of inflexion in Figure 2.2 or the extent of reforestation that might take place. A much broader model of the forest transition has been suggested by Palo and Lehto (2012) that takes into account ecological factors (e.g. the capacity for natural regrowth to actually occur), the local actors involved in deforestation and reforestation (and the factors that influence them), the formal and informal institutions (including markets, regulatory bodies and laws concerning property rights) as well as inter-sectoral factors, such as the role of agricultural development, which affect the opportunity costs of reforestation and, thus, the likelihood it will take place.

## Conclusions

The twentieth century has been marked by extensive deforestation but has also witnessed the beginning of efforts by many countries to undertake reforestation. In most cases these efforts were prompted by a perception that deforestation was leading to a decline in the availability of forest resources.

Government activity rather than private enterprise initiated most initial reforestation. Private enterprise has subsequently become involved once government forestry agencies identified suitable species and silvicultural methods to use and, by doing so, reduced the risks inherent in this new land use activity.

Government have also assisted the involvement of private enterprise in reforestation by offering financial assistance (e.g. grants, subsidies and tax concessions). This assistance has often been conditional and prescribed where reforestation should be carried out and the circumstances under which it should be done.

Most of the new plantings have used a single species chosen from a remarkably small number of exotic species. This is because these have been most productive at the sites available for reforestation. Over time most of these species have been subject to breeding programmes and the productivity of many plantations has increased at the same time as overall genetic diversity has diminished.

Industrial forestry requires that large contiguous plantations be established in order to maximise financial outcomes. This means that many reforestation projects have contributed to landscape simplification. At the same time, they have often had serious adverse impacts on local communities. In the worst cases these communities have been displaced.

Reforestation has usually generated considerable benefits but the distribution of these benefits has often been uneven; reforestation has not always assisted the livelihoods of rural communities nor has it always generated environmental benefits. The clear lesson from the last 100 years is that these matters cannot be left to chance but should be integral parts of the planning process.

State and corporate industrial groups have been engaged in reforestation but smallholders have also become important stakeholders.

The focus of reforestation now appears to be broadening. In the past 100 years most reforestation has been carried out to produce goods such as sawlogs, plywood or pulpwood. This means that the new plantation forests are quite different to the original natural forests. But more recently there has been increased interest in reforestation to also provide ecosystem services.

The silvicultural systems developed to generate timber and other forest products will not necessarily be suitable for plantations established to provide ecosystem services, especially if these are being established by smallholders. This means a much broader range of silvicultural techniques will be needed than have been used until now.

# 3 Natural forest regrowth

## Introduction

The best way of reforesting large areas is to take advantage of the capacity of many forests to recover naturally. Natural regeneration can occur at sites where further disturbances are limited provided the system has not crossed an ecological threshold and moved to a new state condition (where this has occurred more costly interventions such as planting will be needed). The process of natural regeneration has the potential to restore many of the original native species and the emerging forests can be managed to supply a variety of goods as well as services for the benefit of households and communities living in the area. The potential role of natural regeneration is often under-rated but it is especially suitable to use when large areas are to be treated because it is relatively inexpensive. Indeed, it may be a necessary component of all large-scale reforestation schemes.

The approach is not without some risks, however, and is not universally applicable. For example, it is unlikely to be feasible at heavily degraded sites lacking residual trees or old stumps, or at sites without nearby areas of natural forests. Natural regrowth can also include – and even be dominated by – exotic weeds. This means it should be regarded as just one of the approaches that should be considered when planning large-scale reforestation. This chapter reviews situations where natural regrowth has been useful; some of the constraints on its use; and ways in which forests produced by natural regrowth might be managed.

## Ecological basis of natural regrowth

The ecological basis of natural regrowth at deforested sites is reasonably well understood and has been extensively reviewed by Chazdon (2014). Plants can regenerate at cleared sites from either seed or by vegetative regrowth (this assumes that most of the original trees and seedlings of native species have been removed by previous disturbances although this may not always be the case).

## Seed dispersal and seed pools

Seed can originate from that stored in soil seed banks or from seed dispersed into the site from external sources. Soil seed banks usually contain only a small sub-set of the plant species present at a site. In the case of tropical rainforests, for example, they can contain large numbers of seeds of pioneer species but rather fewer seeds of species representative of more mature successional stages. Some of this stored seed can persist for long periods of time and sometimes for periods exceeding several decades (Leck *et al.* 1989, Thompson *et al.* 1997).

The composition of soil seed banks is not necessarily stable and may vary over time depending on the longevity of the stored seed and on the nature of the surrounding land uses. Some stored seed will gradually lose viability and die but new seed can reach the site and be added to the soil seed pool by being blown in by wind or being carried in by animals. Birds and bats are common dispersers but other wildlife and domesticated animals such as cattle can also be effective. The extent to which dispersed seed are likely to contribute to soil seed pools depends on the landscape context, especially: (i) the distance the site is from any residual natural forests; and (ii) the number of dispersers present that are able to cross the intervening landscape (Myers and Harms 2009, Shoo *et al.* 2012). Less seed will arrive at more isolated locations irrespective of whether it is being dispersed by wind or animals. Corlett (2009) suggests seed dispersal in most tropical ecosystems is usually over distances of less than 1,000 m although longer dispersal can occur. Where animals are involved, seed dispersal can also be limited by the fact that many of those remaining after deforestation may be incapable of crossing the new non-forest landscape. Residual trees in an agricultural landscape can often be important bird attractants. These provide perches for birds and can become significant seedling recruitment points (Toh *et al.* 1999, Holl *et al.* 2011).

Germination of seed stored in the soil can be triggered by seasonal changes in temperature, fire or rain. Moisture is necessary to allow germination but follow-up rain is also necessary to allow seedlings to survive and grow tall enough to escape weed competition. These constraints, together with animal predation and herbivory, mean that few of the dispersed seeds germinate and most of the newly germinated seedlings usually die. Nonetheless, nearby seed sources and active populations of seed dispersers are usually enough to ensure a steady supply of potential plant recruits.

This recruitment process is somewhat selective in that not all species are dispersed by animals and the most common animal dispersers usually favour species with small fruit or seeds. In addition, source areas differ in importance and small forest patches are likely to be less effective than larger patches (and have fewer species) while frequently disturbed areas are likely to be dominated by pioneer species (and often by exotic weed species).

### *Remnant forests as seed sources for regrowth forests*

There can be considerable differences in the capacity of different remnant patches of natural forest to act as seed sources. Large patches of remnant forest are likely to have a wider variety of potential colonists and bigger populations of each species than smaller forest remnants. Moreover, the diversity and population sizes of species in small isolated forest remnants usually decline faster than is the case in larger patches. The process has been intensively studied in Singapore where several small forest remnants have been reproductively isolated for more than 150 years. Over that period much wildlife has disappeared from these forests and some of this wildlife may have been important seed dispersers. Most tree species in the remnants have apparently persisted but this may simply be a reflection of their greater longevity and many species may already be functionally extinct (Turner *et al.* 1996, Turner and Corlett 1996). The significance of this is that small patches of natural forest that have been isolated for some time may not be able to supply as diverse a variety of colonists as were once originally present in these forests.

The ultimate composition of the regrowth forest then depends on the outcome of two trends. One is the trend for local species extinction to occur in the natural forest remnants and the other is the trend for species to be dispersed from these forests across the landscape to colonise regrowth forest sites. An illustration of the balance between these two is given in Figure 3.1. When local extinction rates in the residual patches of natural forest are low (e.g. because the patches are large) and colonisation rates in regrowth forest patches are high (e.g. because dispersal distances are short and there are many dispersal agents) there is a high likelihood that the composition of the regrowth forest will come to resemble that of the original forest. But a rather less-diverse regrowth forest will develop if the reverse is true. Delaying the time before regrowth can begin increases the impact of local extinction within the natural forest areas even further.

Once regrowth develops the location and spatial extent of forests across the landscape begin to change. Over time regrowth areas themselves then become seed sources and provide colonists able to move to other locations. This will change the nature of the seed rain across the landscape and may substantially alter successional trajectories.

### *Vegetative regrowth*

Vegetative regrowth is another method by which regeneration can occur (Bond and Midgley 2001, Vesk and Westoby 2004). This can take the form of sprouts or suckers growing from old stumps or from sprouts arising from residual tree roots. Many species are not killed when trees are felled and are capable of producing sprouts from old stumps. This capacity to resprout forms the basis of the well-known coppice management systems that have

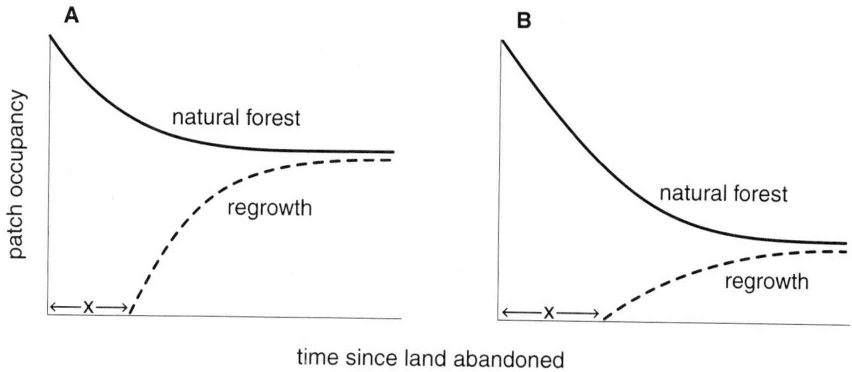


*Figure 3.1* Composition of regrowth forest patches is influenced by the rates of species extinction in residual patches of natural forest (solid lines), the rates at which species colonise newly abandoned land (dotted lines) and the time since the area was originally cleared after which recolonisation can begin (X). (A) Regrowth patches are likely to be more diverse when extinction rates are low and colonisation rates are high. (B) Over the same time period, regrowth patches are likely to be less diverse where extinction rates are high and colonisation rates are low (after Vellend 2004).

been widely practiced in many parts of the world (Smith 1986). Some species are also able to produce suckers from old roots, especially when the root is damaged in some way. The result can be a prolific regeneration of shoots along the length of the root. Indeed, the capacity of the Australian species *Acacia harpophylla* to produce large numbers of root suckers limited the ability of early European farmers to clear otherwise attractive areas with good soils dominated by this species (Dwyer *et al.* 2009).

All of these mechanisms – soil seed banks, dispersed seed, old stumps – may be involved in the recovery of forests in degraded landscape although the relative importance of each will depend on the species present, the history of land use at the area and the landscape context. Subsequent successional development then depends on the relative proportions of the species that have established themselves, their growth rates, their fire and shade tolerance and their final size at maturity. The time before the constituent species can begin to reproduce and their longevity also influences the way the new forest ecosystem develops.

### *Effects of repeated disturbances*

A representation of hypothetical natural regrowth pathways after a disturbance is shown in Figure 3.2. The system loses species and biomass when the natural forest (A) is cleared. Even if cropping then occurs, many (though not all) native woody plant species are likely to persist at the site as stumps and old roots. If cropping ceases and the site is abandoned,

*Figure 3.2* Hypothetical regrowth pathways showing how repeated disturbances can lead to a gradual reduction in the numbers of native species able to regenerate at a site and perhaps also to a reduction in the biomass of the new forest. The original forest (A) is cleared for agriculture and the site is then abandoned. It recovers to B but is cleared a second time. Again it is abandoned and recovers to C. Successive disturbances cause an increasing loss of tree species diversity and biomass.

sprouting will take place and recovery by some woody species can be rapid (to B). But if this regrowth forest is itself cleared before recovery is complete then the legacy of residual stumps is again reduced. This means an even more depauperate flora is left to recover (to C). The capacity of the system to return to the original condition (A) depends on the absence of further disturbances and on the landscape context in which the regrowth is developing. The ability of the original species to recolonise the site will be reduced if there has been extensive clearing across the landscape as a whole. The standing biomass is also likely to be progressively reduced after successive disturbances because of soil erosion following each event. In short, regrowth can occur but the more disturbances that take place and the longer these continue, the less likely it will be that the original ecosystem will be restored.

Once regrowth develops the location and spatial extent of forests across the landscape begin to change. Over time regrowth areas themselves then become seed sources and provide colonists able to move to other locations. This will change the nature of the seed rain across the landscape and may begin to alter the successional trajectories in regrowth forests.

## Examples of natural forest regrowth

There are many examples of natural forest regrowth occurring in a variety of biomes and geographic locations (Benayas *et al.* 2007, Cramer and Hobbs 2007). An early example occurred in Europe when the bubonic plague decimated populations and led to the abandonment of cultivated land and 20–25 per cent of all villages across the region. Williams (2006) suggests many of the present forests of Germany result from this regrowth. Something similar appears to have occurred following the introduction of diseases to America by early European settlers which allowed extensive regrowth to develop (Ruddiman 2007, Dull *et al.* 2010).

A significant amount of natural forest recovery is again underway in many agricultural landscapes of Europe with up to 24 per cent of arable lands projected as being likely to be abandoned by 2050 (Navarro and Pereira 2012). An example of such regrowth in Sweden is shown in Figure 1.4. But there are many other cases across the globe of land abandonment and forest regrowth, especially at the forest frontier. Examples of where it has occurred on a large scale are shown in Table 3.1.

Most natural regrowth occurs when farmland is largely abandoned. There are a variety of possible causes. Sometimes it is because farming is attempted on ecologically unsuitable lands or because good land is so poorly managed it has to be vacated. In other cases it is because socio-economic or political events force changes on landholders. These include changes in markets, changes in government agrarian policy or events such as warfare or widespread disease. Under such circumstances many people chose to leave the land and migrate to urban areas to seek better (or safer) economic opportunities (MacDonald *et al.* 2000, Benayas *et al.* 2007).

However, in some cases natural regrowth has been deliberately encouraged. Contrary to many earlier views, Leach and Fairhead (Fairhead and Leach 1995, Fairhead and Leach 1996, Leach and Fairhead 2000) argue that much of the tree cover in the forest–savannah transition zone of West Africa has arisen from forest regrowth and transplanting. Rather than a progressive Malthusian loss of forests caused by rising populations, forests have often recovered through a neo-Boserupian process of agricultural intensification mediated by such factors as changes in farmers' objectives, tenurial arrangements and environmental changes (cf Boserup 1993). Other examples where natural regeneration has been deliberately encouraged will be discussed further below.

But natural forest regrowth is not always regarded favourably. This is certainly true in parts of the rangelands of inland Australia where the regrowth of woody plants has been of concern to graziers (Noble 1997, Fensham *et al.* 2005). The cause of this resurgence in woody regrowth is a matter of some debate with changes in fire regimes and rainfall being suggested as contributing factors. In Ethiopia extensive bush regrowth occurred as a result of a government ban on fires but this led to the loss of pasture to the considerable disadvantage of local farmers (Angassa and Oba 2008). Those wishing to increase the overall extent of reforestation have to recognise such limitations and find ways of incorporating the regenerative capacity of some ecosystems into existing land management systems.

*Table 3.1* Examples of situations where natural forest regrowth has occurred over large landscape areas

| *Location* | *Initiating factor* | *Reference* |
|---|---|---|
| Australia | Graziers institute a reduction in fire frequency and/or increased rainfall | (Noble 1997, Fensham *et al.* 2005) |
| Ethiopia | Ban on burning | (Angassa and Oba 2008) |
| El Salvador | Remittances from abroad reduce need for as much farmland | (Hecht and Saatchi 2007) |
| Guinea | Population decline caused by warfare | (Fairhead and Leach 1996) |
| Honduras | Marginal agricultural lands abandoned | (Southworth and Tucker 2001) |
| Kenya and Tanzania | Agrarian policy review encourages restoration of traditional management systems | (Barrow and Mlenge 2003) |
| Madagascar | Regrowth found in areas with secure land tenure and reduced population pressure | (Elmqvist *et al.* 2007) |
| Peru | More isolated cropping areas abandoned following loss of government incentives to clear | (Alvarez and Naughton-Treves 2003) |
| Puerto Rico | Farms abandoned because of urban migration | (Aide *et al.* 1996) |
| Spain | Farms abandoned because of urban migration | (Lasanta-Martinez *et al.* 2005) |
| Ukraine | Farming abandoned and migration out of rural areas caused by collapse of Soviet Union | (Kuemmerle *et al.* 2011) |
| USA | Over-grazing and a reduction in fires in semi-arid grassland areas | (Van Auken 2000) |
| USA | Early farming areas in the east abandoned because of better farming opportunities elsewhere | (Williams 2006) |

The case studies which follow illustrate some of the possibilities as well as the constraints involved in relying on this capacity of many forests to recover unaided from disturbances. They also illustrate that 'success' is not necessarily the development of a closed-canopy forest; some landowners may wish to achieve this but others may prefer a more open savannah-like condition (for example, so they can graze cattle). Still others may want to develop forests in specific areas unable to regenerate by natural means. In the discussion that follows the term 'forest' is used even though much regrowth is initially composed of highly scattered shrubs and trees and might not meet the formal FAO definition of a forest (discussed in Chapter 1).

### Case study 1: Forest regrowth in Western Europe

Forest regrowth is taking place in many former agricultural areas in Western Europe abandoned because of changes in economic circumstances and government policies (Gellrich *et al.* 2007, Tasser *et al.* 2007, Améztegui *et al.* 2010, Van Uytvanck *et al.* 2010). The manner in which this is taking place varies with vegetative regrowth being important in some locations and colonisation by species from intact natural forests taking place in others. Examples from Norway and France illustrate some of these patterns.

Government policies in Norway since the 1950s have encouraged a change from general purpose farming to more specialised activities such as dairy, sheep or grain production. This change, together with a greater reliance on mechanisation, has led to some former grass-dominated farmland being abandoned. An example of the type of forest regrowth that can occur is that which took place in southern Norway (60° N) in an area studied by Staaland *et al.* (1998). The area has a long agricultural history with some areas being farmed for up to 6,000 years. But changes in farming practices have occurred over the last hundred years and grazing in the area was finally abandoned in the 1950s. Grass areas and shrublands decreased and patches of forest expanded. Woody plant succession in this area usually involves grey alder (*Alnus incana*) and *Prunus pudus* or *Ulmus glabra* depending on aspect. Upper slopes are colonised by Norway spruce *Picea abies*, birch (*Betula pubescens*) as well as aspen (*Populus tremula*). Norway spruce is likely to be the final dominant in most of these areas. Most of the initial expansion occurred because of aspen and grey alder expanding from existing forest patches through root suckers. Because of this, the amount of woody regrowth decreased sharply with increasing distance from the forest edge. But the rate at which regrowth actually takes place varies with location. Staaland *et al.* (1998) found the rate of regrowth was faster at relatively drier sites than at sites classed as being relatively wet (e.g. at the bottom of ravines). Fifty years after grazing ceased the average age of trees at relatively drier sites was 40–45 years but only 25–30 years at these wetter sites. Over the same period the overall tree cover across the area had expanded by 50 per cent.

A different pattern of reforestation is underway in the Massif Central of France (Prévosto *et al.* 2003, Prévosto *et al.* 2004). In this case mostly former grazing lands have also been abandoned after a long period of use. As in Norway, the absence of grazing has led to a massive colonisation of the area by trees. But, unlike Norway, most of the natural regrowth that occurs on these sites has been by wind-dispersed species such as Scots pine (*Pinus sylvestris*) and silver birch (*Betula pendula*) reaching the sites from external seed sources. Pines tended to dominate the reforestation process in the drier southern areas while birch has been more common in the wetter northern areas. Prior land use influenced successional patterns with forests on more fertile lands previously used for cropping being colonised more slowly than former grazing lands. This appears to be related to the perennial herbs at the cropped areas that have acted as barriers to tree seedling establishment. The former cropping areas also have denser shrub layers of species such as *Corylus.* Most of the new forests have a narrow range of tree size classes suggesting the period of colonisation was short-lived and that little subsequent recruitment of these species is now underway although a diverse shrub and herb layer is developing. Vojta and Drhovská (2012) also report that successional development in abandoned former pastures in central Europe can be slow and that although many forest species had colonised these sites, other species usually found in ungrazed ancient forests were still absent.

### *Conclusion*

Regrowth is developing in many areas of Europe as a result of economic and land use policy changes. The extent of regrowth varies and is dependent upon the land use history, the distance to intact forests and the composition of these forests.

### Case study 2: Forest regrowth in Nepal

Farmers in the middle hills of Nepal grow crops on terraced land and have some livestock. Trees and forests are important to these farmers since tree leaves and grass are used for fodder and animal bedding as well as providing compost for crops such as rice, maize and millet which are grown on hillslope terraces. These terraces cover most of the cultivated land. Trees are also used for fuelwood and construction material. In the 1970s there were claims that deforestation and land degradation was widespread across Nepal. At the time all forest lands in Nepal were claimed by the State and, in theory, under the management control of the Forest Department. The government and many development agencies believed that an expansion of forest cover through large-scale reforestation was the only way to overcome the problem of land degradation.

In fact, as Gilmour (1990, Gilmour and Fisher 1991) showed, the claims of deforestation and degradation were overblown. Many farmers had begun planting trees on their private land and developing indigenous systems of forest management on common land, independent of the government, and the forests were no longer shrinking. On the contrary, there was evidence of careful forest management of many village forests and that forest and tree cover, particularly on private land, was increasing. For example, a survey by Carter and Gilmour (1989) in one area found a near threefold increase in tree cover on private agricultural land over a 24-year period up to 1988, quite independent of formal government and NGO reforestation activities. Similar results at another location were reported by Gilmour and Nurse (1991). This was despite a substantial increase in rural populations and a decrease in average farm size.

The trigger for the development of new forests appeared to occur when supplies of animal fodder and fuel became scarce. An obvious indication of this being when it took several hours to walk to an area of native forest to which they had access and harvesting rights. Once this threshold condition was exceeded, farmers began to engage in tree planting on their own land and supported the development of community forest management activities on their common land.

The increase in forest cover took place in several ways. The first was through planting trees in small community forests, generally with support from government and donor agencies within the framework of a national community forestry programme. The native species *Pinus roxburghii* was commonly used because it is an ecological pioneer and tolerant of most exposed site conditions (unlike most native broad-leaved species which proved difficult to establish in open field conditions). A second method used by individual farmers was to plant wildlings from natural forest across their agricultural terraces. The species used were mostly *Litsea polyantha* and species of *Ficus* because they are good fodder trees. In both cases the planted trees provided an environment that encouraged natural regeneration of broad-leaved species. Natural regrowth also took place along stream banks and in areas between fields that are difficult to use for cropping (Gilmour 1990). These non-cultivated patches within the wider agricultural landscape became available for growing trees because farmers had shifted from open range grazing of large numbers of livestock to stall feeding of small numbers.

The density of trees and the number of species found in regrowth tended to increase with time although the number present varied with altitude and with the identity of initial 'nurse' trees. In each case the exclusion of grazing animals and some initial tree cover appeared to facilitate subsequent natural forest regrowth. Local communities developed rules governing access rights to the new community forests and how harvesting of forest resources might be carried out (Gilmour *et al*. 1990, Gilmour and Nurse 1991).

These findings concerning the extent of natural forest regrowth and the involvement of villagers in initiating its development and managing its use had a major impact on forest policy in Nepal. It was clear that farmers were not ignorant about ways of reforesting both their own land and adjacent common land. It also became evident that the emphasis of the government had to change from using extension officers to promote plantations on deforested lands to, instead, finding ways of supporting indigenous silvicultural approaches to improve both community-based forests and farmer livelihoods. Major government policy changes in the late 1980s opened the way for management rights for community forests to be given to local Forest User Groups, subject to the development and approval of a management plan. This has led to a flourishing of community-based forestry activities covering both planted and natural forests and an increase in forest cover of both common and private land across Nepal (Gilmour 2003). More than 1.2 million ha of community forests have been handed over to local Forest User Groups since 1990. An example of the changes is shown in Figure 3.3.

## *Conclusion*

Despite government efforts to promote reforestation, farmers were not interested until they perceived a shortage of trees and forest products. Indigenous management systems subsequently evolved to overcome this shortage, but only after the government accepted the right of villagers to own and manage trees on their lands. Forest cover in Nepal has significantly expanded following these changes and natural regeneration has been an important component of this increase. It has been largely responsible for restoring considerable plant biodiversity to many areas.

*Figure 3.3* Natural forest regrowth on former farmland in the middle hills of Nepal. Thirty years earlier this hillside had been completely free of trees (Photo: D. G. Gilmour).

### Case study 3: Tree regeneration in Niger

Niger, like much of the African Sahel region has been subject to drought, deforestation and famine. The southern regions of the country were once occupied by open woodland although much of this has been cleared for cropping. Rainfall in this part of the country is 600–800 mm but droughts are common and wind erosion has been widespread. This has meant agricultural productivity can be variable and it is often poor.

During the colonial era the State claimed rights to all vegetation. Following independence, the post-colonial Forest Service still claimed rights to certain commercially important trees beyond forest areas, including those on farms. Landholders could not fell these trees without permits and could be fined for even cutting branches for fuelwood (Stickler 2014). The policy discouraged any interest farmers might have had in retaining trees on their farms. This was re-enforced by government extension staff encouraging farmers to 'modernise' by cutting back other less commercially valuable trees to create larger, more open fields in which to grow crops. But a dramatic change in tree cover has occurred in the last few decades during which some 200 million trees have regenerated across an area of five million ha and made a significant contribution to the livelihoods of people living in the area. The change has been referred to as the 'regreening of the Sahel' (Sendzimir *et al.* 2011).

This increased reforestation was not caused by increased rainfall, although variations in rainfall may have played a part in the response. Nor was it the result of conventional tree planting although a considerable effort has been made over the years by the Niger government to encourage the establishment of tree plantations by farmers, especially with exotics such as *Eucalyptus camaldulensis* and neem (*Azadirachta indica*). In fact most of these conventional plantings failed because of the harsh conditions in the region. Rather, the increase in tree cover was due to a process that has become known as 'farmer managed natural regeneration' (Haglund *et al.* 2011).

The trigger for the development of farmer managed natural regeneration is unclear. Some suggest it was simply a response by farmers to years of drought and it spread because the implementation costs were low while the benefits were high (Haglund *et al.* 2011). Another account argues that the trigger for the transformation was the political vacuum resulting from the death of President Kountché in 1987. This temporarily hobbled the national government and gave more autonomy to local regions (Sendzimir *et al.* 2011). Whatever the case, it is clear that, with the assistance of some NGOs, farmers began exploring ways of using natural regeneration they found on their farms. Although tree clearing had been carried out many years earlier, some trees were still present across the area persisting in the form of old stumps and root systems and, in some cases, tree seed stored in topsoils (Figure 3.4). This has been described as a vast 'underground forest' (Reij *et al.* 2009, Tougiani *et al.* 2009). These plants had the appearance of being shrubs but were, in fact, trees. The most common species include *Bauhinia reticulata*, *Combretum* spp., *Ziziphus spina*, *Adansonia digitata*, *Z. mauritiaca*; and the most highly prized *Faidherbia albida* (a nitrogen fixer which sheds its leaves in the rainy season and so offers only limited competition to surrounding crops when they are actively growing). It is by encouraging this regrowth, rather than cutting it back, that reforestation has taken place. An additional factor is likely to have been a change in the forest law in 1993 that effectively gave farmers ownership

*Figure 3.4* Trees regenerating from old stumps and roots in a farmer's field in Sub-Saharan Africa (Photo: T. Rinaudo).

of their trees and encouraged them to protect and manage those on their own land. This occurred at a time when urban fuelwood prices were rising (Manvell 2001).

The reforestation process has evolved into a three-step process. First, farmers selected stumps to work with based on the perceived usefulness of that species for food, fuel or fodder. Second, a leading shoot on each stump was selected to become the final stem. Then, third, all other shoots and side-branches were removed. The procedure immediately generated some fuelwood and, after a few years, the shoots had grown sufficiently for fuel, poles or building material to be harvested. Some plants also generated local medicines. In the meantime, crops such as millet, sorghum, cowpeas, peanuts, hibiscus, sesame, and cassava could continue to be grown between the regenerating trees. The choice of how many trees to retain is decided by individual farmers. Some farmers were concerned that the growing trees would adversely affect crop growth and limited the number of trees to 40 per ha but further experience has shown that densities of up to 150 trees per ha can be grown without adversely affecting crops.

The advantages of tree regrowth have been self-evident to farmers and it only required a policy change to enable the landscapes to be transformed. Agricultural productivity has increased and, by producing saleable products the system has increased farmers' income as well as diversifying their income base. The new farming systems are less subject to wind erosion and soils have been stabilised. The growing trees have also attracted seed-dispersing birds which have further enriched the pool of trees and species at the site. Overall the system has led to a more ecologically and economically resilient community. Some of the community-based institutions developed to experiment with and manage this natural regeneration are described by Tougiani *et al*. (2009), Reij *et al*. (2009) and Sendzimir *et al*. (2011).

### *Conclusion*

Former government policies encouraging tree clearing in agricultural landscapes caused widespread problems. But many tree species in this region persisted as old stumps, roots and seeds and were able to regenerate once attempts to eradicate trees were abandoned. Regrowth depended on changes in government policies that gave farmers the right to manage, harvest and sell their trees as well as on the development of ways of incorporating these new trees into their farming systems.

## Case study 4: Forest regrowth in Laos and Thailand

Shifting cultivation has been widely practiced throughout the humid tropics although it is becoming less common. The secondary forests it generated were often deemed to have limited commercial or conservation value and shifting cultivators were regularly blamed for 'destroying' forests (e.g. Cramb 1988). But much depends on the methods used. A study of the vegetation developing after recent shifting cultivation in the seasonally dry tropical forests of Laos PDR compared sites cleared just once within a 30-year period and those cleared 3–5 times over the same time (McNamara *et al*. 2012). All disturbed forests have a large component of pioneer or secondary species able to regenerate after clearing. However, the more critical issue is whether these sites are able to retain species representative of more mature successional stages. In this Laotian study the diversity and richness of primary forest species present as seedlings were surprisingly similar in the two sets of secondary forests and were not greatly different from those in relatively undisturbed forests in the district. An ordination showed the fallow sites differed from the nearby undisturbed sites but this difference appeared to be largely due to difference in the relative abundance of species at the various sites rather than the absence of primary forest species from the disturbed sites. The high diversity of species in the two forest fallows appears to be associated with seed dispersal from nearby sources and from the ability of most of the primary forest species to resprout after disturbance. Species found in seasonally dry forest like these are often capable of resprouting (Viera and Scariot 2006). This suggests the system is relatively buffered against change and indicates that recovery after shifting cultivation (in both structure and biomass) should be comparatively rapid if these sites are protected from further disturbances, dispersal distances remain short and seed-dispersing wildlife are able to persist.

But there are limits to the capacity of stumps and old roots to continue to resprout after clearing. If disturbances are too frequent or they continue for too long then the capacity to recover will decline. This was evident from the experience with the shifting cultivation practices once used by the Hmong (or Miao) people in northern Thailand (Geddes 1976, Cooper 1984). The Hmong have sometimes been described as 'pioneer' shifting cultivators because they preferred to move to new sites and clear more forest rather than return to fields left fallow for a few years (a practice referred to as 'established' shifting cultivation). Unlike established shifting cultivators such as those in Laos, the Hmong system involved continuous cropping of an area for several years during which time many old stumps were removed (Poffenberger and McGean 1993). One of the crops favoured by the Hmong was opium poppies and these could be grown on the same site for 10 years or more before an area was abandoned. This long period of cropping accentuated the destructive elements of their agricultural system. Not surprisingly productivity eventually declined to the point where the community was obliged to abandon the site and move elsewhere. Most of these sites became grasslands and some have never recovered (Geddes 1976, Cooper 1984). Where some forest recovery at former poppy sites has taken place the flora is less diverse than that regenerating on other fallow areas used only briefly for upland rice cultivation (Fukushima *et al*. 2008).

### *Conclusion*

The capacity of forests to recover naturally depends on a site's land use history. Some intensive agricultural practices can destroy recovery mechanisms and cause ecosystems to cross an ecological threshold from which recovery is difficult.

### Case study 5: Guanacaste Conservation Area, Costa Rica

Much of the seasonally dry tropical forest of Costa Rica has been cleared for grazing and cropping but several areas remain in the north-west of the country including within Santa Rosa National Park. A project began in the 1980s to acquire the farmland around the National Park and, ultimately, restore forest cover across 110,000 ha to form the Guanacaste Conservation Area. The background to the project is described by Janzen (1986, 2002) and Allen (2001). Much of the farmland surrounding the National Park was pasture and fires in the dry season were common. However, patches of forest remained scattered across the landscape and have formed the basis for most of the restoration effort. This is being tackled by limiting fires and encouraging natural regeneration based on seed derived from these remnants. Fire control is based on fire breaks established around the boundaries of the project area and within grasslands to limit the spread of wildfires. A special fire control group was employed to extinguish any wildfires that did start. Experience showed that fires occur mainly within grassland areas and only rarely in forests. This meant that once grassland areas were reforested the risk of a restoration setback caused by fire usually declined.

Tree regeneration often began within a few years of fires being excluded. Some of the new trees originated as sprouts from stumps and old roots (although the density of these was low and commonly less than one per ha). But the majority of new trees came from seed dispersed into the site from nearby patches of residual forest. Very few seeds have come from old soil seed banks because of the very large number of active seed predators. Two types of seed dispersal have occurred. The earliest colonisers were largely wind-dispersed species (Finegan and Delgado 2000). These were spread in large numbers but mostly over distances of only several hundred metres. Tree seedlings from these began emerging above the 50–150 cm tall grass after about 4–5 years (Janzen 2002). The second dispersal mechanism was via animals. Once the wind-dispersed species had grown into trees a number of animal-dispersed species subsequently colonised the sites and began to grow up under them (Allen 2001). The animal dispersal agents include horses and cattle as well as native wildlife. Through these two processes the smaller areas of pasture have been largely reclaimed by trees within a very short time. Some limited tree planting has also been tested in several areas to complement natural regeneration (i.e. humans have become the dispersal agents) although further experience eventually showed this was unnecessary.

Grasses are a major impediment to tree establishment but competition from grass can be controlled by allowing cattle grazing. This has proved to be controversial because of concerns that the cattle would damage young tree seedlings. However, experience to date has found that cattle were able to control grass growth while causing only slight damage to the young trees. Trials are now determining appropriate stocking regimes and at what point the restoration benefits of maintaining cattle in the area are outweighed by the disadvantages of doing so.

Forest regrowth has been rapid over about 49,000 ha but the successional patterns vary and the outcomes at any single site in such a large and heterogeneous area are uncertain. The species recolonising particular sites depend on the land use history of that site, its distance from seed sources and the nature of the seed sources themselves (e.g. whether wind-dispersed species or wildlife-dispersed species). Other variables include whether there are still remnant trees scattered across some of the grassland areas that might act as perch trees for birds and the duration of any cattle grazing. It is likely to take many years for 'recovery' to occur but there is already evidence that some larger wildlife species are beginning to use areas of regenerating forest (Allen 2001). Managers have taken the view that in restoring forests across large areas it is not possible to restore the original biodiversity to every particular area. Rather, the 235,000+ species thought to be present across the area will be 'allowed to compete, mutualise and commensalise into whatever patterns and habitats they arrive at given the abiotic and biotic heterogeneity intrinsic to the (area)' (Janzen 2002: 561).

Project managers have taken particular care to ensure the project benefits the local community. This is something that Janzen (1988) describes as 'bio-cultural restoration'. The intent is to ensure that the idea of restoration is embedded within the hearts and minds of people in the same way that churches and schools are. This is partly for reasons of equity and to ensure that the benefits of restoration are shared. But it is also to increase the likelihood the new forest will be protected against future man-made disturbances. This 'embedding' process is being done by employing members of the local community (including former farmers whose land has been acquired by the project), by undertaking educational programmes and by seeking to foster economic opportunities such as eco-tourism which will benefit the community. Further details are provided by Holden (1986), Allen (2001) and Calvo-Alvarado *et al*. (2009).

***Conclusion***

Natural regrowth across a large area of former farmland has occurred primarily through seed dispersal from residual patches of natural forest. This is enabling the creation and consolidation of a large new conservation area. The future composition of this forest may not be the same as the forests originally present but will reflect the ecological processes able to operate in this environment now and in the future. Besides encouraging restoration to enhance biodiversity, managers are also ensuring the project generates livelihood benefits for the local community to ensure its long-term viability.

**Case study 6: A new type of forest in Rajasthan, India**

While natural regrowth usually involves mostly indigenous species it is sometimes the case that exotic tree species can spread naturally and also create economically useful new forests. This has happened in the Godwar region of Rajasthan in India. The region can be classed as semi-arid and has an annual rainfall of 400–500 mm. Natural forests of *Anogeissus pendula* in association with *Butea monosperma* and *Ziziphus nummularia* cover the hills while the lowland plains are used for crops and grazing. Little regeneration by these native species has occurred outside the hill area.

In recent years the government has sponsored agricultural intensification across the region by facilitating and promoting crop irrigation. Dams have been built and bores dug to access groundwater. Electrification has enabled constant pumping from these bores. As a result, the area occupied by crops increased by 5,700 ha, or 315 per cent, in the 13 years between 1986 and 1999 (Robbins 2001). Over the same period stall feeding of livestock began to replace open grazing on village commons. Some tree plantations were established within the agricultural zone using the exotic tree species *Prosopis juliflora*, *Acacia nubica*, and *Acacia tortilis*. But this reforestation has been dwarfed by the extent of natural colonisation of grassland areas, largely by *Prosopis juliflora* (Robbins 2001). The extent of this natural regrowth has been rapid. Over a period of 13 years (between 1986 and 1999) the area of forest increased by 9,000 ha, or by an additional 49 per cent, while the area of grassland and fallow shrank by 15,000 ha, or 45 per cent. An indigenous congener of *P. juliflora* is *P. cineraria*. This also grows in the dry regions of India but its growth is not as rapid as the exotic species and it has not been involved in the natural forest regrowth in this region.

*Prosopis juliflora* is well known as an invasive species (Kaur *et al.* 2012). The species is deep-rooted as well as being drought tolerant. It is capable of nitrogen fixation and grows rapidly in environments such as those of Rajasthan. The seeds are probably spread by cattle and other animals but it is unattractive to browsers because of prolific thorns on its branches. The species is thought to have been introduced into the Rajasthan region in 1912 and was heavily promoted as a fuelwood species (in part because it was less subject to browsing damage). However, it is not known to have formed closed-canopy forests in the region until recently. Nonetheless, in the space of 20 years, it has suddenly begun to form a series of forest patches some of which now cover 1,500 ha.

The sudden spread of *Prosopis* coincides with the collapse of the traditional management regimes on village common lands. Forestry authorities stopped allowing herders to graze their animals in the state-owned natural forest in the hills because this had become a wildlife reserve. This meant that the village common lands came under increasing pressure. But, at the same time, the government prompted agricultural intensification meaning that more of these village grasslands were being converted to cropping. The government also declared these common lands were 'waste' lands and began asserting ownership. The collective impact of these various changes appears to have enabled the spread of *Prosopis* (Robbins 2001).

Attitudes to the forest regrowth have differed. Herders and those previously using the grassland areas have been upset by the increase in forest cover because it has shaded out pasture grasses and deprived them of fodder. The new forest may also affect groundwater supplies which could be damaging for agriculturalists now heavily dependent on these resources. On the other hand, many poorer people have welcomed the increase in forest cover because the trees provide high-quality fuelwood and an opportunity to improve their incomes by producing charcoal. A survey found 60 per cent of households used the new *Prosopis juliflora* trees in some way. Foresters belonging to the state forestry agency appear to have mixed feelings. According to Robbins (2001) some welcomed any increase in forest cover because it helps them achieve national reforestation goals. But, at the same time, many are also concerned that *Prosopis juliflora* has also spread into the wildlife reserve and will be difficult to eradicate. There is evidence from elsewhere that, unlike *P. cineraria* growing in its natural habitat, *P. juliflora* can be an aggressive competitor within natural plant communities (Kaur *et al*. 2012).

#### *Conclusion*

Exotic species have become naturalised in many landscapes and can spread by natural regrowth. *Prosopis juliflora* was present in Rajasthan for nearly 100 years before this occurred and the main trigger for its spread appears to have been the break-down in management of village common lands. Once this occurred the exotic species was able to colonise and dominate these sites. These new *Prosopis* forests will benefit some people but disadvantage others. It is likely that the full economic and ecological consequences are yet to be experienced.

## Potential constraints

These examples suggest there can be considerable opportunities to restore trees and forest cover over some very large areas at a relatively modest cost. But there are also some constraints on it developing into a permanent forest cover. These include ecological impediments as well as socio-economic limitations (Box 3.1). The former dictate what is potentially achievable, whereas socio-economic constraints determine what is actually realised.

### *Ecological constraints*

The primary ecological limitation is that there may not be enough residual trees, stumps or seed to initiate the succession (Griscom and Ashton 2011). This is likely to be the case where intensive cropping has been practiced for lengthy periods of time and where intact patches of natural forest are distant. Not all species are able to produce sprouts and species with low

**Box 3.1 Why natural regeneration can sometimes fail to restore forest cover**

There are several reasons why natural regeneration sometimes fails to restore forest cover:

- Sites are too degraded (unfertile soils, weeds) and the density of trees (arising from seedlings or coppice from stumps) is too low.
- There are no residual patches of nearby forest able to supply seed enabling additional trees or species to colonise the site.
- There are no seed dispersers able to carry sufficient seed from residual patches of forest to the site.
- The distribution of regenerating trees is too patchy.
- Managers are unable to protect sites from ecological hazards such as weeds, fires or herbivores.
- Managers are unable to protect the sites from human interference such as clearing or unregulated logging.
- Managers deliberately limit the density and types of species regenerating in order to incorporate tree growing into a farm management system (e.g. also involving crops or grazing animals).

density timbers are likely to rot faster and disappear from the pool of stumps faster than those with heavier wood densities. That is, they may regenerate after single disturbance but be unable to persist in the face of repeated disturbances. In addition, successful coppice is usually found on smaller rather than larger stumps where there is a danger it can be blown or fall off before it becomes established (Jacobs 1955, Smith 1986).

The limitation imposed by a lack of residual trees or old stumps might be overcome if seedlings could regenerate from seed but the capacity of seed to supplement these root stocks may also be limited. Soil seed banks are one possible source but these can sometimes be more of a problem than a benefit. Many seed banks are likely to be dominated by exotic weeds, especially if the site has been subject to recurrent disturbances in the past (or is near land still being used for agriculture). Likewise, exotic weeds may dominate the seed rain, especially if intact natural forest is some distance away. This means that any new woody regrowth able to survive competition and emerge from the grass and herb cover may have a high proportion of exotic species. Sometimes the density and height of these weeds is such that it is very difficult for woody plants to grow and over-top the grasses. In such cases successional development is blocked and further development may require considerable effort to enable woody plants to grow high enough to persist (Cohen *et al.* 1995).

Exotic tree species pose particular problems. In some situations, such as the case of *Prosopis* in Rajasthan described above, they may offer some benefits (although in that case the evidence is equivocal). But in other cases they can be more problematic. Much depends on their competitive interactions with indigenous species. Newton *et al.* (2011) explored the possible outcome of having an exotic species (*Acacia dealbata*) colonise a regenerating dryland forest in Chile. They used the LANDIS-II model to predict what might happen over time. The model suggested *A. dealbata* would disappear from the community after about 100 years in the absence of further disturbances. On the other hand, according to the model, it would persist and could spread if disturbances such as grazing or fire occurred within this period. This posed a significant dilemma because a number of the native species required some degree of disturbance to persist and modelling suggested overall species richness would decline if disturbances were prevented.

Other ecological constraints on forest recovery include the difficulty of protecting regenerating plants from disturbances such as fire or grazing (including domesticated animals as well as wildlife). Both can effectively destroy regeneration and arrest successional recovery although many species are more often able to tolerate both factors as they become older. The damaging effects of herbivores on young tree seedlings has been widely reported in many ecosystems (Van Uytvanck *et al.* 2008, den Herder *et al.* 2009, Lof *et al.* 2010, Van Uytvanck *et al.* 2010).

### *Socio-economic constraints*

The main socio-economic constraint concerns the difficulties of protecting the regenerating forest. Even just preventing wildfire can be expensive, especially if the forest covers a large area and grasses are present. An added risk is that few young forests are able to generate many economic benefits; most benefits are only generated after some decades have passed. This means many landowners may be inclined to clear and replace regrowth forests with something more profitable. This is most likely to be the case in areas where land tenure is uncertain and the regenerating forest is perceived as unused waste land. Or, alternatively, the customary landowners are concerned a state forestry department might now claim the land as part of their state-owned forest. It can also be the case that some government agencies see regrowth forests as temporary 'land banks' that can be kept in reserve and used when needed for infrastructural development, agricultural resettlement schemes or some other purpose.

It may not suit all landholders to have regenerating trees develop into a forest. This was the case in Niger where only certain trees species were favoured and there was a definite limit on the density of trees needed. But in other situations full forest recovery can be welcomed. Much depends on how the trees are to be incorporated into the landholder's management system.

## Establishment and management of natural forest regrowth

The fact that natural regeneration can sometimes occur does not mean it necessarily will occur in locations where new forests are particularly required. Where recovery is unlikely or risky it may be preferable to plant seedlings or directly sow seeds.

Holl and Aide (2011) discuss the general question of when and where to rely on natural regeneration while Dorrough *et al.* (2008) explored it from the perspective of an individual farmer interested in reforestation on part of their land but having to balance costs and risks of failure. They noted the issues are essentially similar to those faced by a regional planner having to decide where and how to invest reforestation funds or incentive payments. Their conclusion was that natural regeneration could be attractive when agricultural productivity is low and the long time periods required for regeneration are acceptable to landholders but it may be less attractive on more productive lands and where farmers are more risk averse.

The most obvious first step to take is to decide if, in fact, there is any potential for natural forest regrowth within the desired time or whether it might be more prudent to undertake other forms of reforestation.

### *Evaluating the regrowth potential*

Sometimes the likelihood that natural forest regrowth will occur unaided is obvious because there are numerous trees or samplings still remaining scattered across a site and these are surrounded by large numbers of seedlings. In such cases, all that may be necessary is to protect the site from further disturbances and allow the trees and new seedlings to develop. But in other situations there may be no residual trees, seedlings may be sparse and there may only be a few old stumps able to produce sprouts. Or sometimes there may be seedlings and sprouts present but they are heavily shaded by a dense grass cover. How can one judge whether to rely on natural regrowth or not?

There are three important indicators. These include: (i) the overall density of well-established seedlings of woody plants; (ii) the evenness of their distribution across the site; and (iii) the identity of these species. It is difficult to be prescriptive about the size a seedling must reach in order to be 'well established'; it is obvious that a seedling only 10 cm tall is less likely to survive than one that is 110 cm tall. But much depends on the height of the surrounding weed cover; even a 50 cm tall seedling may not survive if the surrounding grass competition is 100 cm tall. Nonetheless, 50 cm is probably a useful benchmark, especially if some initial weed control is carried out.

In the case of the actual seedling density, Shono *et al.* (2007) suggest there should be 200–800 seedlings per ha if natural regrowth is to be feasible in the humid tropics. Elliott *et al.* (2013) suggests even higher stockings of around 3,000 seedlings per ha may be desirable in these regions in order to achieve canopy closure within two rainy seasons. This is because the resulting trees will close canopy and exclude grass competition more rapidly than when seedling densities are low. It also means that a greater diversity of species is likely to be present. But it may be unusual to find consistently high seedling densities like these over extensive areas and it may be more feasible to use a more modest standard of, say, 100–200 seedlings per ha depending on the amount of competing vegetation. Lower tree numbers are also likely to be acceptable in drier landscapes, especially where landholders intend to grow crops of graze stock beneath the trees. In short, the most suitable seedling density depends on assessments of risk and the intended future land use (Figure 3.5).

The importance of the species identity is that this determines the attractiveness of the trees to landholders. Much depends on whether the objective is to create an economic resource (e.g. Nepal and Niger case studies above) or to quickly achieve canopy closure for conservation purposes (e.g. the Costa Rica case study). Where economic concerns are paramount a limited variety of economically attractive species may be perfectly acceptable even when some of these are exotic species (provided they are not invasive). But more weight will be placed on having a larger

*Figure 3.5* Naturally regenerated tree seedlings on a hillside in Hong Kong. The density of woody plants is not high but forest recovery will be possible provided the site is not burned.

variety of mostly native species where biodiversity conservation is more important.

Finally, the degree of patchiness matters because this determines whether supplementary plantings will be needed in certain locations within the overall area. Further, large gaps in the distribution of seedlings could enable patches of fire-prone species such as grasses to persist posing a continued fire risk across the site. The definition of a 'large' gap is likely to be governed by judgements such as whether seed are continuing to colonise the site and new seedlings are still appearing, or whether the distribution of seedling colonists is largely finalised. It may be feasible to ignore some small areas currently devoid of seedlings because it is clear that lateral branch growth or future seedling colonists will fill them. But other larger gaps may need to be filled by plantings.

An over-riding problem lies in deciding at what scale to make these assessments. It may not be difficult to do them across a few hectares but rather more difficult to do so on a much larger scale. The Nepal and Niger case studies offer some guidance here. In both these cases the assessment was done by individual landholders who had to decide what would be the best management decision they could make for their own farms; they were less concerned about the broader landscape consequences. But more pragmatic decisions may have to be made where managers are responsible for reforesting much larger areas and have limited budgets. In such cases it may be necessary to simply tolerate slower rates of recovery.

### *Triggers for establishment*

In each of the case studies above regrowth only occurred in the absence of further disturbances and, in the case of drier regions, following a period of good rain. This changed disturbance regime may have been the cessation of cropping, the absence of fires or the control of grazing animals through mechanisms such as enclosures (Mengistu *et al.* 2005, Angassa and Oba 2008). Sometimes these changes can be caused by political events such as wars. But in most cases, the more important underlying trigger concerns the socio-economic factors or policy alterations which induce changes in landowners' practices. Socio-economic factors (especially the shortage of forest products) appear to have been one of the main reasons for the development of community forestry and the encouragement of natural regrowth in Nepal. In this case policy changes lagged behind what was happening in the field. Likewise, socio-economic factors also under-pinned the encouragement of forest regrowth on farms in Niger which initially took place in a policy vacuum. Again, it seems policy settings subsequently changed to reflect what was happening on the ground. But policy changes were responsible for enabling much of the forest regrowth in the case studies from Norway and Costa Rica. Policy changes have also been responsible for the gradual diminution of shifting cultivation in Thailand as well as in many other tropical countries. In Thailand the change was also influenced by population growth that made shifting cultivation unsustainable, political pressure to stop opium cultivation and the expansion of the protected area network to cover many areas where shifting cultivation had been practiced.

A key factor responsible for transforming regrowth from worthless scrub into a valuable resource is land tenure. Where tenure is uncertain or contested there is a high risk that land containing forest regrowth will be cleared for some other purpose. This is much less likely if ownership is clear. Perz and Skole (2003) observed that forest regrowth in Brazil was more likely to be successful in areas with longstanding settlements and stable cropping systems. Conversely, regrowth was difficult to protect when patterns of land ownership were uncertain and when large numbers of people were migrating into an area. Similar observations have been made by Elmqvist *et al.* (2007) in Madagascar. An intriguing additional factor was observed in El Salvador where Hecht and Saatchi (2007) noted that regrowth appeared to be more common where landholders were receiving remittances from overseas family members; this allowed them to purchase food, rather than having to grow it.

### *Catalysing and accelerating successional development in regrowth*

In many cases natural regrowth is followed by rapid successional recovery but in some cases it is not. Successional development can be facilitated in several ways, a process sometimes referred to as Assisted Natural Regeneration (Dugan 2000). The first is where there are already sufficient seedlings or coppice growth present but these are not growing very strongly because of surrounding weeds. In this case the most obvious intervention is to remove or control the weeds (Hardwick *et al.* 2004, Shono *et al.* 2007, Elliott *et al.* 2013). This might be done manually or by the carefully controlled use of herbicide or by grazing (Janzen 2002). By doing this light levels are improved and competition for moisture is reduced. But some caution is needed here because clearing shrubs and grasses can improve the survival of some species but hinder others. In Thailand, Hardwick *et al.* (1997) found the early survival of young seedlings of some tree species was actually enhanced by grass cover, at least in the short term, because they provided shelter without which these seedlings could not have survived. Likewise, in Chile and Argentina, Williams-Linera *et al.* (2011) found surrounding shrubs could protect young tree seedlings from frosts and heat. In Holland, Kuiters and Slim (2003) found brambles protected browse-sensitive oak and other woody species regenerating at abandoned farmland. It is clear that in cases such as these a balance must be struck between leaving some shrub cover to facilitate seedling survival and removing it once the seedlings are established so that, in the longer term, the competition does not prevent the tree seedlings from growing.

A second way of catalysing the successional process is to accelerate the rate at which seeds are dispersed into the site. This can be achieved by planting individual trees or clumps of trees to act as bird perches and nodes of seedling recruitment (Toh *et al.* 1999, Benayas *et al.* 2008, Cole *et al.* 2010, Holl *et al.* 2011). Zahawi *et al.* (2013) found these were as effective as more extensive plantings but cheaper. Some have even had success using artificial structures to accelerate the process (Scott *et al.* 2000). This has not yet been done on a large scale and, while it may be financially advantageous to limit the numbers of trees needing to be planted, it may also be difficult to protect isolated trees like these planted in fire-prone environments. They may also be difficult to relocate in order to control weeds sufficiently for the seedlings to become established.

An alternative to planting individual perch trees is to establish small tree plantations or 'tree islands' of perhaps 100 $m^2$ to exclude weeds and facilitate the commencement of natural successional development (Zahawi *et al.* 2013). Planting enough trees to form a closed-canopy forest is obviously more costly than planting scattered trees or clumps of trees but it is easier to ensure the trees are established. On the other hand, the newly planted forest has some psychological significance because it signals that the land is valued and being used. This can help reduce (though not necessarily prevent) the risk of fires burning though an area (Figure 3.6).

*Figure 3.6* Small forest patches can be exposed to the risk of frequent wildfires, especially in areas with seasonal climates.

A third approach to accelerating the successional process is to deliberately plant seedlings of certain species within the regenerating forest. As noted earlier, the composition of regrowth forests may differ from the original forests (Figure 3.2). Much depends on the disturbance history and the spatial patterns of land use surrounding the particular site. The composition of the new forest can be modified by enriching it with commercially desired species or species having some particular conservation value. This can be done by planting seedlings or sowing seed and has been practiced in many heavily-logged tropical forests (Adjers *et al.* 1995, Lamb 2011). Enrichment plantings have also been used to diversify some temperate forest monocultures (Ammer *et al.* 2008). In most cases the best time to do this is when the forest is relatively young and before the canopy is too closed. If the canopy has closed, canopy gaps large enough to allow light to reach new seedlings on the forest floor will have to be created. Seedlings of the preferred species can then be planted across the site and tended until they are established.

## Management issues

The continued development of regrowth forests depends on how well they are protected and managed. Provided this is done they can become important sources of a variety of goods and ecosystem services (Chazdon 2014). Some goods such as fuelwood or certain medicinal plants may be available when the forests are still comparatively young. By contrast, services such as the provision of habitats for wildlife may take many years to fully develop. The extent to which goods or services are provided can be influenced by the type of management applied to the new forests.

### *Silvicultural systems to generate forest products*

Naturally regenerating forests on former agricultural lands are unlikely to be attractive to industrial forest owners – at least in the short term – but are often of considerable interest to nearby communities and households (who may claim customary ownership of the land). In most situations the forms of management available to such managers are not very well developed. Traditional silvicultural practices used for industrial forestry are usually based on careful measurements of the growth of commercially attractive timber species and the development of prescribed cutting cycles or rotations based on these data. Neither these measurement systems nor the use of fixed cutting cycles are necessarily relevant to many community groups or households, especially those in developing countries. Instead of seeking to maximise timber productivity, these landholders are more likely to manage regrowth forests on a rather more flexible basis and be interested in producing a much wider variety of products other than just timber. These might include fuelwood, fodder and perhaps also medicines and food (Figure 3.7). Some people may also wish to graze animals in the area or

*Figure 3.7* Fuelwood being collected from regrowth forests (Photo: S. Kelleher).

even grow crops beneath the trees (e.g. the Niger case study). In such cases the most appropriate silvicultural system is one necessarily based on the ecological attributes of the species present, but also one that that reflects these various objectives.

There are a variety of alternative approaches that might be used to manage new forest regrowth including periodically coppicing the trees or developing a system of coppice with standards (i.e. seed trees to ensure additional trees are added to replace those lost after successive coppice fellings). Trees might also be heavily pruned or pollarded to allow small branches to be regularly harvested. Alternatively, trees might be harvested on some kind of a selection system which involves longer time periods but produces larger logs.

There are unlikely to be off-the-shelf silvicultural systems available to match the particular circumstances at many regrowth forest areas and in most cases specific silvicultural systems will have to be developed. Assuming regeneration has occurred across the whole site, the key questions are likely to be:

- Are unwanted species present and should these be removed? Unwanted species might include weeds limiting the growth of trees or unwanted tree species likely to compete with more preferred tree species.
- What is the most appropriate tree density? This will depend on environmental conditions (e.g. rainfall) and whether grazing or cropping will be carried out beneath the trees.

- Is the density of the preferred tree species now sufficient or should some kind of enrichment take place?
- What can be harvested from the new forest (timber or non-timber products) and when can this harvesting commence?
- How can over-harvesting (of timber or non-timber products) be prevented?
- How can the occurrence of sufficient tree regeneration be ensured?
- If the area is to be grazed, what stocking rates might be appropriate and how might this be managed to avoid damaging the trees?

### *Management to achieve conservation objectives*

Many regrowth forests quickly acquire considerable biodiversity although, as already noted, much depends on the land use history and landscape context. Considerable biodiversity can be contained in secondary forests surrounded by large areas of primary forest but perhaps rather less in secondary forests surrounded by farmland. Most studies in tropical areas have only investigated forests less than 50 years old. In a major review Chazdon *et al.* (2009) found many volant species (i.e. able to fly), representative of mature natural forests, had recolonised secondary forests by this time but populations of less mobile species were still recovering. In a meta-analysis of data from 600 tropical forest sites Martin *et al.* (2013) found that, in comparison with nearby intact forest, total tree species richness (though not necessarily all the tree flora found in undisturbed forests) had re-established after 50 years of regrowth. Likewise, 83 per cent of the above-ground biomass (and carbon) had recovered after 85 years. They suggested this meant that many younger (<100 years) tropical regrowth forests may have more value as carbon stores than for biodiversity conservation although it is clear that they can achieve both. In such cases, the main management task is simply to protect the regenerating forest and allow successional recovery to continue.

The situation can be different in cultural landscapes that have been used for agriculture over long periods such as those in Western Europe. In these cases a specialised biota evolves to utilise the mosaic of land uses present, such as crop areas, semi-natural pastures, hedgerows and orchards. When these agricultural landscapes are abandoned and forest regrowth occurs there is likely to be a considerable turnover in species. Indeed, Benayas *et al.* (2007) have raised concerns that forest regrowth in European landscapes could lead to an overall reduction in regional biodiversity since landscape heterogeneity declines and species associated with the former cultural landscapes are replaced by those found mostly in forests. They also point to fears that uncontrolled wildfires might become more common and lead to further simplification. It is for this reason that those involved in the European Rewilding Project (Accessed 16 May 2014. Available at www.rewildingeurope.com) have decided that some existing non-forest land will be retained because reforestation of the whole landscape would result in the loss of some biodiversity, especially species preferring open and semi-open landscapes. The issue largely concerns how much weight to give to habitat generalists and habitat specialists (Navarro and Pereira 2012). The topic is discussed further in Chapter 6.

The primary task of those managing regrowth forests destined to become conservation areas is to protect these from further disturbances and monitor the successional process to ensure the desired trajectory is maintained. It may be possible to accelerate the recovery of some wildlife by artificially creating habitat features such as nesting boxes or deliberately translocating certain species from elsewhere. Over time, some wildlife within the new forest may move out and become pests in neighbouring farmlands and ways may be needed to control this (Navarro and Pereira 2012).

### *Some unresolved questions*

There are already large areas of established regrowth or secondary forests in both temperate and tropical areas. In the case of tropical secondary forests remarkably little is known about even the composition of these or the ecological processes that sustain them. There are a variety of questions needing to be resolved to enable the potential value of natural regeneration to be expressed.

- What is the potential for natural regeneration of forests at a particular site? Are there seeds, seedlings or coppice from old stumps present? What is the density of these and how evenly is it distributed? Which species are present and which species predominate?
- What needs to be done to facilitate natural regeneration at sites with limited numbers of residual trees or seedlings? Is the problem one of too few colonists arriving at the site or is it that too high a proportion of colonists fail to germinate or survive and become established?
- What native species are absent from natural regeneration at these sites and how might these be encouraged to recover?
- At what point will various wildlife species be able to recolonise naturally regenerating forests and how might the recovery of these species be assisted?
- Under what circumstances might landholders be interested in promoting natural regrowth? In what way might they benefit? How might landowners be persuaded that natural regrowth is worth conserving? What forms of governance might be needed?
- What needs to be done to ensure government policies and regulations reflect the potential value of natural regrowth to individual landowners, communities and the state?

## Conclusions

Trees are often able to regenerate on cleared land through natural means if they are protected from disturbances. Natural regrowth will not always occur but it can be a powerful means by which large areas of degraded land are reforested at a relatively low cost.

Natural regrowth depends on there being a legacy of old roots or stumps able to resprout, a residual soil seed bank and/or additional seed being dispersed into the site by wind or animals. Its effectiveness depends on there being an appropriate density of trees and a relatively uniform distribution of these. Its effectiveness also depends on the identity of the species regenerating across the area.

The number and diversity of species likely to be dispersed into the site will be a function of the landscape context; large numbers of seeds are likely to be dispersed into sites near intact forest but fewer seeds will reach more isolated sites or those where only small forest remnants remain as seed sources. The tree species most readily dispersed over long distances are those with small fruit or seeds. The rate of colonisation can be increased by establishing perch trees to attract seed dispersers.

Regrowth forests will not necessarily be identical to the original forests: some may have an impoverished flora or be dominated by a relatively small number of species (including weed species).

Naturally regenerating forests are often at risk of further disturbances such as wildfire or clearing. This is less likely if land ownership is clear and a specific management regime is in place. Managers may seek to use natural regrowth to re-establish new forest ecosystems or agroforestry systems involving mixtures of trees with crops or pastures.

Despite its advantages, natural regeneration may not be suitable under all circumstances (e.g. it may be too risky or too slow) and in some situations reforestation may have to be undertaken by planting seedlings or directly sowing seed.

# 4 Types of planted forests

## Introduction

Plantations are usually established because lands are deforested and neither the existing areas of natural forests nor forest regrowth are able to supply the goods and ecosystem services required by the community. Most people associate tree plantations with visions of regimented rows of closely spaced trees, usually of a single exotic species. But there is a much wider variety of possibilities that might be used. The most appropriate of these depends on the manager's objectives and the type of land being reforested. A common objective is to produce sawn timber, pulpwood or some other forest product. But an increasingly important alternative objective is to reforest cleared or degraded areas to protect the environment, conserve biodiversity and restore key ecological functions. This was often a secondary goal of earlier reforestation programmes but is now becoming a primary goal in many places. Different types of reforestation will usually be used for these various objectives (Keenan *et al.* 1999). This chapter explores this variety of silvicultural options and considers the advantages and disadvantages of each for large-scale forest restoration. It includes methods that use seedlings raised in nurseries as well as approaches involving directly sowing seeds in the field. The chapter begins by exploring that most basic of questions – whether to use native or exotic species?

## Species choices and seed sources

A handful of genera have dominated the world's plantations including *Pinus, Eucalyptus, Acacia, Tectona* and *Populus*. Likewise, a relatively small number of species from these are the most widely used. A much debated topic in forestry circles is whether to use such exotic species or whether greater use should be made of native species. There is some ambiguity in the terminology since a 'native' may be indigenous to a country but not normally be found in the area being reforested. In the discussion that follows a 'native' is taken to be a species that previously grew at the site while an 'exotic' is a species not known to have been historically present.

Exotic species are often used in large-scale planting programmes because they are seen as being especially useful, they are well known in the market place, their seed is easy to get and they grow quickly. Some of the other advantages of exotic species are summarised in Table 4.1. There is often some symmetry in these comparisons in the sense that the purported advantage of exotics is often matched by a supposed disadvantage associated with native species (e.g. the relative availability of seed). There are, of course, examples of native species that are widely used in timber plantations because they escape these limitations and it is likely that further research into seed supplies, nursery techniques and silviculture could increase the attractiveness of more native species for plantations.

The consequence of relying on exotic species for large-scale reforestation programmes is that, over time, landscapes previously containing a variety of species are gradually replaced by simplified and homogenised landscapes containing many fewer tree species. But the dilemma for those wishing to avoid this simplification is that much less is usually known about most native tree species. Just where can seed be obtained now that the original forests have been fragmented and degraded? How can one raise large numbers of seedlings from this seed? What types of sites do these species tolerate now that the original site conditions may have been degraded?

The importance of these questions depends on the purpose for which reforestation is being undertaken. If the purpose is to supply, say, a pulp mill requiring large amounts of timber with specific wood properties then it may make perfect sense to use seeds (or clonal material) of an exotic

*Table 4.1* The advantages and disadvantages of using exotic or native species in planting programmes

| *Exotic species* | *Native species* |
|---|---|
| Commercially attractive. | Some have very high commercial value but others have limited appeal. |
| Easy to get seed and store it. | Seed can be difficult to obtain (infrequent seeding, limited knowledge of parent tree locations) or store. |
| Have established nursery techniques. | Knowledge of required nursery techniques often limited. |
| Grow rapidly at a variety of sites including degraded sites. | Often grow more slowly and have more specific site requirements. |
| Many have been subject to considerable tree-breeding. | Have less commonly been subject to tree-breeding. |
| Provide some ecosystem services. | Usually provide more ecosystem services. |
| Indigenous pests and diseases may be absent. | Have associated indigenous pests (with their predators) and diseases. |

species that has been through an extensive breeding programme to produce just that type of wood. But if trees are being grown by landholders who are more interested in producing a variety of products in order to reduce economic or ecological risks then systems using several native species may be a better choice. And those wanting to enhance biodiversity conservation at a landscape scale will certainly wish to use more native species, including those based on local seed provenances. However the decision is not a simple one. Even when the long-term objective is to enhance biodiversity conservation there may be situations where an exotic species should be used as a nurse plant to ameliorate site conditions and facilitate the eventual establishment of the preferred native species. This issue is discussed further below.

### *Invasive exotic species*

A potential disadvantage of exotic species is that they can become invasive (Dodet and Collet 2012). This is more problematic when a species is being introduced to an area for the first time and little is known of its environmental tolerances in the new field situation. There may be fewer risks if the species is already established in the region and its ecological attributes are understood. Richardson and Rejmanek (2011) report that less than one per cent of woody plants are invasive outside their natural range but also warn that woody weeds are now becoming more of a problem in many places. It is clear that a balance must be struck between taking advantage of the benefits of using exotic species and the possible risks involved in doing so. More generally, it is important to recognise that fast growth is not the only factor involved when choosing which species to use and the handful of species favoured for industrial forestry should not become the automatic choices for those undertaking reforestation for other purposes.

### *Seed collection*

Seed collection is an especially critical task for those wishing to use native species. Any large-scale reforestation effort will require significant volumes of seed (Merritt and Dixon 2011). This might not be difficult for some of the more common species but may be harder for species that are less commonly used. Those likely to be most difficult will include species having unknown phenologies, that are hard to store, difficult to germinate and that have low seed viabilities. The problem will be most acute in degraded landscapes that have little natural forest remaining and where seed sources for most native species consist of just a few trees now growing in widely scattered locations.

In order to maintain evolutionary opportunities it is important that seed used in plantings come from a diverse range of parent trees and preferably >50 trees (FAO *et al.* 2001). But genetic diversity may have already been compromised by forest fragmentation and the disruptive effect this will have had on pollination. Reduced pollen flow can lead to inbreeding depression and have a serious effect on future tree performance. In short, it may be difficult to use a wider variety of native species, especially in some tropical regions where diversity is high, without considerable research effort to improve our understanding of their basic seed biology.

## The three main approaches to reforesting cleared land

Broadly speaking, there are three approaches that can be used to reforest cleared land. These are shown conceptually in Figure 4.1. The number of species on the X-axis varies and is obviously lower in boreal forests and higher in tropical forests. The Y-axis is some expression of growth (e.g. biomass) or structural complexity. When a natural forest (A) is cleared, the site loses species and the vegetation biomass and structure at the cleared site is much simplified (B). There are three possible ways of undertaking reforestation at such sites. One is to adopt the industrial forestry model and establish a plantation using a single species (C). With careful site preparation and fertilisation the plantation may develop a greater biomass than the original forest (D). Alternatively, a variety of species could be grown in plantation mixtures (E). These could be grown for production purposes or to rehabilitate degraded lands. Forests at C, D and E lack many of the

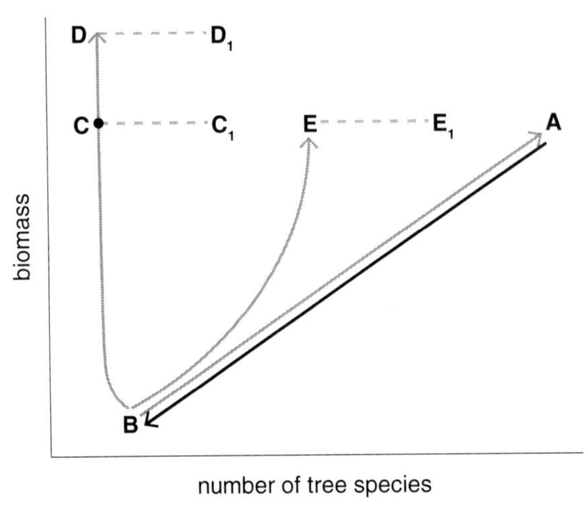


*Figure 4.1* A conceptual outline of alternative ways of restoring forest cover at a deforested site. The original forest is represented by A and the deforested site by B. Different forms of reforestation can be used to create forests represented by C, D or E. As they become older each of these plantations can be colonised by additional species from nearby natural forest thereby increasing overall diversity although not necessarily biomass (C1, D1, E1). Lastly, Ecological Restoration might be used to recreate the forest once present at A. (See text for more details.)

species formerly present in the original forest but, as discussed in Chapter 3, some of these species may recolonise the sites if patches of residual forest remain nearby in the landscape and seed dispersers are able to carry seed from these to the plantations (C1, D1, E1). Finally, an attempt could be made to restore the original forest ecosystem that previously existed at the site (i.e. return to A).

Those undertaking reforestation for production purposes have developed a variety of methods to establish monocultural timber plantations (represented by C or D). In most cases these plantations are grown for a fixed period before being harvested and replanted. But those concerned about conservation and ecological functioning are likely to prefer restoring the more diverse forests represented by A in Figure 4.1. This approach will be referred to here as 'Ecological Restoration' and is a rather more complicated process, especially when it is being done by planting tree seedlings. The third approach (E) lies between these two and involves the use of a variety of species. However, these may be fewer in number than were originally present and may include some exotic species. The advantage of this approach is that it can provide goods like timber but rather more ecosystem services than simple plantation monocultures. At the same time, it is likely to be rather easier to implement than Ecological Restoration and could be the best approach to use in heavily degraded sites where Ecological Restoration might be extremely difficult to achieve. Lamb and Gilmour (2003) referred to this as rehabilitation.

The forms of reforestation used to initiate each of these pathways differ in a number of ways (Table 4.2). Perhaps the most obvious difference lies is in the numbers of species planted. Those with fewer species tend to be managed for production purposes while those with more species tend to be aimed largely at the provision of ecosystem services although forest products might also be harvested. A second difference is in the extent to which the plantings are subsequently disturbed. Forests managed for production (e.g. C and D in Figure 4.1) tend to grow trees for finite periods at which point they are felled and replanted; short rotations (often <7 years) are commonly used for pulpwood production but longer rotations (>20 years) are usually used for trees producing sawn timber. Forests primarily established to generate ecosystem services such as watershed protection may not be harvested at all but, if they are, then this may be done through a modest selective logging operation; no replanting is needed because they are self-sustaining. A third difference between the three approaches concerns the degree of structural complexity allowed to develop: monocultural plantations are usually even aged with only a limited understorey while multi-species plantings often develop into uneven-aged forests and sometimes have very diverse understories. All of these differences affect the capacity of the new forests to supply ecosystem services although even simple monocultures can generate some services.

*Table 4.2* Attributes of the three main types of plantings used to reforest former agricultural or 'degraded' lands (NTFP – non-timber forest products)

| *Planting method* | *Production of goods* | *Tree species used* | *Rotation length* | *Structural complexity* | *Provision of ecosystem services* |
|---|---|---|---|---|---|
| Monoculture – shorter rotation | pulpwood | Single | <7 years | Simple | More limited** |
| Monoculture – longer rotation | sawn timber, ply | Single | >20 years | Simple though can acquire understorey | |
| Mixed-species plantations | Timber and NTFP | Several | >20 years | More complex | |
| Ecological Restoration | Timber and NTFP | Many* | Not harvested | More complex | Higher |

*Actual number depends on geographic location
**Though even plantations grown on short rotations may supply some ecosystem services

## Planting methodologies for production purposes

There is a very wide variety of approaches that can be used to initiate these three forms of reforestation. It is useful to distinguish between methodologies mostly used to establish production forests and those used for conservation and protection purposes while recognising this is not a clear dichotomy and that many production forests are managed for both purposes. The most appropriate form of silviculture to use in production forests depends on the markets available, whether the forest owner is a large industrial-scale organisation or a smallholder and, if the latter, on the socio-economic circumstances of the landholder. Large industrial plantings are likely to be carried out following intensive site preparation techniques such as ploughing or mounding as well as using weedicides and fertilisers to boost tree growth. Smaller growers are less likely to use these methods although they, like large corporate owners, must ensure new plantings are protected from herbivores and wildfire. Some of the common causes of plantation failure are listed in Table 4.3.

### *Monocultures of timber trees grown on short rotations*

These are intensively managed plantations often involving an exotic species. They are commonly grown at densities of around 1,000 trees per ha to produce wood pulp and trees may be clear-felled after as little as 10 years. Site preparation and stand management is intensive because the objective is to maximise timber productivity. Wood volumes exceeding 40 $m^3$ per ha

*Table 4.3* Common causes of plantation failures

| *Issue* | *Expressed as* |
|---|---|
| Land ownership | Land ownership is contested leading to trees being uprooted or the plantation being deliberately burned or grazed |
| Seedling quality | Poor quality seedlings with problems such as being pot-bound or having unbalanced root-shoot ratios |
| Planting | Planting carried out in wrong season (too early? too late?) |
| Protection | Insufficient weed control; unprotected from herbivores; burned in wildfire |
| Soils | Too compacted, too shallow or with nutritional limitations |
| Climatic variability | Droughts and frost |
| Biology | Pests or diseases |
| Silviculture | Inappropriate site for species; lack of thinning so stand stagnates |
| Management | Unprofitable because too small an area planted, the site is too distant from markets or there is no market for products produced by these particular species |

per year have been recorded in some eucalypt plantations, especially in the tropics and sub-tropics, and productivity levels like these are achieved by tree-breeding, heavy fertilising and the intensive use of weedicides. Nutrition can be a particular problem since young trees usually have a high proportion of nutrient-rich sapwood. This means there is a considerable nutrient drain when trees are harvested and this must be replaced at the time of the next planting or second rotation. Pruning or thinning is rarely practiced because the emphasis is on maximising timber volume rather than log quality.

Plantations grown on short rotations can be highly profitable and are used by most large industrial pulpwood plantation owners, especially those operating in tropical and sub-tropical environments. The most profitable plantations are those near processing mills or with good access to ports for export. They can also be profitable for smallholders but only if these growers have land near to mills owned by large timber companies or near to good transport systems. In such locations joint ventures or out-grower schemes can be mutually beneficial to the small landholder (because they provide access to a reliable market) and the large timber corporation (because such arrangements provide access to more land without them needing to buy it). Pulpwood plantations are much less suitable in more distant locations since pulpwood prices are usually relatively low and transport costs quickly reduce profitability.

Most plantations grown on short rotations for industrial timber are replanted after harvesting but some might be managed by relying on coppice from stumps. This may be an especially attractive option in situations where trees are being grown for fuelwood. The number of coppice rotations that can be harvested before replanting is needed depends on the mortality of the stools (i.e. the stumps from which coppice sprouts) but four to five rotations may be possible before replanting is needed (Smith 1986, Evans and Turnbull 2004).

Monocultural plantations of exotic species that are clear-felled at frequent intervals are not particularly attractive for conservation purposes although as Cossalter and Pye-Smith (2003) point out, it depends what type of sites they replace. In addition, fast-growing trees of any species commonly transpire large amounts of water. This can make tree plantations unattractive in landscapes where water resources are limited. On the other hand, this attribute can advantageous in some situations. This has been the case in Western Australia where dryland salinity has become a serious problem and there are advantages in lowering water tables. This is discussed further in Chapter 5.

### *Monocultures of timber trees grown on longer rotations*

Many plantations involve trees grown on much longer rotations with periods of 20–60 years being common. They frequently involve species of well-known genera such as *Pinus, Picea, Populus, Tectona* or *Swietenia*. The advantage of longer rotations is that trees are able to grow to sufficient sizes to be efficiently sawn or peeled for veneer. The higher price for these products, relative to that of pulpwood timbers, is usually sufficient to compensate for the longer growth period. More valuable logs also mean that it is economic to transport these over longer distances to distant markets; that is, plantations grown on longer rotations can often be grown in more isolated locations than can pulpwood plantations. Plantations containing especially high-value timber trees obviously have even more of an advantage.

Unlike pulpwood forests, plantations managed on longer rotations are usually thinned to reduce between-tree competition and so hasten individual tree growth. The initial planting density may be around 1100 trees per ha and these can be thinned to, say, 400 trees per ha at the time of clear-felling. In theory this might seem to offer an opportunity for some early financial returns for growers but, in practice, it is sometimes difficult to find markets for the small trees removed in earlier thinnings. Considerable efforts have been made develop thinning schedules that balance the silvicultural requirements of particular species and local market realities. Trees grown on longer rotations are also usually pruned to increase the value of logs by minimising the occurrence of knots. At the end of the rotation, all remaining trees are normally clear-felled and the site is replanted.

Long rotations offer considerable management flexibility and a large variety of silvicultural systems involving trees grown on longer rotations have developed. These involve differences in rotation lengths, thinning frequencies or tree densities (Smith 1986, Evans and Turnbull 2004). For example, it may suit a grower to use very low tree densities in order to allow grass to grow between the trees and enable cattle to graze amongst the trees.

A particular advantage of plantations using longer rotations is that, unlike most pulpwood plantations, they offer the possibility of supplying some ecosystem services while still producing goods such as timber. For example, most of these types of plantations are able to prevent soil erosion because an understorey of grasses and shrubs usually establishes on the forest floor and this is only disturbed again at the time of harvesting (i.e. at intervals of more than 20–60 years). On the other hand, these structurally simple plantations may not be attractive habitats for many wildlife species and especially species with specialised habitat requirements. But there is scope for manipulating sawlog plantations if management objectives change and greater emphasis on certain ecosystem services is required. For example, some old timber plantations in Europe are being transformed from even-aged forests into uneven-aged forests and some conifer monocultures are being enriched with hardwoods to make them more attractive for recreational purposes and to help conserve biodiversity (Malcolm *et al.* 2001, Lof *et al.* 2010). An example is shown in Figure 4.2.

*Figure 4.2* Enrichment of an older Swedish conifer plantation with seedlings of broad-leaved species. Canopy openings have been created to allow light to reach the new seedlings on the forest floor.

Over time many simple monocultures can acquire very diverse understorey communities, especially if there are nearby natural forests able to supply colonists (Keenan *et al.* 1997, Parrotta *et al.* 1997, Firn *et al.* 2007). To some managers these colonists might be seen simply as weeds competing with the plantation trees. But others will see the developing understorey as a means by which diversity is added to the site and the plantation becomes a stepping stone on the way to restoring a natural forest. An example is given in Box 4.1.

**Box 4.1 The catalytic effect of timber plantations**

The forests of Tam Dao Mountain in northern Vietnam were heavily degraded and some areas were reforested in the 1940s using *Pinus massoniana*. The original intention was that these plantations would be used to produce sawn timber. Subsequently, however, the pine plantations were incorporated within the Tam Dao National Park. Over time a dense understorey community has developed beneath the plantation trees (Figure 4.3). Trees in the understorey are now about 12 m tall. A survey carried out in 2007 found it contained a rich assemblage of species including 122 trees species, 46 shrub species and 33 vine species (Cuong *et al.* 2013). Amongst the species present were 16 species classed as being endangered or vulnerable. The source of all these colonists was the surrounding intact natural forest most of which was within 700 m distance. Most of the species colonising the plantations had seed or fruit less than 3 mm in size and it seems likely these were dispersed into the pine plantations by wildlife. By contrast, plants with larger seed or fruit were not being dispersed.

*Figure 4.3* Understorey of native species developing beneath the canopy of a *Pinus massoniana* plantation now incorporated within Tam Dao National Park, Vietnam. The native species have colonised from nearby undisturbed natural forest.

Park managers now have several options for managing this plantation. One option is to selectively fell some of the pines and sell the logs to improve park finances and then let the understorey trees grow up and take over from the remaining pine trees. This gradual process would maintain tall trees on the site but cause some damage to the new colonists. An alternative option for managers is to girdle or poison all the pine trees and let them die *in situ.* This might be less damaging to the understorey community than logging and the gradual opening of the canopy layer would allow the understorey to flourish and eventually take over. There would be no income to park managers. Finally, managers could do nothing and leave the pines to gradually senesce. Again the native species in the understorey would take over although this might take some time.

### *Multi-species plantings of trees*

The second pathway in Figure 4.1 involves planting mixtures of several species. Multi-species plantations, or polycultures, are stands involving intimate mixtures of trees of several species at a particular site (cf mixtures at a landscape or enterprise level involving a spatial mosaic of monocultures of different species). These intimate mixtures are more difficult to establish and manage than monocultures and this is the reason why industrial plantation owners have tended to ignore them. But there has been a surge of interest in the silviculture of mixed-species plantations in recent years (Nichols *et al.* 2006, Felton *et al.* 2010). Indeed, Knoke *et al.* (2008) argue there is now a clear trend favouring species mixtures in Europe and especially in Germany. Some of the potential advantages of multi-species plantings are summarised in Table 4.4. This illustrates the point that multi-species plantations can have a role in both production systems as well as in situations where new forests are being created to supply ecosystem services. Most of these plantings probably involve fewer than 4–5 species chosen to be complementary in terms of their niche requirements, facilitative capacities, growth rates or economic benefits.

Designs for multi-species plantings have been explored by both ecologists and foresters. Ecologists have focussed on the functional consequences of increased (and decreased) diversity: their studies have addressed such things as species niche complementarity (for nutrients, light, water), nutrient cycling in mixtures and changes in herbivory in multi-species communities compared with those in monocultures (Hooper *et al.* 2005, Jactel and Brockerhoff 2007, Richards *et al.* 2010, Mori *et al.* 2013). These studies have shown that carefully designed mixtures have the potential to

*Table 4.4* Potential advantages of multi-species plantations

| *Advantage* | *Reason* |
|---|---|
| Amelioration of degraded sites | Facilitator species modify site conditions and allow establishment of preferred species (McNamara *et al.* 2006). |
| Enhanced production | Greater complementarity between species with contrasting canopy or root architectures (space) or phenologies (time) allows reduced between-tree competition (Cardinale *et al.* 2007, Paquette and Messier 2010, Forrester and Smith 2012, Zhang *et al.* 2012). |
| Improved nutrition | Forest has greater capacity to access and conserve nutrients, especially if nitrogen fixers involved in mixture (Richards *et al.* 2010). |
| Reduced damage from pests and diseases | Susceptible tree species hidden in space or natural enemies of pests encouraged by habitat conditions (Jactel and Brockerhoff 2007). |
| Improved financial outcomes | Diversity of plant species increases diversity of marketable products. Cash flows improved because some products can be harvested before the end of rotation (Knoke 2008). |
| Improved regional conservation outcomes | Forest contains more tree species and provides a more structurally diverse habitat for wildlife and better watershed protection (Kanowski *et al.* 2005, Bhagwat *et al.* 2008, Benayas *et al.* 2009, Aerts and Honnay 2011). |
| Improved resilience and ability to withstand ecological and economic disturbances | Forest has greater diversity of functional types and of marketable products. Can be more able to withstand environmental shocks (e.g. droughts, windstorms) and market fluctuations (Knoke *et al.* 2008, Roessiger *et al.* 2011). |
| Improved landscape aesthetics and recreational opportunities | Forest has greater visual and structural diversity. |

improve productivity through improved nutrition, better use of site resources and reduced herbivory. But they have also shown these outcomes will not be achieved using random species assemblages and that any benefits are not simply a consequence of having a larger number of species (Nguyen *et al.* 2012). Instead, species must be complementary in their use of niche space or one species must facilitate the growth of others (e.g. by being a nitrogen fixer). That is, knowledge of species traits is critical and care is needed to assemble species mixtures that are truly complementary.

Commercial foresters have tended to consider mixtures from a rather longer term perspective than that used in most ecological studies. They have explored the possible effect of species diversity on commercial productivity over the lifetime of the plantation, the overall financial profitability of plantings and on the timing of cash flows to growers (Knoke *et al.* 2008, Roessiger *et al.* 2011). They have also examined the effect mixtures have on wind firmness of stands and possible responses to climate change (Koca *et al.* 2006, Schutz *et al.* 2006, Knoke *et al.* 2008, Felton *et al.* 2010). One of their objectives has been to establish how plantations containing several species (cf a landscape containing several spatially separated species each growing in a monoculture) might improve the profitability of tree growing and reduce the risk of investing in a long-term land use such as tree growing.

It is important to note here that the task of designing multi-species timber plantations differs from developing methods of assembling multi-species forests for protection or conservation plantings. This is because the competitive interactions in the former are periodically interrupted and eventually truncated by harvesting while they persist indefinitely in the latter allowing for the possibility of competitive exclusion of some species in the longer term.

There are several ways in which multi-species plantations might be established. These differ in the diversity of species used, whether the mixture is temporary or permanent and in the relative proportions of the various species. The various types of multi-species plantings used in production forests are summarised in Table 4.5 and illustrated in Figure 4.4.

*Table 4.5* Different types of multi-species plantings for production purposes

| *Type of planting* | *Rationale* |
|---|---|
| Tree cover with an under-planted crop | The trees provide shelter and the under-planted crop provides an early cash flow (Lamb 2011). |
| Temporary nurse tree used to facilitate the establishment of a preferred tree species | Shelter or additional nutrition is provided by nurse trees or nitrogen fixers which enable the subsequent establishment of species otherwise unable to regenerate or grow at the site (Keenan *et al.* 1995, McNamara *et al.* 2006). |
| Monoculture converted to a mixture | Established monoculture plantations are subsequently enriched with additional species to increase resilience and provide additional ecosystem services (Felton *et al.* 2010, Löf *et al.* 2010). |
| Even-aged mixture with trees having different harvest times | Removal of faster-growing species provides an early cash flow and the reduction in stand density acts as a thinning and improves the growth of slower-growing species (Berqvist 1999, Rothe and Binkley 2001, Haggar *et al.* 2003, Forrester *et al.* 2006, Kelty 2006, Lamb 2011). |
| Even-aged mixture with trees harvested at the same time | Complementary species grown together and finally harvested at the same time (Lamb and Lawrence 1993, Kelty 2006, Knoke *et al.* 2008, Lamb 2011). |

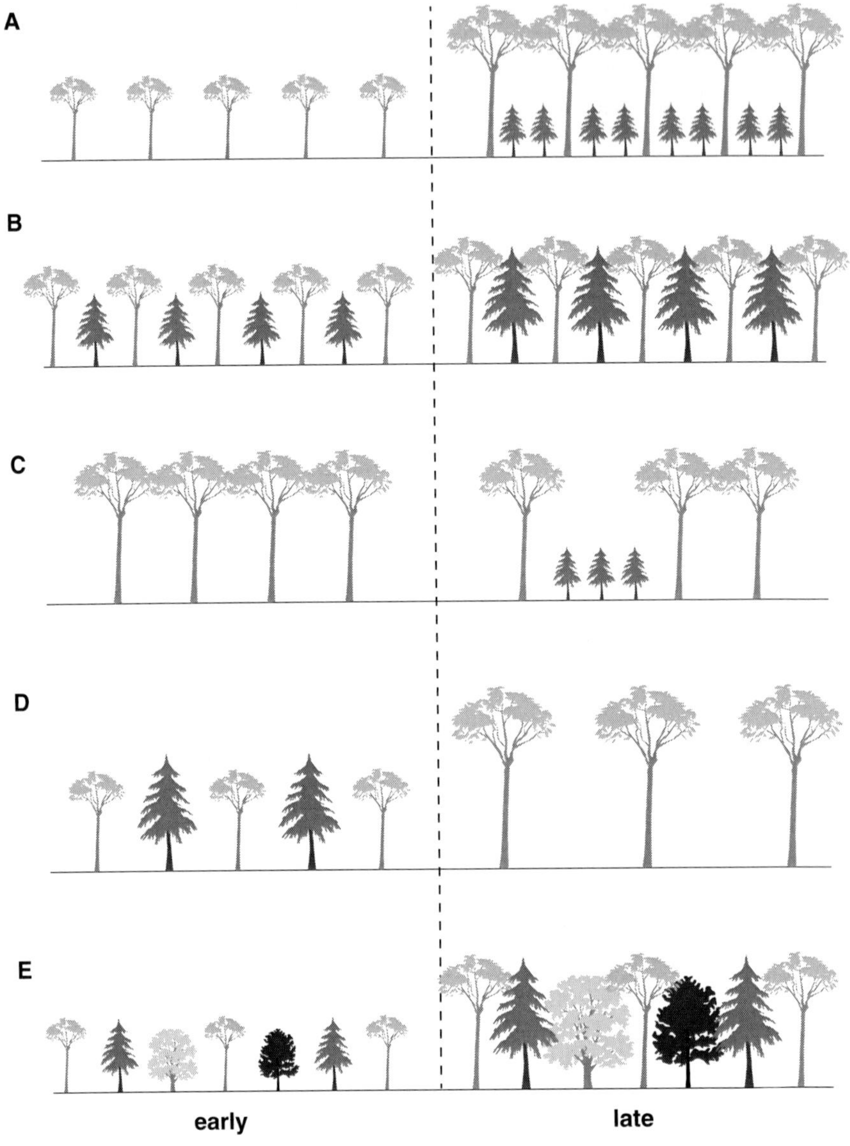


*Figure 4.4* Different types of multi-species plantings established for production purposes. A represents a plantation under-planted with a cash crop; B shows how a nurse tree species can be used to facilitate the establishment of a preferred species unable to be established in the open; C shows a monoculture being converted to a multi-species plantation by creating canopy gaps to allow light to reach the forest floor and establishing additional species below these canopy gaps; D shows an even-aged mixture containing species having different harvest times. Certain species can be harvested at an early time leaving slower-growing species to be felled at a later date; E is an even-aged mixture of species having the same harvest times. (See text for details.)

### *Tree cover with an under-planted crop*

This type of planting is likely to be most useful when landholders are interested in some form of reforestation but cannot afford the long interval before receiving any financial benefit from the planted trees. The mixture involves planting crops in an understorey layer beneath an established tree cover. These species are able to produce a commercially useful product with a year or so and thus provide a cash flow to landholders ahead of any income generated by the trees. The tree cover can involve one or more species and these are grown on a long rotation (>20 years) and harvested for sawlogs. For the system to work the tree density and canopy layer must be manipulated over this period to allow sufficient light to reach understorey layers. Furthermore, the under-planted crop species must be tolerant of these shaded conditions. The approach is common in many tropical agroforestry systems where the understorey plants include food plants, traditional medicinal plants as well as products like rattans (Philpott *et al.* 2008, Lamb 2011). The approach is similar to the well-known taungya system where crops are planted in young tree plantations. It differs from the taungya system because cropping continues after canopy closure while, in the taungya system, farmers must move cropping to another location when tree canopy closure occurs.

### *Temporary nurse tree used to facilitate the establishment of preferred species*

This system can be used when sites are so degraded or the environmental conditions are such that seedlings of a preferred species are not able to grow in the open as a monoculture. The mixture involves a 'nurse' species able to tolerate the site conditions and which facilitates the establishment of another species that may have a higher commercial or conservation value. The first step is to plant the nurse species and create some canopy cover and exclude grasses and weeds. Ideally, this species should have some commercial value and be able to quickly create a canopy cover. If it is a nitrogen fixer it may also be able to improve soil nitrogen levels. Trees of the target species are under-planted once a canopy is formed. Successful under-planting may require that some of the initial nurse trees are removed to open up the canopy and allow more light to reach the forest floor. Growth of the under-planted seedlings must then be monitored to ensure that the benefits and costs of the canopy cover are balanced: if the canopy cover is too dense the under-planted trees will stagnate. But if the canopy cover is removed too rapidly then grasses may re-invade or the shelter required by the under-plants may be too little.

Once the under-planted species are established it may be necessary to thin or remove the remaining nurse trees. While the under-planted species initially require assistance to establish they also require reduced competition to grow strongly. Care is clearly needed to ensure that harvesting the nurse trees does not damage the residual trees. This can be achieved by felling nurse trees along rows and removing the whole row. Eventually all the nurse trees are removed and the preferred target species is able to grow unaided. Provided they have some commercial value, the harvested nurse trees can generate a profit for the landholder. A landholder using this system in Vietnam used these profits from harvesting the nurse species (in that case, *Acacia*) to extend the plantation by planting additional areas with more nurse trees (McNamara *et al.* 2006). It would be possible to use a non-commercial species as a nurse species provided it had a short longevity but this would not generate this intermediate financial benefit.

The approach has been used in Malaysia and the Philippines to facilitate the establishment of dipterocarp species unable to be grown in the open (Anon. 1999). It has also been used to facilitate the establishment of plantations of the valuable timber tree *Toona ciliata* (Keenan *et al.* 1995). This is a member of the Meliaceae and, like most species in that family, is susceptible to attack by a tip borer (*Hypsipyla* sp.) when grown in the open. By under-planting *Toona* beneath a canopy cover, the insect damage attack can be substantially reduced.

### *Transforming existing monocultures into multi-species plantations*

This approach is being used in locations where management objectives change and plantations monocultures must be enriched to become multi-species forests. The approach is being adopted in many areas of Europe because of changing social attitudes and an increasing preference for multi-purpose forests containing broad-leaved species rather than coniferous monocultures (Box 4.2). Depending on the management objectives, the new mixture may be maintained indefinitely (e.g. for conservation purposes) or eventually felled and replanted (e.g. when landholders are also interested in production). A variety of methods have been used to enrich the existing monoculture but all involve creating canopy gaps to allow more light to reach the forest floor and then planting seedlings of the preferred species in these gaps (Felton *et al.* 2010, Löf *et al.* 2010). The size of the gaps required depends on the age and hence the height of the existing plantation as well as its canopy density (Figure 4.3). A common difficulty is that herbivores such as deer can browse the newly planted seedlings which points to the need for temporary fencing to prevent this (den Herder *et al.* 2009, Long *et al.* 2012).

### *Establishing even-aged species mixtures having differing harvest ages*

These mixtures are useful when there is a need to generate an early cash flow to make reforestation attractive to landholders and when there are only limited opportunities to sell thinnings of logs of species being grown on long rotations. Many more valuable timber species are slow-growing

**Box 4.2 Spruce-beech mixtures in Germany**

Around 50 per cent of all new forests planted in Germany over the last 20 years are now mixtures (Knoke *et al.* 2008). Many of these involve spruce (*Picea abies*) and beech (*Fagus sylvatica*) and have been developed by planting the beech beneath older spruce stands although the two species can be planted at the same time when bare ground is being reforested (Thomas Knoke; pers. comm.). This change has been driven by a belief that mixtures offer some significant ecological advantages over monocultures. The yield of mixtures in comparison with that of monocultures of the same species varies with site conditions because these affect the nature of the competitive interactions between species (Pretzsch 2005). Yields also depend on the proportions of each species and the silvicultural treatments applied. However there is increasing empirical field evidence that many forests involving mixtures of spruce and beech are more productive than monocultures of either species (Pretzsch and Schutze 2009). Moreover, there is also evidence that mixtures of these species are able to withstand disturbances such as windstorms or insect damage rather better than monocultures (Knoke *et al* 2008, Roessiger *et al.* 2011). Until recently the economic evaluations of these silvicultural systems was lacking and most simple financial comparisons of conifers and broad-leaved species usually showed that conifer plantation monocultures were more profitable than mixtures. But a different result is found when the risk of disturbances and price fluctuations is taken into account. In this case mixtures provided the superior financial outcome. This was the case when even comparatively simple mixtures involving stands with as little as 7 per cent beech trees mixed with 93 per cent spruce trees (Knoke *et al.* 2008, Griess and Knoke 2011, Roessiger *et al.* 2011).

which makes them unattractive to some landholders who might otherwise be interested in reforestation. The mixture involves planting fast- and slow-growing species together and selectively removing the fast-growing species once it reaches a marketable size. In the humid tropics this might be after about five years. The slower-growing species is then left to mature over a longer rotation (Nguyen *et al.* 2014).

Damage to residual trees during the early harvesting can be reduced by planting each type of species in rows rather than at random. By doing so, the faster-growing trees can be felled along the row which minimises damage to the residual trees. Various designs are possible including having alternate rows of the fast- and slow-growing species or two rows of the fast species alternating with a row of the slower species. The early harvesting generates income for the grower and acts to reduce competition amongst the remaining trees being left to grow until the end of the normal rotation.

The species must be complementary in terms of their growth rates but also in the markets they target. Trees chosen for early harvesting must be sufficiently fast-growing to be able to supply a market at a relatively early age (e.g. for poles, fuelwood). Trees of species chosen to grow for the full rotation must have a high value to justify the long time period. But, in addition, they must be sufficiently tall at the time the first harvesting is done to be able to escape significant damage. A variation of this system is evolving in parts of northern Europe where naturally regenerated *Betula* grows within *Picea* plantations until removed in a late thinning (Valkonen and Valsta 2001, Agestam *et al.* 2006, Hynynen *et al.* 2011). Another variation involves growing a short-lived nitrogen-fixing species with the preferred sawlog species. In this case the benefit comes not as a cash flow but from the improved nutrition. Forrester *et al.* (2006) reviewed experiences when species of the nitrogen fixer *Acacia* are mixed with eucalypts.

*Establishing even-aged species mixtures with similar harvest ages*

The most complex type of timber plantation is that involving several tree species (none of which has more than, say, 75 per cent of the basal area or number of individual trees in the stand) that are planted at the same time and are left to grow through to the final harvest. The advantage of the system is that it is may generate many of the benefits summarised in Table 4.4. The disadvantage is that these types of mixtures are silviculturally challenging since they require the use of species with a high degree of ecological complementarity. Complementary species are those occupying different ecological niches (e.g. they differ in canopy architecture, in growth phenology or in rooting depth). For example, a suitable mixture might involve a shade intolerant species growing in an upper canopy stratum with a shade tolerant species growing in sub-canopy position. Competition is reduced because they occupy different niches (Lamb 2011). Similarly, Richards and Schmidt (2010) report on complementary species based on differences in phenology and nutrient uptake. It is difficult to identify truly complementary species so that most examples of this approach have arisen through a process of trial-and-error and have a relatively small number of species: in many of these, one species tends to dominate the overall stand density (Agestam *et al.* 2006, Petit and Montagnini 2006).

There is a long history of experimentation with long-lived species mixtures in northern Europe (Knoke *et al* 2008) and trials are now underway in many tropical and sub-tropical locations (Piotto *et al.* 2004, Erskine *et al.* 2006, Forrester and Smith 2012). Evidence to date suggests the productivity of these mixtures can exceed those of monocultures provided complementary species are used (and that *ad hoc* combinations are likely to be unsuccessful). There is also some evidence that timber quality can be higher in mixtures than monocultures (Agestam *et al.* 2006) although others have found the reverse may be true (Griess and Knoke 2013). Much depends on the suitability of the particular species to the sites used and the effect this has on their competitive abilities. It also depends on the relative proportions of each species in the mixture (Pretzsch 2005, Forrester *et al.* 2010, Forrester and Smith 2012).

## Planting methodologies for conservation and protection purposes

The primary objective of most of the planting types outlined above is to produce timber and other forest products although many of the landholders using these approaches are likely to also value their capacity to protect watersheds and generate various ecosystem services. However, there is a growing interest in plantings where the primary objective is conservation and environmental protection rather than timber production (Keenleyside *et al.* 2012, Elliott *et al.* 2013). Some of the methods above will have conservation and environmental protection benefits but methods using a greater diversity of species than the 4–5 tree species probably used in most production systems are likely to be better. These plantings usually mix species in intimate tree-by-tree mixtures rather than in alternate row plantings which are more commonly used in production systems.

Two approaches have been used. One is Ecological Restoration: reforesting the site with the intention of restoring the original ecosystem, its original biodiversity and functionality. The second approach also seeks to foster biodiversity and restore functionality but without necessarily attempting to return to the supposed prior historical condition. These types of forests have been referred to as 'novel ecosystems'.

### *Ecological Restoration*

Ecological Restoration has been defined as seeking to assist the recovery of an ecosystem that has been damaged or destroyed (SER 2004). This involves restoring an assemblage of self-sustaining indigenous species characteristic of the original ecosystem and ensuring the system has acquired 'sufficient biotic and abiotic resources to continue its development without further assistance or subsidy'. There has been debate over the feasibility of Ecological Restoration (Diamond 1987, Clewell and Aronson 2013, Shackelford *et al.* 2013). Some argue it can be difficult to achieve because too little is known about the composition and functioning of the original forest. This likely to be the case in heavily modified cultural landscapes that have been used and modified by humans over a long period of time. Under these conditions there may not be any patches of original forest that could act as reference sites that define the target. Another reason why Ecological Restoration could be difficult is because the site has been so severely degraded that the original species are no longer able to tolerate the new environmental conditions. A third reason is because certain species may have become extinct while other exotic species may have been naturalised. Finally, global climate change may have altered the environment in ways that make Ecological Restoration impossible. Clewell and Aronson (2013) seek to resolve some of these problems by arguing that the objective of Ecological Restoration should be the restoration of ecological continuity rather than a historic ecosystem.

Ecological Restoration is an ambitious task even when residual forests remain nearby to act as reference sites and environmental conditions are not too badly degraded. Theoretical ecologists debate whether there are 'rules of assembly' that must be followed in order to achieve this (Weiher and Keddy 1999, Temperton *et al.* 2004). No useful general 'rules' appear to have been developed to date although the importance of a 'founder' effect – the fact that the initial colonists can facilitate or inhibit successional development – is well known (Connell and Slatyer 1977, Firn *et al.* 2007).

In practice, two alternative approaches have been used to restore forests over large areas. Neither seeks to closely replicate normal successional development. Instead, they leapfrog over various stages to accelerate the process. In both cases some site preparation is usually needed although the need for this probably depends on the previous land use history and on the funds available. Fast seedling growth is not as important as in production systems but, nonetheless, early weed control is still crucial for seedling survival and survival is usually also improved by cultivating compacted soils.

### *Framework Species Method*

This approach can be used at sites previously occupied by large numbers of tree and shrub plant species and where it is simply impossible to collect the seed of all of these. The approach depends on there being residual patches of natural forest nearby which are able to supply further colonists as well as on sufficient wildlife able to carry these seeds across the intervening landscape. Reforestation is initiated using only a sub-set of species. Successional development then depends on the additional species colonising the site from the adjoining intact forest. The numbers of species used depend on circumstances but around 20–30 species have been used in some tropical sites (Goosem and Tucker 1995, Lamb 2011).

The species used should include some that are short-lived because they soon die and allow canopy openings to develop which enable seedlings of colonists waiting on the forest floor to grow up and join the canopy. Other longer-lived species that might be used include food plants to attract wildlife to use the site as well as poorly dispersed species such as those with large fruit that might not otherwise be able to reach the site. The trees are closely planted, sometimes at densities of up 2,000 trees per ha with individual species being randomly located. These provide a framework forest which can be colonised by additional species brought in from elsewhere. This builds on observations made in many locations, and described earlier, that even a simple monocultural plantation can facilitate recolonisation of a site by a wide variety of species (e.g. Figure 4.3).

The approach has been used in a number of tropical locations (Tucker and Murphy 1997, Lamb 2011, Elliott *et al.* 2013) but similar approaches involving smaller numbers of tree species and lower tree densities have been used in temperate areas such as in the bottomland hardwood forests in the lower Mississippi Valley in the USA (Gardiner and Oliver 2005, Stanturf *et al.* 2009) and in conifer forests in Finland (Kuuluvainen *et al.* 2005).

### *Maximum Diversity Method*

The second approach involves replanting as much as possible of the original flora and relies less on natural seed dispersal. It is more feasible in less species-rich environments or locations where natural forest remnants are too distant for colonists from these to substantially enrich the site. Those using this species have tended to plant all the tree species at a single time although follow-up plantings may have to be used to ensure the establishment of certain species or to increase the population size of others. The method has been used with apparent success to create native woodlands in the United Kingdom (Rodwell and Patterson 1994), longleaf pine savannah in southeast USA (Brockway *et al.* 2005, Printiss 2013) and restore temperate native forests after mining in Australia (Koch and Hobbs 2007). In these cases not only the tree flora but also the understorey shrubs and herbs have been restored.

The difference between the Maximum Diversity Method and the Framework Species Method is an artificial one. In many temperate forests the numbers of tree species may be low but the understorey flora can be highly diverse. In such cases it may be possible to plant seedlings and restore most of the tree flora but colonists from external sources are needed to restore the shrubs and herbs (Kuuluvainen *et al.* 2005). An example of how different approaches can be used depending on circumstances is described in Box 4.3.

Restoring forests for conservation purposes necessitates assisting wildlife as well as plants to recolonise sites and not merely visit. This will only happen once appropriate habitats have developed and may take some years. In the meantime, it may be necessary to deliberately install certain habitat features such as nesting boxes or bring in old logs to provide habitats for ground dwelling species.

## ***Multi-species plantings or 'novel ecosystems' for conservation and protection***

Ecological Restoration is the preferred option when conservation and environmental protection are the primary goals but it may not always be possible to use this approach. Apart from limitations such as site degradation discussed earlier, it may be simply too expensive. In these circumstances the best and most practical alternative may be to foster the development of a new community made up of as many of the original native species as possible as well as some exotic species. These latter may be useful in replacing native species no longer able to tolerate the site's environmental conditions. Under these circumstances the more modest goal would be to develop a community able restore ecological functioning and, over time, become self-sustaining. A new ecosystem like this is represented by point E in Figure 4.1. Lamb and Gilmour (2003) have referred to this approach as 'rehabilitation'. Others have referred to these as 'novel' ecosystems (Hobbs *et al.* 2009).

**Box 4.3 Restoring species-rich forests in the Atlantic region of Brazil**

The tropical forests of the coastal regions of southern Brazil are referred to as the Atlantic forests. These have been heavily cleared for agriculture and only small fragments of the natural forest now remain. Attempts to restore some kind of forest cover have been made over many years (Rodrigues *et al.* 2009). The earliest attempts used exotic species but indigenous species have been favoured in recent years. Initially, these methods used fast-growing but short-lived indigenous species. These were able to create a closed canopy which excluded grass but the trees were not self-sustaining and all died at a similar time. Subsequently, plantings using up to 140 species were tested.

Over time the number of species used has declined but more attention has been paid to including a greater variety of functional types. These are planted at a density of around 1,600 trees per ha which ensures canopy closure and the exclusion of grasses within a few years. Two groups of species are used in plantings. A 'filling group' including 15–30 fast-growing species with wide canopies able to achieve rapid canopy closure and a 'diversity group' of 70–80 slower-growing species with narrower canopies. By way of context, it is useful to note that many remnant forests in the region have more than twice this number of species (Rodrigues *et al.* 2011). In fact a modified approach is used at sites where natural regrowth is present or recolonisation is probable. In these cases costs can be reduced by decreasing the density of seedlings and the numbers of species planted (Rodrigues *et al* 2011). Reforestation in the Atlantic Forest region now follows a detailed analysis of the site conditions and the surrounding landscape and residual forest areas to ensure that the most cost effective approach is used and that advantage is taken where ever possible of natural recovery processes. Trials suggest faster establishment can be achieved using fertilisers and other methods commonly used in intensively managed timber plantations (Campoe *et al.* 2010).

The process has now been embedded in state legislation which requires, amongst other things, that at least 80 species per ha must be restored within a prescribed time in local reforestation projects (Wuethrich 2007, Aronson *et al.* 2011). Some restorationists approve this because it sets standards and means landholders must meet their obligations to restore deforested land. Others, however, fear this is premature and too specific and will constrain practitioners from adjusting their techniques to suit the variety of circumstances.

The actual number of species established depends on local circumstances and on biogeographic considerations with rather fewer being used in temperate locations than might be used in tropical regions. The number and identity of the species used is likely to depend on local seed supplies and on the confidence of those undertaking the reforestation programme in creating species-rich forests. As in the case of the Framework Species Method, further species are likely to colonise the site if natural forests remain nearby. Exotic species (e.g. nitrogen fixers) can have a role to play in facilitating the establishment of additional native species or as functional analogues of native species no longer able to grow at the site (e.g. because of the reduced soil fertility). There has been some debate over the use of exotics because of the perceived risk that they may become invasive (Peh 2010, Richardson and Rejmanek 2011, Dodet and Collet 2012). However, there may be little alternative in some ecological situations.

This type of reforestation is usually implemented by planting seedlings of all species at the same time. Planting densities of 700–2,500 trees per ha might be used depending on the balance struck between achieving rapid site occupancy and the need to reduce establishment costs. There are no guidelines on the relative proportion of plants of each species that might be used although it would be prudent to not allow any species to exceed, say, 70 per cent of the tree seedling number. Most programmes attempt to randomise the spatial distribution of the species.

Combining novel assemblages of tree species can lead to unexpected successional trajectories. This means careful monitoring is needed so that practices can be corrected if it becomes necessary. An example of this occurring at a site being rehabilitated after mining is given in Box 4.4.

Once establishment has been achieved the main task is to ensure species are able to regenerate. The best way of doing this is to convert an even-aged plantation to an uneven-aged forest. This ensures there are reproductive opportunities for species with differing ecological requirements. If species with differing longevities are used, this conversion process may happen of its own accord. But if all the species planted are long-lived trees then it may be necessary to undertake some selective felling to create canopy openings and initiate the process. The gaps must be large enough to allow light to reach the forest floor so the size of the gaps depends on the height of the trees and the solar elevation. Obviously, any such felling would only be done once the canopy trees were old enough to be producing seed.

Some may be concerned that novel species assemblages of this kind are too risky to use on a large scale. But the empirical evidence is encouraging and there are a number of examples of older, stable and apparently successful reforestation operations of this type. Indeed, there is an argument that these so-called 'novel ecosystems' are likely to be an increasingly common element of many landscapes in the future (Lee *et al.* 2004, Vallejo 2005, Tak *et al.* 2007, Lugo 2009, Newton *et al.* 2011). An example from China of a large-scale reforestation programme of this type is discussed in Box 7.1.

### Box 4.4 Diverted successional development

Revegetation is being undertaken following mining for rutile on South Stradbroke Island in sub-tropical Queensland, Australia (Audet *et al.* 2013). Prior to mining, topsoils are removed and stockpiled. Once mining is complete these topsoils are re-spread over the site and the native forests are restored using a combination of direct seeding and planted seedlings. Soils on the island are infertile sands and in the early days of site rehabilitation, managers included 700 gm per ha of seed of a native *Acacia* species as part of the initial revegetation. It was thought the *Acacia* would act as a facilitator of community development by improving soil nitrogen levels. This did not occur. In fact the *Acacia* diverted the expected successional trajectory by growing more rapidly than the other species and shading them out. Plant biodiversity at the site plummeted. Rather than being a facilitator the *Acacia* acted as an inhibitor of successional development.

A wildfire worsened the problem. Most *Acacia* have hard seed coats and there is often a large dormant seed pool stored in the soil. Figure 4.5 shows the outcome after a fire burned one of these sites and stimulated germination of these stored seed. This event simply accentuated the dominance of the sites by *Acacia*. Since these experiences, managers have progressively reduced the amount of *Acacia* seed added during the rehabilitation process to the point where none is now included in the seed mix. Nitrogen fixation is now carried out solely by *Acacia* seed occurring naturally in the topsoil returned to the site after being stockpiled during mining.

*Figure 4.5* Successional development diverted after a wildfire stimulated germination of *Acacia* seed stored in the topsoil. The large numbers of *Acacia* trees are now growing rapidly and are out-competing other species at the site.

## Direct seeding

Another form of reforestation by which managers can help overcome the failure of natural dispersal mechanisms is through the use of direct seeding. The technique has been successfully used in temperate and tropical regions as well as in low and high rainfall areas (Dalton 1993, Knight *et al.* 1998, Engel and Parrotta 2001, Willoughby *et al.* 2004, Doust *et al.* 2008, Carr *et al.* 2009, Stanturf *et al.* 2009). In some of these cases the areas treated were small but, in others, they were quite large.

Direct seeding is much cheaper than planting seedlings because it avoids the need for a costly nursery stage. Likewise, distributing seed in the field is much cheaper than transporting and planting seedlings. Woodall (2010) found it could be less than 10 per cent of planting seedlings on a 'hectares planted' basis and less than 5 per cent on a 'per seedling established' basis. This makes direct seeding an attractive method of reforesting large areas. But these cost advantages are accompanied by some significant disadvantages (Table 4.6). Thus seeds lying exposed on the soil surface, especially if these soils are compacted, are less able to absorb soil moisture than seeds that are buried. Further, they can be easily washed away by rainstorms. There is also a risk that directly sown seed will be eaten by insects or vertebrates. Those that survive these hazards and actually germinate may be swamped by competition from surrounding grasses and herbs. This means the final number of established seedlings may be a very low proportion of the seed applied (Mergen *et al.* 1981, Doust *et al.* 2006). Of course a low success rate can be overcome by increasing the seed application rate but this is only possible if seed is cheap and easily collected. This will be possible for common species but will be difficult in the case of rare species and those species that only occasionally produce seed.

The variable success rates can also mean that tree establishment rates in the field are highly variable with dense clumps in some locations and an absence of trees in others. Dense clumps may be a problem in production forests and necessitate some form of thinning and Mergen *et al.* (1981) note that the cost of thinning overstocked stands has been a deterrent to direct seeding in some locations. Overall, therefore, direct seeding has some significant potential advantages but is still somewhat risky to use. It has been used to establish production forests but its main value in the future may lie in establishing protection forests.

### *Aerial distribution*

Some of the most difficult sites to reforest are those that are in remote or mountainous areas. These are sites where direct seeding using aircraft could be especially attractive. Not only is the problem of access reduced but, in addition, large areas could be covered in a short time when seasonal conditions (e.g. rainfall, temperature) are most favourable. An account of

*Table 4.6* Some comparative advantages and disadvantages of direct sowing and planted seedlings

| | *Direct sowing* | *Planted seedlings* |
|---|---|---|
| Advantages | No nursery costs<br>Low planting costs<br>Good root development<br>Creates a spatially irregular (more 'normal') forest<br>Can cheaply sow seed amongst existing plants to enrich a site | Higher survival rates<br>More efficient use of seed<br>Can easily weed individual seedlings<br>Can ensure symbiont inoculation |
| Disadvantages | Low success rate (predation, drought etc.)<br>Need large numbers of seed<br>Less suited to rare species, those with limited seed availability or those with special germination requirements<br>Difficult to control seedling density – may have patchy seedling distribution<br>Difficult to control weeds around individual seedlings<br>Limited to species easy to establish from seed | Higher costs (nursery, transport and field establishment)<br>Potted seedlings can have imbalanced root-shoot ratios<br>Seedlings can be pot-bound |

some early experiences in aerial sowing of seed is given by Mergen *et al.* (1981) who describe trials and operational flights in a number of countries involving a number of species. These used fixed-wing aircraft as well as helicopters. In some cases seeds were sown directly while in others the seed had been pelletised to reduce drift and mixed with insecticides and repellents to reduce predation. Some large areas have been treated although results have been variable because of predation and weed competition.

One area where aerial seeding has been successful has been its use to regenerate native eucalypt forests in Australia after logging (Mergen *et al.* 1981, Florence 2004). These sites were burned before seeding so there was limited early weed competition. In initial trials the very small eucalypt seeds were embedded in clay pellets to ensure they did not drift too far from the aircraft flight lines (many eucalypt species have 100 or more seeds per gm). Insecticide, fungicide and a dye were also used to reduce insect predation and ensure that gaps in the distribution could be clearly seen on the ground (Hodgson and McGhee 1992). Further development work and the use of helicopters led to the pelleting being abandoned and seed being directly dropped from the aircraft. The technique has become routine for regeneration in these southern forests (Florence 2004). Aerial seeding has also been successfully used in some mining operations. Again direct seeding is attractive because the sites are free of weeds if this is done immediately after mining and the soils can be ripped to roughen the surface and ensure they are not compacted.

Perhaps the largest use of aerial direct seeding has been in China. Xinhua and Jingchun (1988) report that 15 million ha were reforested in this way between 1956 and 1985. More recently, Cao (2008) reports a total of 30 million ha of aerial seeding in China. These early treatments mainly used pines although broad-leaved species were subsequently used as well as mixtures of species. Seed was treated with repellents to reduce predation and pelletised to reduce drift during distribution, and existing vegetation was burned or cleared ahead of seed distribution to reduce weed competition. Although very large areas are still being treated, Sannai (2006) reports that the quality of the forests is low. They may, however, be suitable for protection purposes.

### *Ground applications*

The major disadvantages of aerial seeding are that seed is shed onto the soil surface and that existing vegetation will out-compete those seedlings able to become established at the site. It is much easier to deal with these problems if seed is distributed at ground level. For example, seed distributed at ground level can be planted and this is known to substantially improve success rates (Doust 2004). Even large seed is more successfully established if buried (Doust *et al.* 2008, Tunjai and Elliott 2012). In addition, weed control can be carried out before seeding and this can also improve success rates (Willoughby *et al.* 2004).

Seed can be planted by hand or using some form of mechanised sowing. Some farm machinery developed to prepare seed beds and plant the seed of cereal and vegetable crops are now being modified to use with woody plant species (Schmidt 2008, Carr *et al.* 2009, Jonson 2010, Woodall 2010, Campos-Filho *et al.* 2013). Some large areas have now been revegetated using these methods including in low rainfall areas where direct seeding might appear to be too risky to use (Knight *et al.* 1998, Schneemann and McElhinny 2012). A problem with mechanised sowing is that there can sometimes be bare spaces between the sown strips. Jonson (2010) discussed ways of overcoming this and modifying sowing equipment to avoid the problem (Figure 4.6). The outcome, five years after direct seeding using this machine, is shown in Figure 4.7.

Much direct sowing has involved sowing seed of a single species or relatively simple species mixtures. But there is no reason why more diverse mixtures cannot be used if the seed is available. Koch (2007) describes sowing a mix of 78–113 species (most of which were understorey species) during the rehabilitation after bauxite mining. Topsoil was spread over the site and seed was applied using a seeder attached to a bulldozer with a ripper. Fertiliser was subsequently applied from a helicopter to avoid driving over the site and compacting the soil.

*Figure 4.6* A row seeder used to sow a mixture of seed of tree and understorey species on former farmland in Western Australia. Blades on the machine remove weeds in five separate rows and seed is sown below the soil surface in each weed-free strip (Photo: Justin Jonson).

*Figure 4.7* A large area of newly restored *Eucalyptus* woodland containing a variety of species established on former farmland in the southern region of Western Australia. This was established by direct seeding using the machine illustrated in Figure 4.6. The new forest is now five years old (Photo: Justin Jonson).

There are relatively few reports of successional development in directly seeded forests over the longer term. One study by Schneemann and McElhinny (2012) found that species diversity subsequently declined over a period of 17 years but this may have had more to do with the very high seedling densities established (17,000 stems of trees and shrubs per ha) than with the fact the site had been established by direct seeding. A different outcome would probably have been found with a lower seed application rate. They recommended that sites be monitored and that periodic disturbances be introduced to increase further regeneration opportunities.

### *Improving outcomes*

In a major review Schmidt (2008) concluded that direct seeding is a promising technique but is still not used as extensively as it might be. If direct seeding is to become a routine method of reforestation then ways must be found to reduce the amounts of seed used and to increase the reliability of seedling establishment. Key issues are:

- *Optimise timing:* sow seed when soil moisture and temperature conditions are favourable for germination of the particular species and before weed competition is too great. Optimum conditions will obviously depend on the species being used but timing will be especially important in seasonally dry locations (Knight *et al.* 1998, Woodall 2010). In these situations there may be only a limited number of days each year in which conditions are suitable.
- *Pre-treat seed:* germination can be hastened by pre-treating seed before it is dispersed. Pre-treatments can include soaking in water to trigger the germination process, breaking dormancy through well-known treatments such as heat or physical abrasion or stratification. Schmidt (2008) warns that imbibed seed can be more difficult to disperse.
- *Pelletise small seed:* seed predation and fungal attack on newly emerged seedlings can be reduced by pelletising the seed using a clay slurry containing fungicides and insecticides. The pellet may also make it easier for the aerial dispersal of species having very small seed. If using this technique, care must be taken to ensure poisonous chemicals do not affect wildlife.
- *Add microbial symbionts:* mycorrhiza and nitrogen-fixing *Bradyrhizobium* can be added to seed to ensure inoculation. Thrall *et al.* (2005) found the latter substantially increased the survival and seedling growth rate of *Acacia* sown at sites with harsher climatic conditions.
- *Improve seed beds:* the roughened soil surface left after cultivation makes for a more favourable micro-site for many seeds enabling them to absorb moisture, avoid being blown away and also escape predation. These operations will also reduce weeds.

- *Bury seed:* burying seed reduces seed predation and improves the likelihood of seedling establishment especially in sites with compact soils because moisture and temperature conditions are more stable (Woods and Elliott 2004, Doust *et al.* 2006, Woodall 2010). On the other hand, small seeds need to be close to the surface to enable their cotyledons to emerge. Trials need to be carried out to find the optimum planting depth for particular species and the conditions under which these might be varied.
- *Reduce weed competition:* remove competition before seed dispersal with fire, physical removal of the surface layer containing seeds of weed species or by using weedicides. Carr *et al.* (2009) argued that finding the right herbicides and rates of application to overspray directly seeded plants is a major research priority.
- *Sow seed beneath established canopies:* an alternative approach to dealing with aggressive competitors such as grasses can be to sow seed beneath an established tree canopy (Cole *et al.* 2011, Wang *et al.* 2011). This might be of a short-lived cover crop or a plantation with longer-lived tree species where the added seed act to enrich the species diversity. The approach obviously requires that only seed able to tolerate some degree of shade is used.

## Choosing a reforestation method

These various approaches summarise some of the main methods that might be used to reforest presently deforested landscapes. Large-scale reforestation programmes will almost certainly involve using more than one of these methods in different parts of a landscape. One reason is that landscapes are not homogenous and the capacity for natural regeneration varies. Rodrigues *et al.* (2011) describe how a portfolio of reforestation approaches could be used at various sites in the Atlantic Forest region of Brazil depending on the capacity for natural regrowth to develop from old stumps or from dispersed seed. Relatively simple plantings could be needed to initiate a forest succession in some places but rather more complex techniques will be needed elsewhere.

A second reason why a combination of different approaches might be used is that economic opportunities and landholders' objectives are also likely to vary across the landscape. Some locations might be favoured for production purposes while others may be critical for the provision of certain ecosystem services. The importance of where reforestation takes place for the provision of various services is discussed in the next chapter.

Despite the diversity of techniques outlined above, there is still much to be learned about silvicultural techniques needed to implement reforestation on a large scale.

Considerable research has been undertaken to develop silvicultural methods able to generate industrial timbers but there has been far less research into techniques more suitable for large-scale reforestation primarily aimed at producing ecosystem services at such sites. Indeed, Puettmann *et al.* (2009: 59) argue: 'The adoption and dominance of the agricultural research model (by silvicultural researchers) has not led to a culture of trial, innovation, or examination of trade-offs amongst practicing silviculturalists, but has supported a conservative culture of implementing standardized prescriptions'.

In short, the task of reforesting of degraded landscapes on a large scale and doing this in ways that focus on outcomes other than just timber production represents an important challenge for silvicultural researchers.

## Conclusion

There are a great many ways deforested land may be reforested. Apart from natural regrowth, these include planting seedlings and sowing seed. It is likely that large-scale reforestation will involve all three approaches rather than simply relying on the techniques developed to create industrial timber plantations.

Simple plantation monocultures involving native or exotic species have been widely used in the past because they can be efficient producers of timber and can be easily implemented over large areas.

Carefully designed multi-species plantations have the potential to produce as much, if not more, timber and other forest products while also generating a wider range of ecosystem services. They may also have some potential financial advantages (especially to smallholders) in the form of early income and reduced risk from ecological or economic hazards.

These potential advantages are offset by the necessarily more complex management regimes. This has made multi-species plantings less attractive to industrial growers and for anyone wishing to implement these systems over large areas. However, this is not necessarily a disincentive for smaller individual growers.

Considerable silvicultural research has been undertaken to identify appropriate species and develop industrial monoculture plantations. By comparison, much less research has been undertaken to develop robust multi-species plantations.

Mixed-species plantings primarily established to produce timber will often supply more ecosystem services than plantation monocultures but may not be the most appropriate ways of reforesting deforested lands to conserve biodiversity, protect watersheds or supply a wider variety of ecosystem services. These objectives are likely to be better served by plantings containing a more diverse variety of tree and understorey species.

Ecological Restoration is usually seen as the benchmark for such reforestation but may be difficult to implement. This is because the original condition is unknown, there are various environmental constraints or the cost of doing so is too high.

Diverse, multi-species plantings of species (including, if necessary, exotic species) capable of growing at the present site and being functional analogues of the original species assemblage may be the only alternative. There is some risk attached to such novel ecosystems because it is difficult to predict the way the various species will interact over the longer term. Monitoring and adaptive management will be important to ensure appropriate successional trajectories are maintained.

The capacity of various types of reforestation to supply goods and ecosystem services depends on the landscape context in which they are carried out. The importance of this context and the spatial location of plantings are discussed in the following chapter.

# 5 Where in the landscape should forest restoration take place?

## Introduction

The type of new forest established affects both its commercial and ecological value. But the location and spatial distribution of any new forest is also important. Just where in the landscape should forest restoration take place? The question is one that industrial plantation owners have always considered but in recent years it has also attracted the attention of conservation biologists. This is because of a debate about how to conserve biological diversity while increasing agricultural production? One way of doing so would be to partition landscapes into large, intensively managed production areas and large conservation areas. Highly intensive production would take place on one and strict conservation on the other. An alternative approach would be to integrate both production and conservation and create landscapes having a patchwork of wildlife habitats scattered across agricultural areas. The former approach has been labelled 'land sparing' and the later has been referred to as 'land sharing' or 'wildlife friendly farming' (Fischer *et al.* 2008, Phalan *et al.* 2011).

The advantage of partitioning landscapes into separate zones is that it helps ensure food security while allowing the creation of large conservation reserves in which the world's remaining biodiversity would have a better chance of being protected. The advantage of the shared landscape approach is that it achieves many (though not necessarily all) of the same goals. But, in addition, it also provides other environmental benefits such as watershed protection across the broader landscape. Further, the 'sharing' approach recognises that the task of conserving the world's biota will be difficult to achieve by relying solely on scattered and reproductively isolated conservation reserves and ignoring the matrix in which they are located (Bengtsson *et al.* 2003, Benayas *et al.* 2008, Nagendra and Southworth 2009, Perfecto and Vandermeer 2010, Tscharntke *et al.* 2012). Finally, the division of lands into intensive production or large and strictly protected conservation areas may be possible in some areas but in many parts of the world the patterns of land ownership are such that the more integrated landscape approach is more politically realistic.

This debate amongst conservationists is essentially another version of the debate amongst agriculturalists described in Chapter 1 over whether large-scale and intensive industrial agricultural production should be favoured over smaller-scale farming systems (Collier 2008, De Schutter 2011b). In both debates the dichotomy is a false one and, in most cases, elements of each approach will probably be needed (a fact implicitly recognised by many of the protagonists).

Forest restoration is likely to have a role to play in developing these future landscapes, irrespective of the choice that is made. In the case of the partitioned landscape it could be used to rehabilitate degraded areas within or around the large conservation reserves. Part of the 'production' area of the landscape might also be used for intensively managed industrial timber plantations. In the case of the multi-purpose landscapes, reforestation could be used by landowners to diversify income sources and make better use of marginal lands within their farms. In addition, areas within the agricultural landscape that are especially prone to erosion might also be reforested in ways that also provided wildlife habitats.

Whatever the choice that is made, there will be ecological consequences. This is because many ecological processes largely operate on a landscape scale (e.g. hydrological processes, maintenance of species populations) and will be affected by the removal or restoration of forest areas. The actual consequences will be affected by just where any reforestation is actually done. This means strategically located new areas of forest are likely to be more effective than randomly placed plantations. The question, then, is how these benefits might be maximised. Just how might reforestation be done in a way that improves ecological functioning at a landscape scale?

Changing land use practices is always difficult but redesigning landscapes to achieve certain strategic purposes is even more so. What might seem rational planning to some may appear to others to be a form of 'land grabbing'. This chapter explores this question and examines where reforestation should be carried out to achieve the greatest economic and functional benefits while maintaining (or even enhancing) agricultural productivity. In particular, it examines the preferred locations for timber production, to manage hydrological flows, limit erosion and conserve biodiversity.

## The landscape mosaic

The term 'landscape' will be used here rather loosely to describe a broad area of interest. Several alternative definitions are given in Box 5.1.

Most agricultural landscapes are a changing mosaic of vegetation types and land uses. The vegetation might include crops, pastures and various kinds of natural vegetation. The areas of natural vegetation may include

**Box 5.1 What is a landscape?**

The term 'landscape' is widely used but its meaning usually depends on the context in which it is being used. Humans may think of landscapes as areas varying in size from 100s to 1,000s of hectares. But, from a conservation biology perspective, it may be better to think of a landscape as a function of the scale over which a particular species moves and how this species perceives the environment (Lindenmayer and Fischer 2006). On the other hand, from a planning perspective, a landscape might be thought of not as a planning unit, but as the scale at which it is necessary to intervene if one is to balance trade-offs and optimise conservation and livelihood benefits (Boedhihartono and Sayer 2012).

intact forests as well as regrowth forest and be present in patches of differing size and connectedness. Some of the remaining natural forest areas may retain most of their original biota but others may have lost many of their original species and have been colonised by invasive exotics. Consequently, these patches will differ in their conservation significance and ability to supply colonists to help reforest new areas (see Figure 3.1). The agricultural areas may be represented by productive and high-quality farmland as well as more marginal lands. Lands in the mosaic may be owned by the state, by corporations or by smallholders. The cadastral boundaries may be settled or contested. All of these patterns are affected by differences in topography and soil fertility. Finally, this complexity is being constantly modified because some existing forest areas are still being cleared while some agricultural lands are being abandoned and reverting to shrubland or forest. These changes are caused by changing land values and agricultural market prices as well as human population movements (onto as well as off the land).

Despite the complexity outlined above there has been a tendency in many locations for landscapes to become simplified and homogenised over time as agriculture has intensified. The reduction in habitat diversity usually leads to biodiversity losses. One way of countering this is to undertake some form of reforestation. This can diversify the range of habitats present and influence the spatial movement of certain plants and wildlife species across the landscape. It is also likely to have other environmental benefits and improve the economic circumstances of landholders. But the need and the opportunities for reforestation vary with location and some places will be more important in this respect than others. Some of the attributes of particular elements of the landscape mosaic that may be influential in reforestation planning are shown in Table 5.1.

*Table 5.1* Elements of the landscape matrix of importance for forest restoration planning

| *Landscape component* | *Significance for forest restoration planning* |
|---|---|
| High-quality agricultural land | Unlikely to be available for reforestation since the opportunity costs are too high |
| Marginal agricultural land | May be available for reforestation, especially if the current use is only possible because of agricultural subsidies; the need for reforestation is enhanced if erosion is present |
| Former agricultural land (now abandoned or under-used) | Possibly available for reforestation but depends on extent, spatial distribution and whether it already contains some regrowth; the need for reforestation is enhanced if erosion is present |
| Undisturbed residual forest | A potential source of plant and wildlife colonists; significance depends on extent, spatial distribution and composition; also depends on the amount that is protected from further clearing and the amount that remains at risk |
| Rivers and wetlands | Some likely to be more at risk of sedimentation than others |
| Hills and mountains | Possible erosion areas depending on steepness and present vegetative cover |
| Road and rail networks | May affect the commercial attractiveness of adjoining land; can have both a positive and a negative influence on the relative attractiveness of the site for reforestation |

## Choosing locations for timber production

Like agricultural crops, the best sites for timber plantations are those with good soils, level topography and good road access. In practice such sites are usually taken by crops and plantations are mostly established on more marginal areas although the need for road access remains critical, especially for large industrial timber growers.

Sometimes owners of timber or pulp mills are unable to buy enough land sufficiently close to their mill for their operations to be fully viable. In such cases it is common for them to try to form joint ventures or out-grower partnerships with small landowners in areas surrounding the mill in order to assemble a sufficiently large plantation area. The prices they can offer such landholders often make commercial tree growing more attractive to them than other alternatives. Such agreements can provide significant benefits to both parties and help improve rural livelihoods. But there can be tension in these relationships if small landowners are not willing partners. And problems will develop if landholders are overcharged for items such as seedlings supplied by the company or for harvesting costs and do not get an equitable share of profits (Mayers and Vermeulen 2002, Nawir *et al.* 2003).

Large and homogenous timber plantations can be commercially profitable but may diminish regional biodiversity. Some have referred to them as 'biological deserts' (Cossalter and Pye-Smith 2003). The criticism is valid if they have been established at the expense of natural forest but perhaps less so if they are planted on degraded land; it depends what they replace. However there are ways in which large industrial plantations can contribute to regional biodiversity conservation. One way is to break the plantation into compartments and encourage the development of regrowth forests in these intervening strips. This practice is now being adopted by a number of large industrial plantation owners. For example, a plantation owner in Sabah has a concession covering 288,000 ha. Plantations of *Acacia mangium* or *Eucalyptus grandis* are being established on previously logged land and these will cover 38 per cent with the remainder being regrowth forest retained on steeper lands or along riparian areas (Lamb 2011). Similar spatial arrangements involving plantation being embedded within a matrix of regrowth or restored natural forest are being developed elsewhere in Africa, Asia and South America (Cyranoski 2007, Nasi *et al.* 2008, Samways *et al.* 2010, Mesquita and Passamani 2012). The restored forests or regrowth areas between the plantation compartments provide multiple connections for biota to move across the landscape and link populations remaining in natural forest fragments. Some of this biota may also be able to utilise the plantation areas themselves. The plantations effectively pay for the corridor system.

Creating these heterogeneous landscape mosaics is a potentially attractive way of balancing the need for both production and conservation but care is needed to establish what proportion of the landscape should be retained as natural forests (or restored forest). Likewise, care is needed to ensure that areas critical for watershed protection or biodiversity conservation reasons are protected.

Spatial considerations also influence the location of timber plantations established by smallholders. Some might plant trees on good soils to define their farm boundaries or to assert ownership. However, farmers growing trees for commercial purposes are likely to use more marginal lands unless they are near a large mill and can form a relationship of the kind referred to above. These areas may be lands with infertile soils or in remote or steep locations. Of course some will also grow trees to achieve certain functional outcomes such as erosion or salinity control and this is discussed further below.

Location also influences the type of reforestation undertaken. Landholders located near plantations owned by large corporations are likely to use the same species and silvicultural methods as these companies. This is in order to benefit from their technical expertise as well as to be eligible as suppliers to their processing mills. But landholders in locations more distant from such markets must use higher-value species that can bear the greater transport costs or use types of plantations able to produce a wider variety of products for a more diverse set of markets. Pulpwood plantations grown on short rotations are unlikely to be suitable in these circumstances.

## Choosing locations to modify hydrological flows

New forests will usually increase evapo-transpiration and alter hydrological flows. Their capacity to do so depends on the types of trees planted but also on the location within a watershed. Most studies find trees established in lower topographic positions and along riparian areas are likely to have access to, and therefore use, more water than those growing in upper slope positions (Van Dijk and Keenan 2007). In most cases the task facing silviculturalists is to find ways of minimising this water use in order to maintain river flows. However, there are some situations where the capacity of trees to use water and cause water tables to decline is, in fact, advantageous.

### *Reforestation and dryland salinity*

Dryland salinity is a form of land degradation that affects large areas of farmland worldwide, especially in areas with <900 m rainfall. It is often caused by a change in the hydrological flow through landscapes arising from clearing forests to create farmland. Replacing deep-rooted perennial plants by shallow-rooted annual plants reduces evapo-transpiration causing an increase in water flowing below the rooting zone. This causes the water table to rise and mobilises any salt stored deep in the soil profile (originating from old marine sediments or brought in as cyclic salt from coastal areas). These salts either rise to reach the rooting zone of the new crops or move into streams. Reforestation can delay, and possibly reverse, this process. But scale is important. In some locations it may be possible to abandon agriculture entirely and reforest the whole of a catchment. In this case the economic value of the forest products allows commercial forestry to replace agriculture as the dominant land use. But, in many other situations, timber yields and market prices make this change economically unattractive. The issue, then, is how to integrate reforestation with current farming practices.

Some of the ways in which reforestation might be done are described by Hatton and George (2001), Harper *et al.* (2012) and Stirzaker *et al.* (2002). The four main options are as follows.

#### *Woodlots in rotation with agriculture*

In this case trees are planted on farmland where relatively fresh (i.e. non-saline) water has accumulated in the upper soil profile because of the reduced evapo-transpiration. The trees can utilise this water (preventing it reaching the groundwater) as well as any nutrients added to these farmlands in fertilisers. This may result in higher tree growth than would normally be expected in this particular rainfall zone. But once this water is used and the water table is lowered, the trees are then entirely dependent on water from further rainfall and some may die. At this point the trees can be removed, their hydrological task being complete. Agriculture can then be recommenced. Alternatively, if agriculture is not to be re-established then thinning of trees might be done to recover the appropriate balance between rainfall and tree cover.

A land use system cycling between farming and plantations has been called 'phase farming with trees' (Harper *et al.* 2009) or 'kamikaze forestry' (Harper *et al.* 2013). This involves using a series of short rotations of densely-planted woodlots which are rotated around the landscape. This clearly requires a market for the timber produced by these trees. The best outcome is likely to occur at sites with deep, heavy textured soils and where there is a market for trees grown on a short rotation (3–10 years). Poorer outcomes are likely when water tables are more saline and salt continues to build up in the soil.

### *Tree belts on hillslopes*

Rather than planting trees in blocks, they are established in belts across slopes to intercept water moving across the soil surface and through the sub-soil. This option is available when the moving water is mostly fresh and within reach of tree roots. The extent to which this is the case depends on the slope and the hydraulic conductivity. When the hydraulic conductivity is low or the slope is gentle, some water may drain vertically into deeper profile layers. The width of the belts and the distance between them depends on how much, and how quickly, water is moving across the slope. Closely spaced and wide belts could use more water than is available leading to tree deaths. By contrast, more widely spaced and narrow belts may allow additional water to reach lower parts of the landscape causing the water table there to rise. Stirzaker *et al.* (2002) outline methods of assessing both factors.

### *Mixing tree belts with agriculture on flatter land*

This approach might be used when there is less lateral movement of water. Trees can be planted in plantation blocks or as belts of trees separated by alleys of crops or pasture land. The difference between the two lies in the length of the tree-crop interface. Planting 25 per cent of a 100 ha field using a single woodlot would generate an average tree-crop interface of 20 m per ha. Planting the same area using belts 12.5 m wide at 50 m spacings would generate an average tree-crop interface of 400 m per ha. The latter will obviously have a greater competitive impact on crops than the former.

Much depends on how far into the crops this competition extends and on the capacity of the trees in the belts or woodlots to drop water tables and improve crop or pasture productivity. More widely spaced trees (such as in belts) are likely to use more water than closely spaced trees (such as in woodlots) so that belts may be especially useful in preventing groundwater recharge if the width of the competitive zone is limited. But the economic balance is also affected by the commercial value of the trees themselves (Stirzaker *et al* 2002). In situations where trees also generate a commercial yield the net benefit is described by:

$$\text{Net benefit} = T + (Y_2 - 2Y_1) - D$$

Where T = value of tree products, $Y_2$ = gain in productivity of crop or pasture caused by the trees, $Y_1$ = crop or pasture loss caused by competitive effects of the trees and D = value of crop products displaced by the trees (Figure 5.1).

*Planting trees over shallow saline water tables*

In this case trees are planted in discharge areas in lower parts of the landscape subject to periodic waterlogging and rising salt levels. The approach depends on the species being used being sufficiently tolerant of salinity to be able to use the additional water that is available. Although this option may initially succeed in dropping the water table, the longer-term outcome is less certain except where soils are sandy and salts can be leached away. Otherwise, salts continue to accumulate in the soils eventually leading to a reduction in water use, a rising saline water table and tree deaths. In such cases other more tolerant species may be needed or engineering works must be used, such as drainage channels, to remove the salt. Tree planting in discharge areas may be thought of as a way of 'buying time' while complementary plantings in recharge areas are undertaken to reduce the amount of water reaching the area.

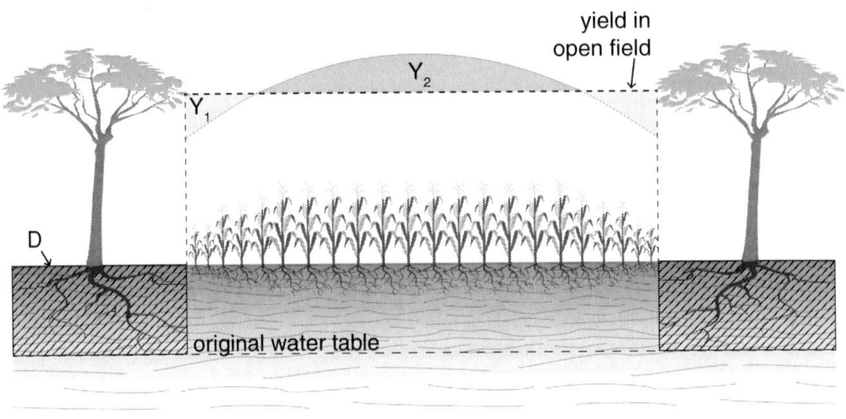


*Figure 5.1* The net benefit of planting strips of trees amongst crops or pastures in dryland salinity areas depends on the value of tree products and the benefits from enhanced crop or pasture productivity less the value of crop losses induced by the trees replacing or competing with crops or pasture. D is the value of crops displaced from this area by trees, $Y_1$ is the crop loss induced by tree roots competing for water and $Y_2$ is the crop gain induced by the trees (after Stirzaker *et al.* 2002).

These four models were developed in a situation where salinity was induced by clearing natural woodlands and replacing it by crops and grasslands. The situation may be different when reforestation (i.e. afforestation) takes place in naturally occurring grasslands. Nosetto *et al.* (2008) describe such a situation when small areas of plantation were established in grasslands in some drier areas of Argentina. Prior to reforestation precipitation exceeded evapo-transpiration but this was reversed and the water balance became negative once trees were established. This meant solutes accumulated in groundwater rather than being leached out. The trees made increasing use of these groundwater resources and eventually began to be affected by the rising salinity. The situation resembles that described in the fourth and final reforestation model described above although the hydrological events leading up to the salinisation process in each case were different. These contrasting causes of salinity point to the need for detailed understanding of the impact of vegetation changes (both positive and negative) on water balances and, specifically, on vegetation-groundwater relationships.

In summary, reforestation can assist controlling dryland salinity but the location of any plantations depends on the particular attributes of the landscape, including rates of leakage below the rooting zone, rates of discharge, the hydraulic conductivities of soils and topography. The choice facing a landholder is to decide whether they can afford to simply abandon certain areas of their land to salinity or whether the functional benefits of spatially targeted forms of reforestation might be achieved by using the trees as an economic resource as well as an ecological tool (see Box 5.2).

**Box 5.2 Dryland salinity in Western Australia**

The agricultural areas in the south-west of Western Australia have a Mediterranean climate with much of the region receiving <900 mm of rain. Large areas of this farmland have been affected by dryland salinity as a consequence of forest clearing. The affected areas were estimated to cover 4.4 million ha of the state in 2000 and these are continuing to expand (Harper *et al.* 2012). Some farmers have simply abandoned the affected areas and concentrate their agricultural activities on unaffected lands. This is because the cost of reforestation is high and any functional benefits from reforestation may develop outside their own farm area. The problem of opportunity costs is now being addressed by farmers leasing land to private timber companies who grow mostly the non-native *Eucalyptus globulus* trees on short rotations (<10 years) for pulpwood. This species is not especially tolerant of salinity but grows well in these soil and rainfall conditions. Since 1990 over 289,000 ha have been planted in this way. Many of these broad scale plantings have been established in the 600–900 mm rainfall zone where timber production is most profitable. Financial modelling suggests reforestation that covers more than 70 per cent of the watershed can be financially attractive if payments are made for carbon sequestration as well as timber (Townsend *et al.* 2012). Payments might also be made to landholders if, following reforestation, water quality could reach certain prescribed thresholds.

These types of broad scale plantations are less profitable in drier areas and methods of undertaking reforestation in the 300–600 mm zone are now being explored (Harper *et al.* 2012). There are several strategies being examined. The purpose of these is to determine whether landscape hydrology can be restored to something approaching its pre-agriculture condition without reforesting the whole area. The first strategy involves seeking species able to provide a wider range of products other than just pulpwood. The species being tested include native and exotic woody species able to produce timber, biomass for energy, activated carbon, eucalyptus oil, sandalwood; or services such as carbon sequestration and habitats for biodiversity conservation. Markets exist for many tree products but markets for ecosystem services such as clean water, carbon or biodiversity are still evolving and, in many locations, the identity of buyers is unclear. Governments will have a significant role to play in establishing or structuring these markets.

The second strategy seeks to find ways of integrating tree growing into farming systems across the region. Trials involve matching species to sites; belts of trees interspersed amongst croplands; blocks of trees targeting areas of greatest recharge and water accumulation; and alternating short phases with trees and periods with crops (Sudmeyer and Simons 2008, George *et al.* 2012, Harper *et al.* 2012).

## Choosing locations for erosion control and watershed protection

Reforestation has been used to control erosion caused by water and wind. In both cases strategically located reforestation is likely to be more effective than random plantings.

### *Reforestation and erosion caused by water movement*

Erosion is increased when soils are directly exposed to rainfall, overland water flow or wind. This is especially true in locations with heavy rainfall or that are tectonically active. Erosion lowers soil fertility and increases river sediments. Trees can help limit erosion by breaking the impact of raindrops and slowing overland flow. This enables water to infiltrate into groundwater stores and conserves soil. Erosion is especially common in steep areas. Studies in the Andes have found that the extent of vegetative cover was a primary determinant of erosion and sediment yield, at least in watersheds $<100$ km$^2$ (Vanacker *et al.* 2007, Molina *et al.* 2009). The relationship is exponential and Molina *et al.* 2008 found that in small watersheds (<25 km$^2$) a small (10–25 per cent) increase in vegetation cover could lead to a significant decrease (60 per cent) in sediment yield.

It is not just the total amount of vegetation that is important, but also its spatial distribution. This is because this affects the distribution of sediment sources and sinks. Vegetative buffer zones on hillslopes or along drainage lines can be particularly effective at controlling sedimentation but in degraded hilly areas the most effective sediment trapping is likely to occur in gullies (Rey 2003).

Trees are usually assumed to reduce erosion but the most effective protection comes from herbaceous and shrubby vegetation close to the soil surface (Rey 2003, Molina *et al.* 2009). This means that the effectiveness of reforestation depends on the type of reforestation that is carried out. Plantation forests with dense canopies that limit the amount of light reaching the forest floor and reducing the development of understorey vegetation may not be effective at preventing erosion. This is often the case in teak plantations which are well known to offer only poor soil protection. On the other hand, monocultures with shrubby understories or mixed-species plantings having a variety of vegetation layers are likely to offer more soil protection.

Some have argued that grasslands should not be reforested since these already provide good soil protection and the reforestation may lead to increased rather than decreased erosion. But much depends on the environmental conditions and the nature of the sites being reforested. Tree cover can have a particular role in stabilising steep slopes to prevent mass wasting and landslips (Sidle *et al.* 2006, Forbes and Broadbent 2013). Tree roots help to dry out soils and structurally re-enforce the upper soil horizons. They are most effective in preventing shallow landslides but less effective in limiting deep-seated landslips.

Steep slopes are less attractive for agriculture and the opportunity costs of reforesting these may be low. The question then becomes one of defining the point at which forests replace agriculture. Slopes on which agriculture is usually deemed to be safe are those less than 11 degrees (20 per cent) while steeper slopes are more problematic (El-Swaify 1997). This suggests these might be locations where active reforestation is done. But steep slopes can be both difficult and costly to replant (Figure 5.2). An approach used in planning reforestation across the drier parts of the Loess Plateau region of China recommended that slopes greater than 15 degrees should be left to regenerate naturally and that active reforestation should be focussed on areas at the break of slope below these steeper hills where the gradient became more gentle (McVicar *et al.* 2007, McVicar *et al.* 2010). This is because the soils on these steep slopes are often dry and it can be difficult to reforest them without accelerating erosion. The design is shown in Figure 5.3. This maximises the interception of soil and nutrients moving down the hill into streams but limits the extent of water losses caused by evapo-transpiration by trees. Such losses might be minimised further by only reforesting priority areas. More details of the large-scale programme to reforest agricultural areas on steep slopes being currently implemented more broadly across China are given in Box 5.3.

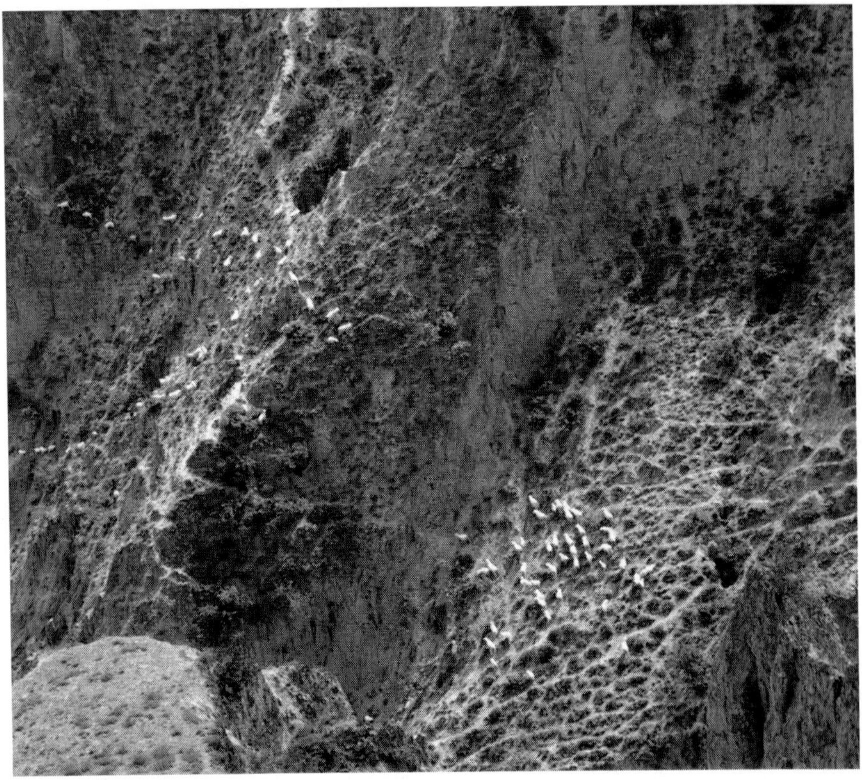

*Figure 5.2* Some steep slopes such as these in the Loess Plateau of China may be difficult to deliberately reforest and may have to be left to regenerate naturally. This is only likely to happen if herbivores like these goats are removed (Photo: J. Ernstberger).

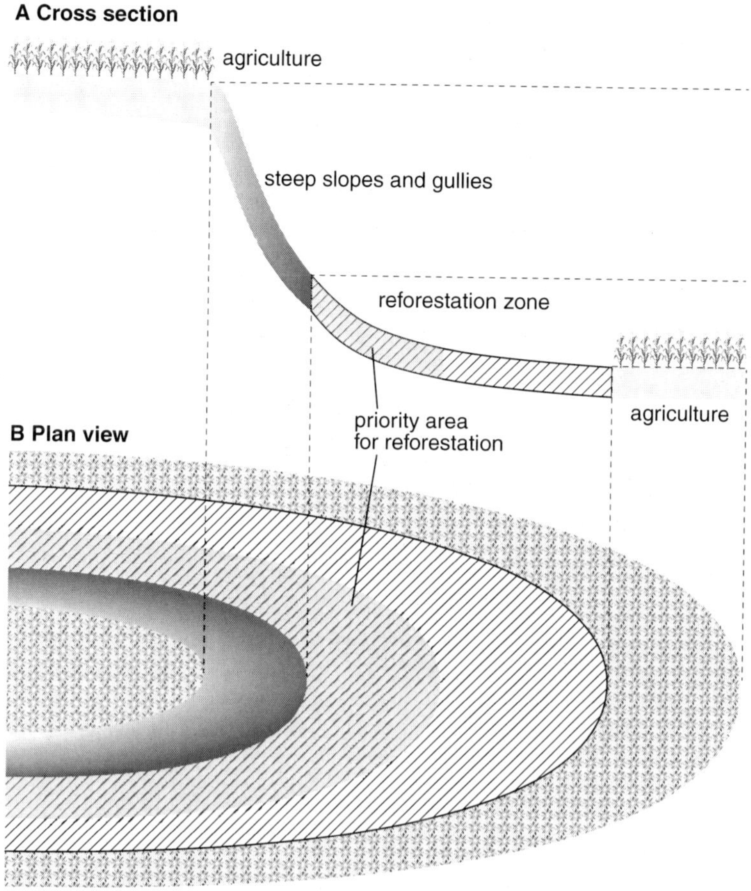


*Figure 5.3* Relationship between topography and priority areas for reforestation in the Loess Plateau region of China showing (A) sectional view and (B) plan view. Reforestation is very difficult on the steep slopes and gullies so is concentrated at the break of slope at the base of the steep hills and in adjoining areas. Reforesting only the priority area within this zone limits the impact of reforestation on streamflow while still reducing soil and nutrient movement (after McVicar *et al.* 2010).

**Box 5.3 The Grain for Green Program, China**

The Chinese Grain for Green Program, or Sloping Land Conversion Program, is one of the world's largest reforestation schemes (Xu *et al.* 2004, Bennett 2008, Liu *et al.* 2008). It was triggered by major floods in the Yangtze River basin in 1998 but the emphasis of the programme is on preventing erosion in mountainous areas and reducing rural poverty. It is also intended to restructure traditional farming practices to make them more environmentally and economically sustainable. The programme is being implemented across nearly 30 million ha of croplands and marginal agricultural lands including

4.4 million ha of land with slopes exceeding 25 degrees. The areas involved are mainly in the north and south-west of China and involve many of China's poorer rural households.

Apart from its sheer scale the programme has several unique elements. One is the emphasis given to the provision of ecosystem services rather than the production of forest products (although some harvesting will be allowed). The scheme relies on tree planting to achieve this rather than natural regeneration and most plantings to date have been monocultures that use commercial timber species rather than mixed-species plantings. Many of the trees planted are exotic species (and some are fruit trees). It has often been difficult to match species with sites, especially when sites are degraded. This is one of the reasons why the survival rates of planted seedlings has been variable. A minimum survival threshold of 75 per cent was set for the programme but many planted areas appear to have fallen below this level and, not surprisingly, some of the plantings have failed (Trac *et al.* 2007). Liu *et al.* (2008) and Cao (2008) argue that more emphasis should be placed on using native species and mixed-species plantings in future stages and that more technical support should be given to landholders.

A second unique feature of the programme lies in the way reforestation is being implemented. Unlike most previous reforestation schemes in China, the programme uses a public payment scheme and directly engages rural households inviting them to volunteer their land and labour. Participants are given free seedlings, a cash subsidy and an annual in-kind grain subsidy to compensate for forgone income from crops. These subsidies are paid for eight years. Areas targeted for reforestation are selected by the government although the final decision may be made at the county, town or village level. Xu *et al.* (2006) and Bennett (2008) report that, in practice, the programme has been excessively top-down although supposedly reliant on volunteerism. Surveys suggest there was limited consultation with communities and people often had restricted opportunities to decide whether or not to participate, which land would be reforested or which species would be used. Some communities have been unhappy about being forced to reforest some of their farmland (Weyerhaeuser *et al.* 2005, Trac *et al.* 2007). On the other hand, others appear to have had a generally positive experience although it is unclear what will happen to the planted trees once the payments stop (Uchida *et al.* 2005, Xu *et al.* 2006). Interestingly, many rural people have now moved to find work in the cities and the programme has probably assisted some to do so (Démurger and Wan 2012). These landowners are now probably less concerned about generating income from these lands so that their forests are likely to remain undisturbed, at least until they return.

This immense project has transformed a large area of rural China. Given the scale of the programme and the variety of ecological zones in which it has been carried out, it is not surprising that the degree of success has varied. However there are already signs that reforestation may be beginning to control erosion in some locations (Liu *et al.* 2008, Deng *et al.* 2012). Concerns that the conservation benefits of the programme may be at the expense of crop production and food security appear to be unfounded since only marginal lands are being reforested and production is being intensified on other agricultural lands (Xu *et al.* 2006).

Riverine areas are a second focal point for reforestation. This is because riparian plantings can act as filters that trap sediments and nutrients and safeguard stream water quality. But the opportunity costs of reforestation may be high since these lands can be attractive for cropping. In this case the question is how wide must any filter strip be to be effective? The answer depends on the steepness of the slope of land adjacent to the river and the type of soil present; sites with steep slopes and heavy clay soils are likely to need wider filter strips than flatter areas with more porous soils. Experience with reforested riparian strips over 20 m wide suggests they can be effective but they may need to be complemented by additional erosion controls elsewhere within a watershed (Parkyn *et al.* 2005, Verstraeten *et al.* 2006).

Some landholders on floodplains are opposed to the use of riparian plantings, especially in flood prone areas. They argue that riparian tree plantings slow down water movement during floods and, therefore, prolong flood duration. On the other hand, the absence of vegetation during such periods means water velocities are higher and the capacity for stream bank erosion is much greater (Hubble *et al.* 2010). This means a trade-off must be made between flood duration and erosion losses. The nature of the trade-off depends where along a river the particular site is located with the most important locations for soil stability being where bank slumping is most common. This may be where rivers emerge from steep areas and enter broader flood plains (Abernethy and Rutherfurd 1998).

### *Reforestation to control wind erosion*

Windbreaks and shelter belts have been widely used to boost crop production and control wind erosion in a variety of environments and climate zones. They act by cutting wind velocity and, hence, the capacity of the wind to carry soil material. In most cases they involve multi-storey plantings of various woody species in belts three or more rows wide and oriented perpendicular to the wind. Foliage should be present from crowns to the ground but semi-permeable windbreaks are preferred to dense windbreaks to reduce turbulent wind movement. The downwind benefit can persist for up to 15–20 times the windbreak height and the upwind benefit can be 2–5 times the windbreak height. A checker-board spatial arrangement of windbreaks is useful where wind directions change during the year. In drier regions the capacity of trees to control erosion (and provide fuelwood and other tree products) has to be balanced against their ability to compete with crops for water.

### *Combating desertification with 'green walls'*

Erosion is often especially significant in drier landscapes. Some large-scale attempts to combat desertification through reforestation are being undertaken in Africa and China. Both were originally described as 'green walls' to 'hold back deserts'. But, over time these have evolved into more complex rural development schemes involving trees as well as other forms of land stabilisation to combat desertification.

The Great Green Wall for the Sahel and the Sahara Initiative is being developed by the African Union and appears to be a sequel to a variety of earlier efforts to construct barriers such as the earlier Green Dam in Algeria described in Chapter 2 (OSS 2008). The original concept was for the establishment of a belt of trees 15 km wide and stretching across the African continent from Dakar in the west to Djibouti in the east. It is now seen as a linear mosaic of related interventions involving reforestation and various cross-sectoral development programmes aimed at improving rural livelihoods. The project is still in the planning stage but will probably be far more decentralised than the original Green Wall concepts.

The Green Wall of China was launched in 1978 with the aim of holding back the Gobi Desert. It was planned to be about 4,500 km long and stretch across much of northern China when completed in 2050. It is now more commonly referred to as the 'Three Norths Shelterbelt' (Cao 2011, Cao *et al.* 2011a). The scheme involves planting trees and shrubs across the north-western, northern and north-eastern provinces. Erosion is to be limited by developing windbreaks in agricultural areas and stabilising sand dunes. Reforestation has been done using aerial seeding as well as planting of seedlings. Various other forms of soil stabilisation such as straw checker-boards are also being used in areas with bare soils. A relatively small number of fast-growing exotic species have been used for reforestation including commercially attractive species such as *Robinia pseudoacacia* and species of *Populus* and *Pinus*.

The scale of the project is huge (27.5 million ha) and, given the sometimes harsh environmental conditions, it is not surprising that results have been variable. In some more arid locations the survival rates of planted trees have been low although there is some dispute over the actual rates (Cao 2008, Yang and Ci 2008, Cao *et al.* 2011a). There have also been unintended consequences. One widely discussed problem has been the fact that the planted trees have made heavy use of ground water resources, especially in more arid areas and water tables have dropped. In many cases the trees have subsequently died (Cao *et al.* 2011b). Critics have argued that these arid and semi-arid areas cannot support dense plantings of timber trees, even for wind breaks, and that more ecologically appropriate plantings should be used (Xu 2011). More shrubs are now being planted but, in areas which are not too degraded, it could be better to fence off areas to exclude stock and foster natural regeneration of grasses and other native species that are not necessarily trees (Jiang *et al.* 2006). These would not form windbreaks but would reflect the ecological realities of these sites.

The functional consequence of the programme to date is a matter of debate. There is some evidence that the frequency and duration of dust storms being reduced (Parungo *et al.* 1994) although Cao (2011a) and

Wang *et al.* (2010) argue there has been a tendency to overstate the extent of these gains.

## Choosing locations to conserve biodiversity

Large-scale forest restoration will increase the heterogeneity of agricultural landscapes and help soften the boundaries around protected areas. The benefits will obviously be greater when more structurally complex forms of reforestation are used instead of simple plantation monocultures. New forests will improve the ability of species, including seed dispersers, to move around the landscapes thereby helping conserve regional biodiversity (Vandermeer and Perfecto 2006, Oliver *et al.* 2010, Tscharntke *et al.* 2012). Although protected areas are commonly seen as the primary means by which biodiversity will be protected, evidence is mounting that the effectiveness of many protected areas depends on activities in the surrounding landscape. Even small clusters of trees or patches of forest within agricultural areas can provide substantial support for biodiversity (Mendenhall *et al.* 2011).

There are two basic strategies by which reforestation can help conserve existing biodiversity. One is to protect remnant patches of natural forest by using reforestation to build buffer zones around them and the other is to overcome the effects of fragmentation by using reforestation to link these remnants. A series of best-worst choices are illustrated in Figure 5.4 based upon general conservation principles. The most basic recommendation concerns size: for a given type of new forest, reforesting a single large area is usually better than reforesting a series of small areas, even when these cover the same overall area (Figure 5.4A). This is because forest margins are less attractive habitats for many species (the 'edge effect') and the proportion of the forest area uninfluenced by the forest edge is greater in larger patches. Other choices all concern the relationship between new reforestation areas within the landscape and existing patches of natural forest. These show that buffer strips entirely surrounding natural forest patches are better than new blocks of adjoining forest because they are more effective at protecting them from external disturbances and minimising the edge effect within the natural forest (Figure 5.4B); that wide buffer strips are better than narrow buffer strips because they offer greater protection (Figure 5.4C); that having corridors linking forest patches is better than reforestation that does not improve connectivity (Figure 5.4D); and that broad corridors are better than narrow corridors, again because it reduces the impact of the edge effect (Figure 5.4E).

### *Protecting and supporting existing patches of natural forest using buffer strips*

Many landscapes contain a variety of residual patches of natural forest. Is it better to use reforestation to create buffer zones around smaller or larger

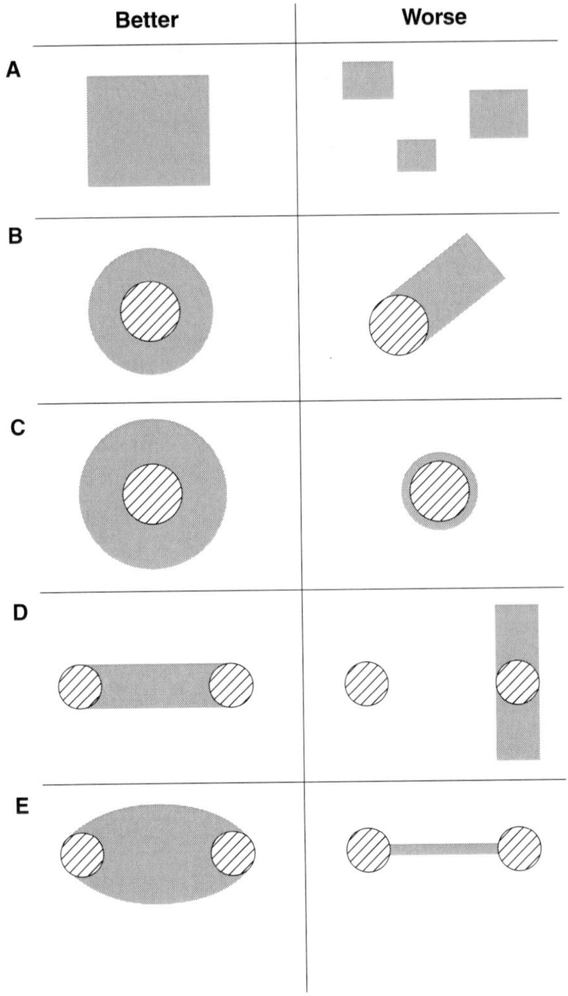


*Figure 5.4* Better and worse reforestation options for improving biodiversity conservation. Residual forest is shown as being hatched and new plantings areas within the landscape are shaded. (See text for explanation.)

remnant forest patches? Empirical evidence as well as that from experimental studies suggests isolated forest remnants gradually lose some of their original biota and that smaller remnants lose species more rapidly than larger remnants. The process can take some time depending on the particular biota. The pattern is shown in Figure 5.5. Smaller remnants are also proportionally more affected by edge effects than larger remnants (Laurance and Bierregaard 1997). This suggests it might make more sense to concentrate efforts on establishing planted buffer zones to protect large remnants than smaller remnants simply because this will protect more species.

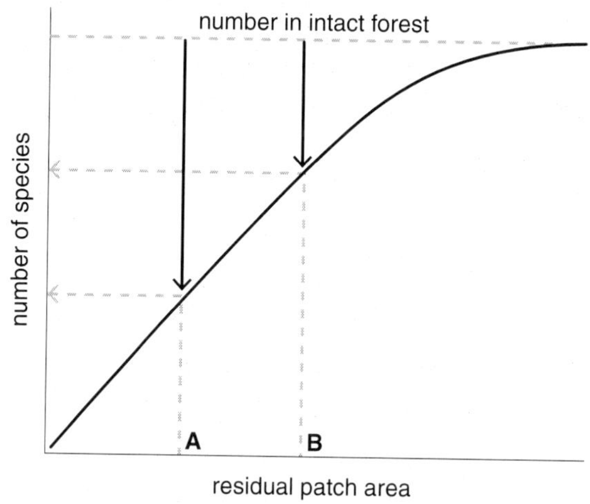


*Figure 5.5* Species are lost when forests are fragmented. Species-area relationships suggest that a smaller forest remnant (A) will eventually have fewer species than a larger patch (B). Restoring forest cover around a remnant can effectively increase its size and slow this extinction process.

But there are several reasons for qualifying this conclusion. One is that it takes time for the extinction process to be complete and newly isolated patches may still have much of their original biodiversity (indeed patch A in Figure 5.5 might conceivably have more species than patch B because it has only recently been created while patch B was isolated many years earlier). A second qualification is that at a landscape scale, protecting a number of smaller remnants might, collectively, be able to capture more species overall than a few large remnants, especially in heterogeneous and highly degraded landscapes (Arroyo-Rodriguez *et al.* 2009, Tscharntke *et al.* 2012). Third, it may not be the total number of species that is important but the identity of the particular species that are retained in a patch. A patch with rare or endangered species should attract more support than one that is heavily degraded and occupied by weeds. Similarly, Sgro *et al.* (2011) also argue for the importance of sites that may have acted as evolutionary refugia in the past and may do so in the future. These sites are likely to be identifiable on the basis of the phylogenetic diversity of the species present and forest patches at sites like these deserve special consideration. Fourth, a patch regularly exposed to disturbances such as wildfires or human activities (e.g. ones near roads or settlements) might be given a higher priority than one where such threats are lower. All of these differences in patch attributes mean that while area remains an important indicator of which remnant forest patch to prioritise when choosing where to invest in reforestation for conservation purposes, other factors may also need consideration.

An approach to dealing with these issues is outlined in Table 5.2. This is based on the risk of disturbance and the overall conservation value. The

*Table 5.2* Prioritising the choice of different types of natural forest remnants for protection using reforested buffer strips

| *Conservation value* | *Lower disturbance risk* | *Higher disturbance risk* |
|---|---|---|
| High | Moderate priority<br>Plant adjoining area or surround with a buffer strip<br>Preferentially establish a buffer using mixed-species plantings or Ecological Restoration | Highest priority<br>Surround with a wide reforested buffer strip<br>Preferentially establish a buffer using Ecological Restoration |
| Low | Lower priority<br>Establish a buffer strip even if it is only a plantation monoculture | Lowest priority<br>If possible establish a buffer strip |

forest patches with the highest priority for protection via reforestation are those that still have high conservation value but which are threatened by continuing internal species losses (because of the patch size), wildfire or an illegal activity such as grazing, hunting, logging or agricultural clearing. Surrounding such patches with wide buffer strips established as mixed-species plantings or Ecologically Restored forests enlarges the habitat areas and helps protect the forest from future disturbances. Forest patches that are equally valuable from a conservation viewpoint but less exposed to threat might be treated in the same way. In these cases there is still need to increase the overall habitat area but less need to have a wide buffer or perhaps even to protect the complete boundary.

Forest patches with a lower conservation value will have a lower priority. Nonetheless these may still be worth targeting because of their location within the landscape. For example, an isolated patch of such forest in an otherwise homogenous and extensive agricultural landscape could still prove valuable as a source of tree seeds and as a wildlife habitat. These, too, could be protected by buffer strips even if these new forests were only monocultures. In all cases the quality of the natural forest patches might be improved by removing invasive weeds and enriching the sites with species of conservation significance.

### *Areas to improve connectivity*

Forest fragmentation caused by agricultural clearing means that many forest-dependent species become reproductively isolated because the landscape has become impermeable to them. Local extinction is a common outcome (Turner and Corlett 1996, Laurance and Bierregaard 1997). The most obvious way of overcoming the effects of fragmentation is to establish strips or forest corridors to link these isolated areas (Figure 5.4 D and E).

Where that is not possible, an alternative might be to establish patches of new forest between the remaining areas of natural forest to create, in effect, a series of habitat 'stepping stones'. Provided the quality of the newly created habitat is sufficient for the species involved, these corridors or stepping stones may enable individuals to move across the agricultural matrix between meta-populations. Corridors aligned along altitudinal, latitudinal or rainfall gradients may also enable the seasonal migration of species or allow species populations to move and adjust to future climate changes. Strategic investments in improving connectivity are likely to generate greater benefits than a similar investment that does not make connectivity a target.

Corridors can be developed at a variety of scales ranging from local through to continental or transnational. An example of an early large-scale corridor is given in Box 5.4 (although this is better described as a barrier than a corridor).

Corridors built to achieve local conservation objectives are likely to have a relatively uniform composition because environmental conditions are essentially similar along its route. The relatively small area needed probably also means that only a limited number of landholders will be involved in negotiations over the corridor's establishment. Designing corridors that are spread over larger areas in order to conserve populations of wide-ranging species or to help multiple taxa adjust to climate change are more problematic. Landholders will have to agree to reforest some of their land or sell land to the corridor proponents. Not all landholders on the route may want to do either of these things so that it could be difficult to achieve continuity. In addition, a much greater number of species and reforestation methods may be needed because of the wider variety of environmental

**Box 5.4 The great Indian hedge**

In late nineteenth century a large vegetative strip in the form of a thorny hedge was built across India. The purpose of this curious structure was to form a physical barrier for the customs authority, allowing it to collect a tax on salt. It began in what is now the Punjab Province of Pakistan and reached 1,300 km across India to Burhanpur in Madhya Pradesh, about 500 km north-east of Mumbai. The route was determined by political boundaries and geographic barriers such as rivers. The hedge was around 3 m wide and 4 m tall and covered an overall distance of perhaps 2,000 km although the best quality 'live' component was probably around 660 km at its peak in 1878 (the remainder being made up of 'live' and 'dry' hedge formed by heaped dead branches). The live hedge was composed of *Acacia catecha, Ziziphus jujuba, Carissa curonda* as well as various *Euphorbia and Opuntia* species. The hedge was very dense but probably had little conservation value because of its narrow width. Once established the hedge required a staff of 12,000 men for its maintenance. Both the customs line and the hedge were abandoned in 1879 and barely a trace now remains (Moxham 2001).

conditions likely to be encountered. This variety of reforestation methods and the mixture of corridor and stepping stones mean these new forest areas might be thought of as pathways rather than corridors, at least in the short term. Some pragmatism may be needed. Parris *et al.* (2011) described these large corridors as: 'Managed and physically interlinked areas of land that are structured around existing natural, intact and functioning areas of vegetation and existing ecological processes. These are managed for landscape resilience, and are supported through appropriate socio-economic institutions'.

This description goes beyond just conservation biology considerations and recognises that large-scale corridors will inevitably have socio-economic consequences and are more likely to generate differences of opinion within the communities through which the corridor passes. Those designing such corridors must take this reality into account.

The concept of landscape corridors has captured public imagination and has attracted some political support. Some recently proposed large corridor projects, including several large transnational corridors, include the Mesoamerican Biological Corridor in Central America (Muller and Barborak 2010), the Yellowstone to Yukon corridor linking USA and Canada (Locke 2010), the Greater Mekong Subregion corridors (Moinuddin *et al.* 2011) and the European Greenbelt running from the Barents Sea near the Norwegian-Russian Border to the Balkans and the Black Sea (Accessed 16 May 2014. Available at www.europeangreenbelt.org). Many of these are more properly described as corridor networks. A large number of other large national and transnational corridor initiatives are described in Worboys *et al.* (2010) and Fitzsimons *et al.* (2013).

The idea behind corridors seems intuitively attractive but not everyone supports it (Simberloff *et al.* 1992, Hodgson *et al.* 2009, Baguette *et al.* 2013). Some say that corridors encourage the spread of invasive species or diseases. Others argue that species differ in their habitat requirements and perceive landscapes – and hence corridors – differently to humans. If landscape linkages are to be created then the habitat requirements of species must be understood. Further, we need to understand what causes species to disperse from their existing locations and the means by which they search for new habitats. The danger is that a landscape that is functionally connected for one species may not be so for another. Finally, some critics have argued against corridors on economic rather than ecological grounds and say that the considerable investment in funds and resources needed could be more profitably spent in other ways such as creating buffers around specifically threatened forest patches.

Despite the scepticism, empirical evidence to date based on a meta-analysis of a large number of studies suggests many corridors are functioning in the ways hoped for. Corridors have been found to be especially effective for plants, invertebrates and non-avian vertebrates although not for birds (Gilbert-Norton *et al.* 2010).

### *Guidelines for corridor design*

There are four important elements of corridor design. The first involves the type of reforestation carried out. Ecological Restoration is the preferred reforestation method to use when creating a corridor because it aims to recreate the original habitat found at the site. But, as discussed earlier, it may be difficult to implement, especially if the corridor is being established along a rainfall, altitudinal or latitudinal gradient having a variety of environmental conditions. Little might be known of the original forests at these sites and reference sites may be lacking. This means other forms of reforestation may be needed. This will certainly be the case if the intent of the corridor is to anticipate environmental changes induced by global warming. Other forms of reforestation may also be required because some landholders need to obtain a commercial benefit from any trees they plant.

The second element concerns the width of the corridor. Narrow strips may suffer from edge effects that penetrate from both sides and influence the whole strip. Corridors probably need to exceed several hundred metres width to avoid this although dense protective plantings on margins may effectively 'seal' the boundary and minimise these effects.

The third element concerns continuity. Continuous and unbroken corridors are obviously preferable but may not always be achievable. There may also be variations in the quality of the forest cover. This means the corridor may come to resemble a set of forest patches or stepping stones across the agricultural landscape. Whether these are functionally effective or not will obviously depend upon the species using the area.

Finally, the spatial layout of the corridor will influence both its effectiveness and its cost. Recommendations are to:

- Use the shortest or most direct route between intact core forest areas.
- Use the least-cost pathway. That is, take account of not only the distance to be covered but also the cost of any necessary land acquisition. The total cost will also include that needed to reforest cleared land.
- Use existing forest remnants and regrowth areas. Include even those that are heavily degraded and are currently of marginal conservation interest (if necessary these can be enriched).
- Use the known seasonal movement or migration routes of key wildlife species to plan the corridor layout.
- Use riparian strips because these areas are attractive to wildlife, especially in drier areas.
- Take into account the possible impacts of global climate change on the spatial distribution of local biota and biological communities. This may mean favouring latitudinal and altitudinal corridors as well as rainfall gradients.
- Take account the attitudes and preferences of local communities.

Different weight might be given to each of these guidelines at different points in a landscape over which a corridor passes. Many are intended to facilitate species movement across the landscape but McRae *et al.* (2012) argue that attention should also be paid to identifying locations within the landscape that *impede* rather than facilitate movement. They suggest this perspective can sometimes generate new ways of improving connectivity. Irrespective of the method used, most corridor designs place considerable emphasis on the use of existing patches of natural forest and look for ways of filling gaps with either natural regrowth or by some form of reforestation.

It is important to keep in mind that the purpose of corridors is to maintain biodiversity and allow for evolutionary development over the longer term. Present day markets and opportunity costs will change over time and current landowners will be replaced by others. Present day limitations on the quality of the corridor or concerning its location need not be final and there may be opportunities for managers to overcome these impediments in the longer term provided appropriate institutional settings are in place. Much of the early conservation literature concerning corridors emphasised the biodiversity benefits of various designs and paid rather less attention to the socio-economic implications that their proposals would have for landholders and communities. But, more recently, corridor design has now become a blend of conservation biology and land use planning (Worboys *et al.* 2010, Whitten *et al.* 2011). An example of a large-scale corridor seeking to integrate conservation objectives and farming practices is described in Box 5.5.

**Box 5.5 Example of a very large-scale corridor**

Gondwana Link is a large corridor project in the south-west of Western Australia spread over a distance of 1,000 km (Accessed 16 May 2014. Available at www.gondwanalink.org). It will connect the ecosystems of the tall moist eucalypt forests in the far south-west of the state with those of the drier woodlands and heaths of more arid inland areas further to the east (Bradby 2013). Much of the area has been cleared for agriculture but 27 per cent of native vegetation still remains (Jonson 2010). The corridor links Fitzgerald River, Stirling Range and Porongurup Range National Parks and several smaller nature reserves but extends to the west and east of these core areas. The region is a global biodiversity hotspot with a very large number of native species (Hopper and Gioia 2004) and the corridor is aligned across a major rainfall gradient that is expected to change with global warming. The aim of the project is to build resilience by strengthening the ability of the region's biota to withstand these changes.

The project is being managed by a consortium of NGO partners and other stakeholders. The group was organised to the stage where it could begin spending funds in 2002 although planning began before then. Considerable inputs to the project are also being made by landholders and volunteers. The corridor is being built by buying properties in key areas and by assisting landholders to revegetate parts of their farms. These areas will target priority vegetation associations and sub-catchments. More than 2,300 ha have been revegetated through planting and sowing using biodiverse mixtures of native species (Jonson 2010). Directly sown seed has been used for most sites (e.g. Figure 4.7) but these have been complemented by strategically located clumps of planted seedlings. These mixtures have been designed to establish defined species assemblages that reflect specific soil types and landscape positions. Vegetation is being established in thickets as well as more open configurations in order to provide habitats for a variety of wildlife. It is expected that 16,000 ha will have been restored by 2015. Areas of existing forest or woodland are also being fenced off and protected. It may not be possible to establish a complete wide linkage across the whole landscape and it may be that the final outcome will consist of a series of 'conservation nodes'. This will depend on the opportunities that become available in future.

Large-scale projects like Gondwana Link involving many stakeholders and actors require unusual degrees of organisation and very flexible governance structures to make them work. The project has been guided by several elements (Whitten *et al.* 2011):

- articulating a compelling vision of the future ('audacious but not too audacious');
- demonstrating empathy and compassion for both people and nature;
- recognising the 'power of place' and building a decentralised conservation approach that takes account of local differences;
- building astuteness and common sense approaches through open and rigorous discussion amongst stakeholders;
- taking advantage of the power of good will and good work.

## Priority areas for reforestation

Given the range of options described above, are there areas that deserve priority and should be reforested first? A common way of choosing between alternatives is to carry out a cost/benefit analysis. This is relatively straightforward when the costs and benefits can be easily quantified and financially discounted to the present time. It is more difficult when these benefits come in forms that are difficult to measure in strictly financial terms and where any benefits take time to develop. For example, how might one compare reforestation at a site where the primary benefit will be to reduce erosion with reforestation at another site where the primary benefit will be to improve the habitat of a particular wildlife species? In such cases the choice is likely to be made on the basis of stakeholder preferences rather than on the supposed maximum net financial benefits.

In recent years a number of models have been developed to optimise land use decision making (Polasky *et al.* 2008, Wilson *et al.* 2011, Drielsma *et al.* 2014). Many of these have been primarily concerned with finding locations where a particular outcome (e.g. habitat restoration) can be achieved most efficiently in terms of areas treated or funds spent. Others have begun to attempting to locate reforestation areas to achieve more general environmental and economic benefits (Orsi *et al.* 2011, Booth 2012, Bateman *et al.* 2013) These models have become increasingly sophisticated but they require large amounts of data describing topography, soil conditions and species distributions as well as the costs of reforestation and information about spatially distributed opportunity costs. Such data may be available in western countries but is often unavailable in many developing countries.

An alternative approach is to identify priorities based on general principles and use these as the basis for initiating negotiations amongst stakeholders. Some suggested priority locations for reforestation are shown in Table 5.3. The relative importance of these will depend, of course, on the views of individual landholders and other stakeholders. In most cases priority should also be given to areas where natural forest regrowth has already commenced since this can offer economic as well as environmental benefits.

*Table 5.3* Preferred locations for reforestation within the agricultural matrix depending on the primary objective

| *Outcome needed* | *Preferred location* |
|---|---|
| Economic benefit | Sites close to roads or markets (to reduce transport costs)<br>Sites distant from roads (to reduce opportunity costs)<br>Sites marginal for agriculture (low fertility, steep slopes) since these have lower land prices and opportunity costs<br>Areas with naturally occurring regeneration |
| Hydrological benefit | Lands with compacted soils (to improve infiltration rates)<br>Water-logged areas (induced by prior deforestation)<br>Recharge areas and areas within farms to combat salinity<br>Areas with naturally occurring regeneration |
| Erosion and sedimentation benefit | Areas with active water or wind erosion<br>Areas with compacted soils<br>Areas with steep slopes (to prevent future erosion)<br>Riparian areas (to act as filters preventing sediments reaching waterways)<br>Areas with naturally occurring regeneration |
| Biodiversity benefit | Areas around natural forest remnants (to create buffers and to increase the overall area)<br>Areas between natural forest remnants especially along altitudinal and latitudinal gradients (to create corridors or stepping stones)<br>Riparian strips (especially in seasonally dry areas)<br>Areas with naturally occurring regeneration |

Especially degraded sites pose a particular challenge. Should these be automatically seen as priority locations where reforestation must be implemented? Some may deserve treatment because the consequences of not doing so are too great. Examples might be a severely eroding hillslope or an old, abandoned mine site where chemicals are leaching from tailings and the effects of degradation extend far into neighbouring areas or downstream. But, otherwise, the issue may be less clear. Suppose it cost, say, $1,000 to rehabilitate 10 ha of severely degraded land but the same amount of money could reforest 1,000 ha of less degraded land. Which is the best choice? Many might say it is better to treat the most critical sites first. But in some circumstances it is possible that the less degraded area might be more deserving. This might be the case, for example, if a reforestation programme was being developed in a particular region and needed demonstration sites in easily accessible locations to publicise the new land use option. In such a situation reforesting a 1,000 ha area near a main road might be more attractive than reforesting a 10 ha area that is hidden away in a more remote location.

The point is that objective criteria can sometimes be used to make decisions but stakeholder preferences and local politics are often the primary determinants of reforestation priorities. It is also the case that externally supplied funds tied to specific projects and locations will determine where reforestation is carried out. But the situation can change over time so that it may be possible to revisit these decisions at a later date if the preferred reforestation choices are not immediately possible.

## Planning arrangements to achieve strategic reforestation

The choice of where to undertake any forest restoration will usually be made by individual landowners based on their perception of where reforestation is likely to generate the greatest benefit. Timber plantations may be found on flatter land with good road networks while hills and riverine areas may have more complex forms of reforestation to reduce erosion. More complex plantings for conservation purposes may be found forming buffer areas around residual patches of natural forests or corridors between them. Small plantings of various types may be scattered on farms across the agricultural landscape. The collective impact of all these choices will not necessarily be optimal for the delivery of conservation benefits or the delivery of ecosystem services if the plantings are randomly located. Undertaking more strategically located reforestation at a landscape scale has been referred to as Forest Landscape Restoration. This can be described as a process that aims to regain ecological integrity and enhance human livelihoods in deforested or degraded landscapes (Maginnis and Jackson 2007). Further details of the planning arrangements needed to achieve Forest Landscape Restoration are described in Chapters 7 and 8. Much, too, depends on the extent of plantings at particular locations. This is considered in the following chapter

## Conclusions

The spatial location of forest restoration within the landscape mosaic determines the economic profitability of new plantings and their functional outcomes. Carefully located plantings will generate better outcomes than randomly located plantings.

The preferred locations for commercial timber plantations are usually those with good soils and low land prices, and those located close to markets or transport although more distant locations can be used if high-priced timbers are being grown.

The preferred location for restoring ecosystem services depends on the particular service required although there may be some opportunities for reforestation at particular locations to restore multiple services (e.g. control erosion and conserve biodiversity).

Many ecosystem services are likely to require that specific types of reforestation are undertaken at specific sites (e.g. belts of tolerant species planted at certain sites for dryland salinity control; structurally complex and multi-species plantings used to create wildlife corridors).

It may be difficult to reconcile broad ecological objectives with the more focussed economic objectives of landholders. This will require careful planning to ensure the costs and benefits of doing so are shared.

# 6 How much forest restoration is needed?

## Introduction

Previous chapters have shown that the type of reforestation and its spatial location are both important in determining the economic benefits and functional consequences of extending forest cover. But the scale of reforestation also matters and large areas of reforestation will often be more effective than small areas. However, this may not always be the case. Indeed, large-scale forest restoration may not always be possible. So what is the nature of the relationship between the area reforested and the benefits generated? In some cases, such as carbon sequestration and storage, there appears to be a relatively simple linear pattern with increasing forest area causing a proportional increase in the amount of carbon captured, provided environmental conditions are similar. But in other cases, such as hydrological impacts or biodiversity conservation, the relationship is more complex. Further, the nature of the relationships is usually dependent on factors other than just the area reforested. One of the most critical of these is the time since forest regeneration began or planting took place and, consequently, it can take time for functional outcomes or ecosystem services to be generated. This chapter explores some of the economic and ecological consequences of large-scale reforestation and the some of the factors that are responsible for these outcomes.

## Reforestation area and timber production

Scale matters for commercial plantation owners. Most industrial timber growers prefer to establish relatively large plantation areas (say, 50,000 ha) to generate an annual yield of a uniform product sufficient to sustain a mill or plant and to achieve economies of scale when doing so. Ideally these plantations surround, or are near, the mill so that transport costs are minimised. Alternatively, they will be close to transport able to link the plantations with a more distant market. In these cases the regular supply and large volume of forest product help cover the higher transport costs.

Owners of smaller plantations may have to adopt a different approach. Those with land and plantations located near large company-owned plantations growing the same types of timbers may be able to arrange contracts to sell this timber to the bigger entity. In this way they avoid the problem of having a small plantation. But landholders not located near big industrial plantations cannot achieve economies of scale unless they can join with a number of other small growers to exceed a scale threshold. The size of this threshold obviously depends on the types of plantations and the nature of the market (e.g. firewood, high-quality timbers). Once their plantations collectively exceed such a threshold they can then supply a market on a sufficiently regular basis to sustain it over the longer term. But this can be a difficult transition to make; owners with older plantations must defer harvesting until other owner's plantations also become old enough and the collective harvestable resource reaches a sufficient size. If an individual sells before the group reaches this size threshold they all risk becoming episodic market participants having only a small volume to sell at any one time. This is likely to be commercially unattractive and may even cause the local market to disappear.

Those unable to form such arrangements must seek an alternative strategy. In these cases success is likely to be less dependent on scale and rather more dependent on the types of new forests established as well as on the nature of the goods and services they produce. Many small growers find it is more profitable to diversify the species planted in order to diversify the markets they can supply. Likewise these plantation mixtures can mean the financial returns are more frequent and risks are diversified. These types of reforestation will involve more time-consuming management systems but smaller growers may be willing to make this investment for the sake of what they see as a preferred outcome.

## Area reforested and hydrological outcomes

Reforestation scale also affects the hydrological cycle. Large-scale deforestation is commonly blamed for flooding as well as water shortages (Bruijnzeel 2004). As a consequence, many believe reforestation will reduce flooding and restore river flows during dry periods. For example, the cessation of flow in the Yellow River in 1997 and the flooding of the Yangtze River in 1998 in China are often quoted as the triggers for some of the very large reforestation programmes carried out in China after these dates (e.g. Yin *et al.* 2005). But reforestation is not necessarily a 'cure' for either of these problems and the effects of scale are complex.

### *Water yields*

Rivers and streams are supplied directly by overland flow (e.g. from storms) and/or indirectly by sub-surface flow from soils and groundwater. This latter source is often referred to as baseflow and is usually responsible for most dry season streamflow. Both pathways can be affected by vegetative cover. Overland flow is affected by changes at or near the soil surface such as litter layers and undergrowth. These impede surface water movement and enable more water to infiltrate into the soil. Base flow is affected by the density of trees with deep roots able to access and transpire more soil water and, sometimes, groundwater.

Although deforestation is often blamed for floods and declines in dry season flows the the hydrological effects of deforestation and reforestation depend on a variety of factors (Andréassian 2004, Bruijnzeel 2004). In the case of reforestation, these include the condition of the watershed prior to reforestation, soil depth and the climatic conditions, especially the amount and seasonality of rainfall. Flow rates are also influenced by the type of new forest and the proportion of the watershed that is reforested.

Most experimental watershed studies have shown that reforestation of cleared land causes a reduction and *not* an increase in annual streamflow totals or water yields (Table 6.1). Jackson *et al.* (2005) reported a global study of 504 annual catchment observations involving reforested watersheds and found an annual decrease of 38 per cent or 180 mm of streamflow, on average, across all ages (although most studies involved reforestation areas of less than 40 years). There was a tendency for greater decreases to be found in younger forests and some indication that the impact might lessen in older (>25 years) plantations. Streams draining reforested areas dried up completely in 13 per cent of cases, especially in drier regions. This decline in streamflow is one of the reasons why there has been much discussion about the merits of using eucalypts in agroforestry settings (Poore and Fries 1985). It is also the reason why certain types of reforestation can be useful in countering dryland salinity (discussed in Chapter 5).

The reason why water yields decrease after reforestation is because evapo-transpiration is greater with trees than with crops, grasses or shrubs. More rainfall is intercepted by tree canopies and tree roots are often deeper than those of most grasses or agricultural crops. These allow trees to access and use soil water, and perhaps groundwater, even when upper soil profiles are dry.

Although larger forest areas mean more water is transpired, the actual reduction in water yield is dependent on the proportion of a watershed that is reforested and this usually increases linearly as the proportion increases. Small areas of reforestation may have little impact; Van Dijk and Keenan (2007) suggest declines in yield only occur when more than 15–20 per cent of a small watershed is reforested (Figure 6.1A).

But different types of forest and silvicultural systems generate different hydrological responses (Scott *et al.* 2005). Evapo-transpiration is greater in dense, young forests than in less dense, older forests largely because of differences in leaf area. Likewise, differences in canopy and root architecture (especially depth) may also influence responses to reforestation. An

*Table 6.1* Consequences of deforestation and reforestation on hydrological flows (based on Bruijnzeel (2004), Scott *et al.* (2005), Van Dijk and Keenan (2007))

| | *Total water yield* | *Local flooding* | *Dry season flow* |
|---|---|---|---|
| Deforestation | Water yield increases in proportion to the extent of watershed clearing. | Flooding may increase but rainfall intensity and watershed characteristics (size, extent of road network, degree of land degradation) are usually more important. | Flow may increase if infiltration capacity is maintained but reduce if topsoils become degraded and infiltration capacity is severely reduced. |
| Reforestation | Mean stream flow is reduced in proportion to the extent that a watershed is reforested; this is also influenced by tree density and age. Reduction may be less at larger scales. | Flooding from moderate storms may be reduced but effects are less clear when heavily degraded sites are reforested. Any impact is likely to diminish as watershed size increases due to 'dilution' effects and the size of rainfall fields relative to watershed size. | Flows usually decrease but may increase at badly degraded sites if reforestation improves hydraulic conductivity of soil sufficiently to overcome the higher water use of the trees. |

illustration of the way different types of forests can produce different outcomes is shown in the study reported by Zhou *et al.* (2002) which compared changes in stormflow run-off from two types of plantations at a severely degraded site in southern China and an adjacent bare area. In this case a degraded site was reforested with (i) a eucalypt plantation and (ii) a mixed-species plantation created by under-planting a eucalypt plantation with a variety of native tree species. Trees in the eucalypt plantation intercepted less rain and had more stemflow than the mixed-species forest. Of water reaching the soil, annual run-off from the bare land was 51 per cent of precipitation, 23 per cent at the eucalypt site but only 3 per cent at the site containing the species mixture. Similar differences were noted in other studies (reviewed by Scott *et al.* (2005), Van Dijk and Keenan (2007)). More research is needed to fully understand the implications of different types of reforestation (species, tree densities, harvesting schedules, etc.) on patterns of water usage and movement but the study is sufficient to indicate the potential importance for water yields of forest type as well as the forest area.

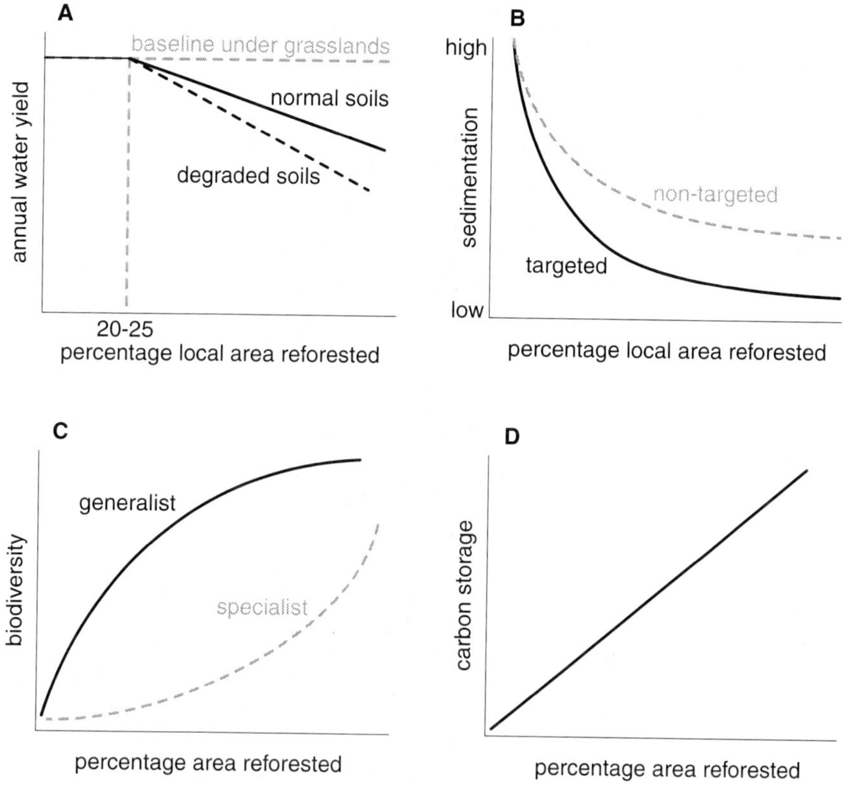


*Figure 6.1* The hypothetical effect of increasing the area reforested on (A) water yield, (B) sedimentation, (C) biodiversity and (D) carbon stores. The effect on water yield depends on whether the soils being reforested are degraded and usually occurs after at least 20–25 per cent of the area is reforested. The effect on sedimentation rates depends on whether erosion-prone areas are targeted or not. The effect of increasing habitat area on biodiversity depends on the responses of particular biota with habitat generalists usually responding more quickly than habitat specialists.

Hydrological processes become more complex as the areas affected by reforestation increase. Most research activity has been carried out in small paired watersheds of less than 1,000 ha (Brown *et al.* 2005) and the question is whether these studies also reflect changes going on at larger scales. One study of large-scale reforestation was carried out in Uruguay where reforestation was carried out across an area of natural grasslands (Silveira and Alonso 2009). In this case the new plantations occupied 540 km$^2$ of the 2,097 km$^2$ watershed. The study found replacing grasslands by trees caused a reduction in annual run-off which was especially noticeable in spring and summer. It also reduced peak flows following storms. A second study of the impact of large-scale reforestation was carried out in the Mediterranean climatic region of Chile (Little *et al.* 2009). In this case logged-over *Nothofagus* forests were cleared and replaced by *Pinus radiata* plantations in two large (>25,000 ha) watersheds. This study also found a reduction in water yield following reforestation with summer flow declining from 13.1 mm to 7.5 mm in one watershed and from 7.3 mm to 5 mm in a second.

In both these cases the areas being reforested were relatively homogenous and a different outcome was observed in a more heterogeneous landscape. Beck *et al.* (2013) assessed changes in forest cover across 12 middle-sized watersheds in Puerto Rico. These differed in the type and extent of reforestation (some were patchy while some were contiguous), soil conditions, degree of urbanisation and in the prevailing climatic regimes. They also varied in size from 23 $km^2$ to 346 $km^2$. In each case natural regrowth since the 1950s had caused the forest cover to increase but the extent of the increase in forest cover averaged 26 per cent and ranged from 2–55 per cent across the 12 watersheds. Despite this increase in forest cover Beck *et al.* (2013) found there was no significant relationship between changes in forest cover in the watersheds and water discharge.

The issue of scale is clearly one deserving further study in a wider range of climatic and soil conditions. The evidence to date suggests reforestation can significantly alter the timing of hydrological flows and commonly reduces water yields. Manipulative studies undertaken in small experimental watersheds have provided useful insights into the processes involved but these may not necessarily represent the hydrological consequences of reforestation at a larger scale, especially when this reforestation is not contiguous or complete.

### *Floods*

Deforestation (especially if followed by soil degradation or urbanisation) can cause an increase in the frequency and intensity of flooding. But much depends on the size of the watershed and the impact of localised rain storms can be diluted at larger scales where other land uses also affect overall landscape hydrology (Table 6.1). Reforestation may reduce local flooding arising from small to medium storms, especially in locations with deep soils, but even large-scale reforestation will not prevent floods being caused by severe and long-duration storms. In such cases flooding is less influenced by the type of vegetation and rather more by rainfall intensity which exceeds the infiltration capacity of soils, the extent to which soils are already saturated by prior rainfall and by steep slopes.

### *Dry season flows*

The amount of streamflow during dry periods of the year is an important issue for most land users. This type of flow occurs during rainless periods and is the result of water gradually moving from soil and groundwater stores into streams. Larger groundwater stores mean that water is able to keep moving into streams over a longer period. As long as surface soils are not damaged, deforestation usually leads to an increase in dry season streamflow since crops or grasses that replace forests typically use less water and baseflows tend to be greater. When the site is reforested evapo-transpiration increases again and groundwater recharge may be reduced. In many cases, dry season flows then decline (Jackson *et al.* 2005).

However, this decrease may not occur if the enhanced evapo-transpiration caused by reforestation is compensated for by an increased rate of groundwater recharge (Bruijnzeel 2004, Scott *et al.* 2005). Such might be the case if soils were highly degraded following deforestation and reforestation was able to improve the soils' rainwater absorbing capacity over a large area. For example, deforestation may have led to topsoils being compacted or eroded leaving a former sub-soil as the new ground surface. Under these circumstances, much less rainfall is able to infiltrate the soil and reach the groundwater. Instead, it is quickly lost as overland flow. But reforestation can improve soils by enriching them with more organic matter and improving soil structure and porosity. Greenwood and Buttle (2014) found the development of root channels formed by understorey shrubs was also important. Over time the infiltration capacity of the soils is gradually increased and more precipitation should then be able to reach groundwater stores. Ilstedt *et al.* (2007) reviewed what evidence was available for the tropics and concluded that, on average, reforestation probably caused a threefold improvement in infiltration rates. There is still only limited evidence showing the benefits of this on actual dry season flows although improvements in dry season flows have been reported by Bruijnzeel (2004) and Scott *et al.* (2005). Of course, soil physical properties may also be improved by mechanical site preparation techniques like ripping or ploughing which are often used to establish commercial plantations but are less commonly employed by smallholders.

There appears to be little known about how big an area must be treated to generate an improvement. Nor is it clear how long it might take. The evidence is mixed and is complicated by differences in the species used in reforestation and management histories. A study by Bonell *et al.* (2010) in south-west India found evidence that young *Acacia* plantations had begun to improve soil hydraulic conductivity but not to the point where they matched those in less disturbed natural forests. The *Acacia* plantations were less than 12 years old and the recovery process was clearly still in an early stage. There were no signs of improvement in a 30-year-old teak (*Tectona grandis*) plantation growing in the same region. This may have been because of actual differences in the capacity of the two species to change soil physical properties or, alternatively, because the teak was growing on dense vertisols while the *Acacia* plantations were on lateritic soils. Interestingly, another study in Costa Rica of changes in soil conductivity following reforestation using teak also found only modest improvements in topsoil conductivity in

20-year-old teak plantations in comparison with secondary forest (Fernández-Moya *et al.* 2013).

Evidence of soil conductivity recovering after several decades was reported in studies in regrowth forests undertaken in Vietnam (Giambelluca 2002, Ziegler *et al.* 2004). In these sites, conductivity of surface soils increased from a little over 20 mm per hour not long after regrowth commenced to 60–70 mm per hour in the oldest sites (25–30 years after agriculture was abandoned). Estimates of recovery times in other studies range from more than eight years to ninety-five years (Hassler *et al.* 2011, Greenwood and Buttle 2014, Hevia *et al.* 2014).

The significance of all of these changes in conductivity depends, of course, on the prevailing rainfall regimes and whether the improved conductivities exceed commonly occurring rainfall intensities. But it also depends on how the regenerating forest is managed. Studies of changes after 25 years in topsoil hydraulic conductivity under pines growing in Nepal found that changes in the top 25 cm of soil were not nearly as great as had been supposed even after this comparatively long period. The hydraulic conductivities of the 36-year-old pines were less than they had been when originally measured 25 years earlier. This lack of response was attributed to the harvesting of leaf litter and shrubs and the fact that animals were allowed to graze amongst the trees. Presumably this helped keep the soils compacted (Ghimire *et al.* 2013). The finding highlights the need to consider just what management regimes are appropriate when degraded lands are being rehabilitated. Should grazing be temporarily or permanently excluded in key locations? What if this cannot be achieved? And what role does the choice of species and species diversity (of canopy trees as well as shrub layers) have in modifying hydraulic conductivity?

One of the most interesting studies on the effect of large-scale reforestation on dry season flows was one carried out in the humid sub-tropics of southern China where grasslands and shrublands have been reforested over a period of 50 years using various types of plantings (Zhou *et al.* 2010). Forest cover over this period increased from 20 per cent in 1956 to a current value of 56 per cent. Although these authors were unable to link this change to changes in overall river run-off patterns, they concluded that large-scale reforestation had helped sustain river flows in the dry season. They thought their results were due to the fact that evapo-transpiration in this region was much less than precipitation (in contrast to many drier areas). They also suggested that groundwater recharge was likely to be greater in more extensively forested areas because the pathway for water to reach rivers is longer. A longer pathway increases the chance that more water will infiltrate into the soil. Overall they concluded that water yields may be less sensitive to vegetation change at this scale than has been previously assumed (cf studies by Little *et al.* (2009) and Silveira and Alonso (2009) described earlier).

In summary, large-scale reforestation may reduce dry season flows unless the increased evapo-transpiration is compensated for by the trees improving the infiltration rates in surface soils allowing more rainfall to reach groundwater. The time it takes for any such improvement is likely to be measured in decades. Large contiguous areas will obviously have the greatest impact but, within an agricultural landscape, even small scattered patches of new forest downhill from an area of impacted soil will impede overland flow and increase the likelihood of more water infiltrating into the soil.

### *Reforestation and precipitation*

There has been considerable interest over a very long time period concerning the relationship between forest cover and rainfall. The so-called 'desiccation theory' was major topic of discussion amongst the scientific community during the eighteenth and nineteenth century and many were concerned about the implications of deforestation, especially on some of the smaller oceanic islands then being acquired during the era of colonisation (Grove 1995).

The issue continues to receive attention because of the present scale of deforestation and its potential impact on food security. The least contentious relationship is that between forests and fog capture. It is well known that trees in certain areas are able to intercept low cloud and thereby increase the amount of precipitation reaching the ground as drip (Bruijnzeel *et al.* 2011). The loss of forest, especially in montane areas where fog is more common, means such trapping ceases thereby causing a reduction in the overall precipitation. It follows that large-scale reforestation in such areas could increase the amount of water received. Little is known about whether this actually occurs sufficiently to affect streamflows. Nor is much known of the effectiveness of different types of reforestation or of the scale at which reforestation should be implemented in order to make a practical difference.

A more controversial relationship is that between overall forest area and rainfall. Nobody has suggested that small-scale changes in forest cover influence local rainfall but there is now some uncertainty about the relationship at a continental scale. Forests are known to influence the exchange of energy and moisture between the ground and the adjoining atmosphere. However, recent work has suggested the need for a fundamental rethink of some of these ideas (Sheil and Murdiyarso 2009). The conventional view is that pressure gradients driven by temperature and convection are the primary drivers of air movement. A new theory argues other factors may also be involved. It suggests that evaporation rates are higher in forests and this leads to a reduction in atmospheric volume over the forest canopy as water vapor rises and condenses (Makarieva *et al.* 2006). The low atmospheric pressure then draws in additional moist air from areas with weaker evaporation. According to this theory forests act as 'biotic pumps' and drive atmospheric ocean-to-land moisture movement. Under some conditions deforestation may act to switch off the pump leading to increased aridity. But if true, the same mechanism could presumably work in reverse provided the scale of reforestation is large enough. The theory has attracted criticism and debate (Makarieva and Gorshkov 2007, Meesters *et al.* 2009) and the debate appears to be far from over (Ellison *et al.* 2012).

The nature of the physical processes involved means that most studies of the relationship have involved complex atmospheric models. But there is some intriguing empirical evidence suggesting that reforestation on a large scale across the northern and north-western provinces of China may be beginning to have observable effects on atmospheric water vapour across this region (Jiang and Liang 2013). Some 27 million ha may have been reforested across the region and, although survival rates were sometimes low (Cao *et al.* 2011a), the vegetative cover increased from 8 per cent in 1949 to 18 per cent in 2005. Changes in vegetative cover were measured as changes in the 'greenness' in satellite imagery and concurrent changes in atmospheric water content were expressed as 'precipitable water'. The two were linked over space and time in much of northern China and the authors of this study concluded that reforestation was having a significant effect on the moisture cycle in these arid and semi-arid areas. It remains to be seen whether further study can clarify these relationships and how they may be influenced by ongoing global climate changes.

## Large-scale forest restoration, erosion and watershed protection

Much of the land available to be reforested is likely to be marginal for agriculture or degraded in some way. Some is likely to be actively eroding. This can occur through more-or-less constant surface erosion across slopes, by occasional landslides on steep slopes or through periodic riverbank slumping at times when river levels are high. Large-scale reforestation can help reduce most of these forms of soil erosion although trees alone are no guarantee that some erosion will not still occur (e.g. it is well known that some erosion can occur even in undisturbed natural forests (Douglas and Guyot 2005)).

### *Erosion caused by water*

Forests reduce erosion by reducing the direct impact of rainfall on soils. In addition, understorey vegetation and litter help slow down water moving across the soil surface, intercepting and trapping any sediments being carried. Water moves most rapidly on steeper slopes and erosion tends to be greater in hill or mountain areas (where rainfall is also often higher). There is no slope threshold above which erosion is sharply increased but it can be appreciable at slopes greater than 25 degrees and this is the nominal threshold used by the Chinese Grain for Green program (discussed in Chapter 5) to define mountain areas where agriculture should be phased out in favour of forest cover. Deep roots with high tensile strength can help stabilise soils on steeper slopes (Sidle *et al.* 2006, Forbes and Broadbent 2013). But forests can only limit soil erosion and not prevent it entirely. This is especially the case when soil layers are shallow and lie above an impermeable sub-surface. In such situations heavy rain and seismic activity can trigger landslips. Heavy rains may be an outcome of future climate changes in some locations in which case erosion losses may become more frequent (Nearing *et al.* 2004).

Large-scale reforestation is likely to reduce erosion in steep areas but the consequences of reforestation are heavily dependent on the type of reforestation employed. The most effective types of forest are those with complex canopy structures with multiple leaf layers and/or dense understoreys of shrubs or grasses. This is because crown drip from tree canopies alone can be larger than rain drops and more erosive (Hall and Calder 1993). Forests with thick litter layers are also effective (Hevia *et al.* 2014). On the other hand, those such as teak plantations which have limited litter layers at certain times of the year and sparse undergrowth are often prone to severe erosion (Valentin *et al.* 2008). Plantations managed on short rotations or subject to grazing or frequent fires that remove litter and vegetation will also be more erosion-prone.

Grasses can also provide an extensive ground cover that shelters soils from high kinetic energy rainfall and impedes overland flow. In such cases it might seem unnecessary to reforest grassland if the primary objective is simply to limit soil erosion. But the effectiveness of grass cover depends on its density and on the condition of the soil between the tufts; surface crusts or seals between grass tufts can substantially reduce infiltration and increase overland flow. The effectiveness also depends on how the grass cover is managed and substantial erosion can occur at heavily grazed sites and after wildfires (Zuazo and Pleguezuelo 2009). Some comparative studies have shown grasses alone are less effective at limiting erosion than woody vegetation (Chirino *et al.* 2006, El Kateb *et al.* 2013) but in many situations, especially in drier environments, a mixture of grasses and trees may be an optimal choice (Wang *et al.* 2012).

Assessing the collective impact of large-scale reforestation on erosion and sedimentation is difficult. Just how much of a watershed must be reforested to make a difference? It is relatively simple to measure the effectiveness of different types of revegetation on erosion on small areas but these results are not necessarily indicative of the effectiveness of any rehabilitation activity at a larger landscape scale. There are three particular problems in assessing the effectiveness of any kind of reforestation. One is that it can take time for soil to move down a slope and into a stream. Much eroding soil can be temporarily stored on flatter areas or behind litter or vegetation on lower slopes before being remobilised by heavy rain. Likewise, sediments in rivers can be temporarily stored in sand banks before being remobilised by rising river levels. A second problem is that much of the sediment in rivers can originate from a few episodic landslips or river bank collapses rather than from more widespread and slow, continuous erosion on hillslopes. That is, the precise location of reforestation matters. Finally, it may take time for trees in any reforestation programme to grow old enough to become functionally effective and limit erosion.

Nonetheless, as noted earlier in Chapter 5, there is evidence that overall vegetative cover is a primary determinant of sediment yield in watersheds <100 km$^2$ (Vanacker *et al.* 2007, Molina *et al.* 2009). Evidence is now also emerging of large-scale reforestation programmes that have reduced erosion within a decade or so. A review by Deng *et al.* (2012) reviewed outcomes of large-scale revegetation initiatives within 11 river basins covered by the Chinese Grain for Green program (see Chapter 5). This scheme includes reforestation on steeper lands and establishment of grasslands in drier regions with maximum rates of reforestation taking place in 2002 and 2003. By 2010 the programme had converted 14.7 million ha of farmland to forest and grassland and a further 17.3 million ha of barren ground had also been reforested. The overall increase in forest and new grassland cover across the region surveyed was 4.5 per cent of the landscape. It took some time before the scheme began to have an impact on erosion losses but, by the end of the programme, erosion had apparently decreased by an average of 45 per cent across the 11 watersheds. Deng *et al.* 2012 attributed this decrease to changes caused by the reforestation strategy. A similar finding within a smaller area of the Loess Plateau had been reported by Zhou *et al.* (2009) who concluded that recorded changes in river sedimentation were due to reforestation and not simply because of variations in rainfall. In this case the area reforested covered 5–7 per cent of farmland in Sichuan and Shaanxi provinces. Although the areas reforested are large in absolute terms, the proportion of the landscape covered is comparatively small. This highlights the strategic importance of location as well as just the extent of the area treated (Figure 6.1B).

### *Wind erosion*

In drier and more seasonal landscapes erosion is also caused by wind that picks up and mobilises soil particles. Trees or shrubs planted perpendicular to the wind in strips or shelterbelts two or three rows wide are a well-known method of reducing wind erosion in agricultural areas with the distance between belts usually being around 20–30 times tree height (Kort *et al.* 1998). If winds are variable in direction a checker-board pattern may be used with crops grown in the space between the strips of trees. Care is needed in designing more extensive belts of forest in drier regions because of the capacity of the trees to increase evapo-transpiration. Cao *et al.* (2011a) describe situations in sub-humid China where fast-growing exotic species have initially flourished but then died because of water stress. More successful outcomes are likely where mixtures of native species adapted to particular environmental conditions are used, especially grasses and shrubs.

## Large-scale forest restoration and biodiversity outcomes

In principle, increasing the area of forests at the expense of degraded land is likely to increase regional biodiversity. This follows from the well-known species-area relationship which is one of the more widely accepted concepts in ecology (see Figure 5.5). It proposes that as habitat area increases (or decreases) the number of species present also increases (or decreases). As a rule-of-thumb, a tenfold increase in area doubles the number of species present. Although earlier studies of the relationship were mostly based on the numbers of terrestrial species found on isolated islands of different sizes the relationship has been widely used to explore the impact of habitat loss on biodiversity (e.g. Brooks *et al.* 1997). In principle it might equally be used to anticipate the effects of large-scale reforestation. Increasing the area of forest increases the area of habitats for wildlife and may begin to reconnect isolated fragments of natural forest. This would then allow plant and wildlife species to move across the landscape and enable genetic exchange (Lemenih and Bongers 2010).

But it is obvious that the value of the increasing forest area also depends on the types of reforestation carried out with natural regrowth and multi-species plantings having more value as habitats than simple plantation monocultures. Similarly, the significance of the increased area depends on the landscape context. It may be impossible to restore forest-dependent wildlife species to landscapes from which they have disappeared. This may be because clearing has been too extensive and no forests containing these species remain or because hunting has removed these particular species from the remaining patches of natural forests (i.e. the 'silent forest' syndrome). In short, the species-area relationship provides a model describing what might happen to biodiversity when large-scale habitat restoration takes place but it may not be very useful in landscapes where many fauna have already become locally extinct.

### *Thresholds*

But suppose various biota are able to recolonise an area – must a certain amount of habitat (restored and as residual forest) be present before this occurs? Although there have been attempts to specify minimum threshold areas, the task is difficult because of the variety of habitat requirements and environmental tolerances amongst species (Swift and Hannon 2010). Species experience or perceive the environment in different ways and there is simply not a single threshold that applies to all taxa (Figure 6.1C). This means restoration managers cannot specify a minimum target area that must be reached to conserve regional biodiversity (although more specific prescriptions may be possible when the objective is to conserve a population of a particular species with known habitat requirements). In practice, many managers are likely to adopt a pragmatic more-is-better approach. Upper limits might then be imposed by patterns of land ownership and the availability of financial resources to undertake reforestation of the preferred type. But there are some species that require especially large areas where further consideration is needed.

### *Large-bodied forest species with broad home ranges*

Large-bodied and mobile forest species with large home ranges pose a particular dilemma for conservation biologists. These species include top-order carnivores, such as wolves and tigers, as well as large herbivores and frugivores, such as bison and hornbills. Each individual may require a large habitat area meaning that very large areas will be needed to support (or restore) a viable population of the species. The importance of such species lies not just in their existence value, but in the fact they can have an important role in improving ecosystem functioning. The role of large frugivores in distributing large seeded fruit is well known but in recent years there has been increasing debate about the importance of large-bodied carnivores for the functioning of ecological systems. The removal of such species can cause consequences throughout the food chain. This has been referred to as a trophic cascade (Estes *et al.* 2011, Ritchie *et al.* 2012). Thus the disappearance of large carnivores has been found to cause an explosion of herbivores and a consequent decline in plant regeneration in a wide variety of environments including temperate and tropical forests (Estes *et al.* 2011).

On the face of it, this evidence suggests such species must be re-introduced into all degraded landscapes in order to restore ecological integrity and improve functioning. But Sergio *et al.* (2008) take a rather more cautious view and argue that large carnivores act as structuring agents in some ecosystems but not others. Much depends on the context in which particular species are found. For example, in parts of Asia, factors such as hunting that caused the local extinction of top-order predators have also led to the loss of their meso-fauna prey. In such cases the loss of the top-order predators has not necessarily led to a loss of seedling regeneration (although, it must be said, that the exact functional consequences of these local extinctions are still mostly unknown).

Nor is it clear what happens when top-order predators are re-introduced to previous habitats since there are relatively few long-term studies of where this has happened. In the USA the re-introduction of wolves in Yellowstone National Park appears to have enhanced the populations of some species and improved plant regeneration (Ripple and Beschta 2012). On the other hand, the re-introduction of wolves to Finland caused a decline in threatened forest reindeer (Ritchie *et al.* 2012). It is clear that generalisations can be misleading and much depends on the particular circumstances.

But supposing large fauna do play a structuring role, do they need large areas of specialised habitats to be established before they can be re-introduced? If so, the task facing those having to restore their habitats over large areas could be formidable. On the other hand, the task is a little easier if they are mostly habitat generalists because the current landscape mosaic may be sufficient provided it is supplemented by some strategically located areas of reforestation. In fact most evidence suggests these larger-bodied species are habitat generalists meaning a heterogeneous landscape containing at least some forest cover will probably be sufficient to support them (Linnell *et al.* 2000, Tscharntke *et al.* 2005). Those species more properly described as habitat specialists (e.g. wildlife species restricted to forest interiors) appear to have much smaller home ranges or dispersal distances so that the scale of the reforestation effort needed to sustain them will be less (Kisdi 2002, Phalan *et al.* 2011). This means the problem associated with restoring large-bodied fauna is not so much about the forms of reforestation that might assist their populations to recover but, rather, how they might then interact with rural human populations – either by attacking humans (e.g. wolves, tigers) or by killing their livestock or damaging their crops (e.g. pigs, elephants). That is, the key issue is not one of finding a solution to an ecological problem but of finding a solution to a political problem. This dilemma is discussed further by Linnell (2000) and Ritchie *et al.* (2012).

### *When is too much?*

Restorationists sometimes assume that the higher the proportion of the landscape that can be reforested, the better the outcome for conserving biodiversity. After all, this is the logical implication of the species-area relationship discussed earlier. But this might not always be the case and much depends on the nature of the landscape being reforested, its history and the type of species already present. Reforestation of a recently formed agricultural landscape containing patches of forest scattered amongst crops and pastures may lead to the gradual recovery of the former forest biota. At the same time any of the non-forest species that had taken advantage of the more open landscapes will be lost. Large-scale reforestation leading to complete reforestation might favour the recovery of species preferring forests but it will do so at the expense of those species preferring more open areas or ecotone strips. The likely trend is shown in Figure 6.2. The number of species in the new habitat is likely to increase asymptotically as reforestation progresses and restores the original forest habitats. But at some point the non-forest species begin to disappear from the landscape and so the overall regional biodiversity declines.

This pattern may be even more complex in agricultural landscapes that have been extensively deforested for many years and where land use practices maintained over a long period have allowed the development of relatively stable species assemblages in the farmland areas. The cultural or 'heritage' landscapes of Western Europe provide an example of such a situation and several studies carried out at contrasting scale have explored the effect of large-scale natural forest regeneration on biodiversity when agricultural lands have been abandoned. The first by Navarro *et al.* (2012) represents a Europe-wide study. These authors examined the habitat preferences of species across the whole of Western Europe and concluded that increasing forest cover and increased connectivity made some species 'winners' while others were 'losers'. The former were species favoured by the creation of forest habitats while the latter were species preferring the previous agricultural landscape (Table 6.2). Many of these losers are from farmland areas classified as having high natural value. Interestingly, they found some species could be 'winners' in some localities but 'losers' in others, depending on circumstances.

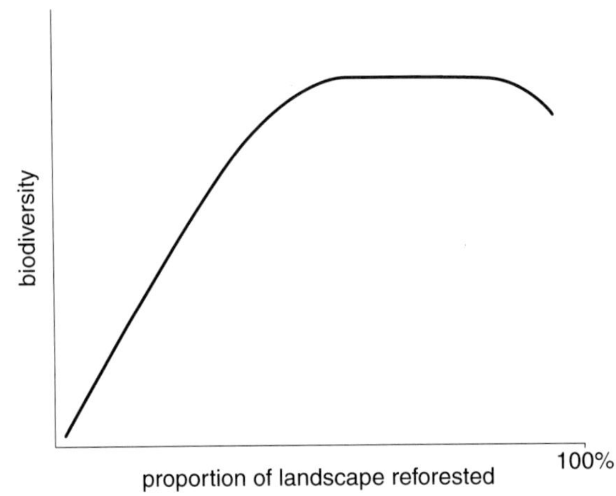


*Figure 6.2* The overall number of species present increases as an agricultural landscape is reforested because more of the original species requiring a forest habitat are able to recolonise the area. But, eventually, some non-forest species or ecotone specialists disappear as more of the landscape becomes forest and their habitats are replaced.

*Table 6.2* The consequence of large-scale forest recovery in Western Europe for species of vertebrate fauna (Source: Navarro *et al* 2012).

| *Type of species* | *Species likely to be advantaged* | *Species likely to be disadvantaged* |
|---|---|---|
| Invertebrates | 26 | 25 |
| Birds | 60 | 64 |
| Mammals | 24 | 3 |
| Reptiles | | 6 |
| Amphibians | | 3 |
| Total | 110 | 101 |

A second study undertaken in the Mediterranean region also examined the effect of large-scale reforestation following the abandonment of agricultural areas but focussed specifically on vertebrate fauna (Moreira and Russo 2007). It concluded that forests would generally support more species than the more open agricultural areas now being replaced. However they also calculated that the overall cumulative change would be negative for all groups except amphibians because the identity of many of the species found in more open areas varies across the region while the identity of the forest species still present in the region is broadly similar irrespective of locality. That is, heterogeneity would be replaced by homogeneity. Moreover, most of these forest species are distributed over Western Europe as a whole in contrast to the wildlife in the agricultural landscapes which are largely Mediterranean species. In short, the more homogeneous landscape created in this region by forest regeneration will support less overall biodiversity.

Finally, a study of bird species in the north-west of the Mediterranean region also found increasing forest cover on a large scale was having a dramatic effect on biodiversity (Sirami *et al.* 2008). Again, birds that were resident woodland species were favoured by the change while species found in farmland were less favoured and migratory species using farmland were also negatively affected. Interestingly, there was evidence that changes occurred more rapidly in landscapes containing greater proportions of forest, presumably because there were more forest species present that were able to recolonise than in landscapes dominated by farmland. The species present in a site changed according to the successional status with species using farmland habitats being replaced initially by shrub habitat specialists and then forest habitat specialists.

So, the answer to the question of how much reforestation is enough depends on the context. There are three possibilities:

- *Recently deforested agricultural landscapes:* large-scale reforestation is likely to allow the populations of species still persisting in natural forest remnants to increase and spread across the landscape improving local biodiversity levels.
- *Agricultural landscapes with 'silent' remnant forests:* large-scale reforestation of landscapes in which much of the forest wildlife has been removed (by hunting or because of habitat loss) may have only a limited impact on regional or local biodiversity levels unless some of the indigenous species can be re-introduced from elsewhere.
- *Cultural landscapes that have evolved over a long period of human use:* large-scale reforestation may have a limited impact on overall biodiversity but will change the composition of the species present in these landscapes. The significance of this will depend on the types of ecosystems being replaced and on the particular species involved. Reforestation of intensively farmed areas will probably improve

biodiversity but may not necessarily do so in more complex agricultural landscapes. Care should be taken to ensure species classified as threatened or vulnerable are not impacted by increasing forest cover but preference should also be given to protecting habitat specialists over habitat generalists. In the United Kingdom guidelines suggest certain areas having particular conservation significance should be excluded from reforestation plans. Examples include heaths, wetlands and mires (Rodwell and Patterson 1994). In some cases, more modest and strategically located areas of reforestation may be preferable to extensive reforestation (see Chapter 5).

**Box 6.1 Rewilding Europe**

The idea of 'rewilding' humanised landscapes is to recreate large wilderness areas with sufficient space to enable the restoration of larger populations of more indigenous species including larger carnivores and herbivores. This might be achieved by restoring large core areas and providing connectivity between these. Rewilding Europe (Accessed 16 May 2014. Available at www.rewildingeurope.com) is one example of the idea being put into practice. Rewilding Europe says 30 million ha of former farmland are being abandoned across Europe. It seeks to take advantage of these lands by restoring habitats in ten areas scattered across Europe and covering over one million ha. They aim to achieve this by 2020. Species of particular interest include large carnivores like wolves, brown bear, Iberian lynx, and large herbivores such as roe deer, moose, chamois, and ibex. This vision necessarily involves making certain critical decisions. One is whether the benchmark should be the Pleistocene era or at some more recent period during the Holocene (Donlan 2005, Caro 2007). Another point of contention concerns how complete the forest cover should be. Most new forests will regenerate by natural means and some have argued this should be allowed to progress as far as possible while others argue that landscapes should still retain substantial areas of open pasture (Fyfe 2007, Sommer *et al.* 2011). Rewilding Europe has decided their objective will be to restore landscapes containing areas of pasture as well as forests since this will allow the development of natural grazing in wooded pastures with primitive cattle (Aurochs; *Bos primeginius*), horses (Tarpan; *Equus ferus ferus*) as well as European bison.

## Carbon capture

Natural forests, including managed forests subject to logging, can contain very large stores of carbon. The amount at any particular location depends on the type of forest present and the environment in which it grows. It will also be affected by the management regime and the impact of events such as fires. Changes in land cover (e.g. deforestation) and land use have contributed about 33 per cent of global carbon emission over the last 150 years. These currently contribute only 10–13 per cent because of the rising use of fossil fuels (Houghton 2013). Large-scale reforestation of cleared or degraded lands could remove substantial amounts of atmospheric carbon and contribute to efforts to slow global climate change. Some of this carbon would be stored in above-ground biomass while some would be contained in soil organic matter. The magnitude of this contribution depends on the amount of carbon already stored in existing vegetation on the lands being reforested as well as on the type of new forest and the way these new forests are managed.

### *Carbon in above-ground biomass*

The impact of large-scale reforestation on carbon capture is greatest if the amount stored on any particular areas can be maximised and if the new forest is then protected from subsequent disturbances such as logging or wildfire (Hunt 2009). Carbon will be released if these forests are harvested although the impact of harvesting depends on how the timber is used. Timber used for fuel or paper will release carbon to the atmosphere relatively quickly but timber used for buildings, furniture or biochar may immobilise carbon for some time.

There is already considerable information on the capacity of different types of tree species and plantings to grow and sequester carbon at various sites (Winjum and Schroeder 1997, Böttcher and Lindner 2010) although rather less is known about how this storage capacity will be affected by future global warming. However, a distinction must be made between the rate at which carbon is taken up by plants and the amount actually stored once it has been absorbed. Some refer to these as carbon sequestering and carbon storage respectively. A high initial growth rate means that carbon is rapidly sequestered but most plantations using fast-growing species (e.g. *Eucalyptus*) are grown on short rotations and so store this carbon only briefly (and produce products such as paper which usually represent only an ephemeral carbon store). Species with higher wood densities and grown on long rotations for sawn timber production sequester carbon more slowly but store this in the standing forest over a longer time (Böttcher and Lindner 2010). The difference is illustrated in Figure 6.3.

Those undertaking large-scale forest restoration to maximise carbon storage may prefer not to undertake any harvesting but manage the new forest as a permanent carbon store. In such cases, the rate of sequestration or the size of the carbon store may be less important than establishing new types of forest that are more resilient in the face of future disturbances such as fires and which will be self-sustaining. This could be done by direct seeding, Assisted Natural Regeneration or various forms of multi-species plantings including Ecological Restoration. The long-term objective would be to convert any even-aged forests established in this way into uneven-aged forests containing a variety of species that are able to regenerate without

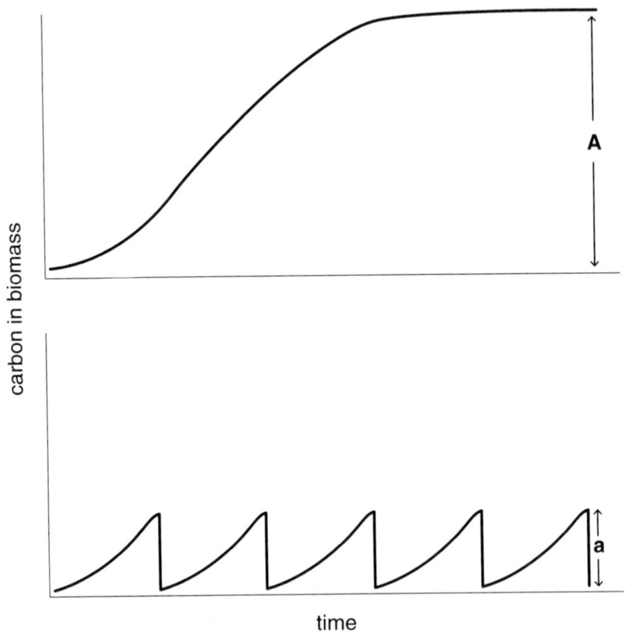


*Figure 6.3* More carbon is initially sequestered in biomass using plantations grown on successive short rotations than in a single long rotation occupying the same time period (that is, 5a>A). However this carbon is removed from the atmosphere as pulpwood and is unlikely to be stored for a long time. More is likely to be stored and immobilised in a forest grown over a longer period.

additional management inputs. Once a forest has reached maturity the biomass stabilises, net primary productivity falls close to zero and carbon gains are largely matched by respirational losses. That is, even with Ecological Restoration there is a finite carbon storage capacity. The magnitude of this depends on the biogeographic location of the new forest and the level of soil fertility.

### *Below-ground soil organic carbon*

Carbon can also be stored in soil organic matter. This carbon store can be increased as litter on the forest floor and roots in the soil decompose and are incorporated within soil material. There has been considerable interest in just how much additional carbon can be stored in soils following reforestation since the benefits accrue not only to carbon storage but to overall fertility and soil health. An emerging consensus is that it initially depends on the previous land use and that there is a difference between sites previous dominated by pastures and those used for cropping (Paul *et al.* 2002, Laganiere *et al.* 2010). Soil carbon often initially decreases in the first few years following site preparation prior to reforestation. The decrease is accentuated at sites previously occupied by grasses as old grass roots decompose before organic matter from forest litter is able to replace it. Recovery appears to be much faster at sites previously used for crops which lack a large permanent root biomass. Recovery is presumably aided by new inputs of carbon from litterfall which usually accelerates after canopy closure. In many plantations canopy closure begins 3–5 years after planting. It may then take some decades for soil carbon to recover to the levels present before planting.

Besides land use history, the rate at which soil carbon increases depends on the tree species used. Carbon appears to be most rapidly gained when broad-leaved species are planted but appears to be gained more slowly, or not at all, when softwoods are planted (Paul *et al.* 2002, Laganiere *et al.* 2010) although this was not as strongly expressed in an analysis of reforestation impacts in the USA (Nave *et al.* 2013). Laganiere *et al.* (2010) summarised improvements in soil carbon after an average of 22 years following reforestation as being 57 per cent for *Pinus* spp., 52 per cent for broadleaf species (excluding *Eucalyptus* spp.), 46 per cent for coniferous species (excluding *Pinus* spp.), and only 16 per cent for *Eucalyptus* species. The differences are probably associated with differences in rooting depths and differences in the ways species affect the formation of organo-metallic complexes in the soil. Jandl *et al.* (2007) concluded that mixed-species forests can be more effective at storing soil carbon than monocultures. Higher rates of soil carbon accretion are also found at sites with higher clay contents again because of the formation of stable organo-mineral complexes.

Soil carbon accretion is also be influenced by climate. Soil carbon increased more rapidly after reforestation in drier grassland areas (<1,100 mm) in Argentina and Uruguay after reforestation but initially declined in wetter grassland areas (>1,100 mm) before recovering after several decades (Berthrong *et al.* 2012). These authors concluded the differences were due to alterations to the nitrogen cycle caused by leaching in wetter areas. Primary productivity is greater in tropical areas with warm moist conditions but so is decomposition and opportunities for soil carbon accretion are more likely in boreal regions although lower forest productivities there mean it may take some time (Laganiere *et al.* 2010). Indeed a meta-analysis of mainly tropical sites by Marin-Spiotta and Sharma (2013) found changes in soil carbon were not strongly related to time since reforestation but were more linked with climatic variables.

Most studies suggest soil carbon accretion after reforestation is a very slow process and largely occurs in topsoils. Similarly, most researchers find that periods in excess of 30 years are needed before significant increases in soil carbon will be observed (Nave *et al.* 2013). A synthesis of a number of studies made in various locations and climatic zones across the USA found the overall increase in soil carbon was only 15 per cent after 100 years following reforestation. This is not a trivial amount but indicates that most carbon will be stored in above-ground biomass after reforestation and not in the soil.

### *Management implications*

The capacity of reforestation to store large amounts of atmospheric carbon means it can be one of the responses that will be needed to counter current trends towards a global climate change. Unlike the effect of reforestation area on services such as water yield, sedimentation and biodiversity conservation there is a largely linear relationship between area reforested and carbon stored although the rate depends on the type of reforestation (Figure 6.1D). The obvious implication of past studies is that plantations managed on long rotations for timber production (or managed as unharvested carbon forests) will conserve more above and below-ground carbon than those managed on short rotations. Minimising disturbances to forest structures and soils reduces unintended carbon losses (Jandl *et al.* 2007). Indeed successive harvests of short rotation pulpwood plantings are likely to lead to a progressive decrease in soil carbon since the time between harvests is too short for recovery to take place (Sang *et al.* 2013).

Many timber species can be grown on long rotations but there are likely to be advantages in establishing multi-species plantations and assisting these to make the transition from being even-aged to becoming a mixture of age classes. This is not so much because such forests store more carbon, but because they are more likely to be stable and self-sustaining over time. Fostering natural regrowth or undertaking Ecological Restoration may be especially attractive particularly if there is no intention to harvest trees and if consideration is given to generating other ecosystem services. The most appropriate locations for forests primarily managed for carbon storage will be abandoned cropping areas where the opportunity costs for landholders are low (e.g. distant from agricultural or timber markets and having infertile soils) or on steep hillsides in watershed protection areas where the need to limit erosion makes timber harvesting unattractive. Finally, care should be taken to minimise soil disturbances during site preparation ahead of planting to reduce losses of initial soil carbon.

Although considerable work has been done to understand the biology underpinning carbon storage in forests, there is still some uncertainty just how much will be stored following different types of reforestation in various environments. Part of this is because of our current inability to forecast how much is likely to be stored in particular soils over time. But there is also great uncertainty over the methods to be used to assess and monitor carbon accretion across landscapes following reforestation (and to monitor the consequences of future climate changes on these storage capacities). There are well-known techniques to quantify and model above-ground carbon stores but measurement of changes in below-ground carbon are more problematic not least because of spatial variability, uncertainties over carbon storage in soil horizons below the top 30 cm and the occurrence of erosion. The problems are reviewed by Sanderman and Chappell (2013).

## Implementing forest restoration on a large scale

There are two final issues that deserve some comment. One concerns the area that must be reforested to generate a specific ecosystem service in a particular target area. This ratio of area-treated-to-area-benefited can be quite high for some ecosystem services. For example, Pannell *et al.* (2001) considered it could be more than 1,000:1 in the case of some dryland salinity situations in Western Australia, meaning in that case that reforestation might not be worth doing. But the ratio could be rather lower in other cases. Thus only a relatively small area of corridor might be needed to enable interbreeding between members of a meta-population of wildlife in some large and isolated remnant forest patches. Likewise, only a small area of reforestation might be needed to establish enough trees with flowering times that complement those of trees in patches of residual forest and that sustain insects required to pollinate agricultural crops.

A second issue concerns the way any large-scale reforestation is actually implemented. Large landscapes often have many landowners who are likely to hold contrasting views about how their land should be used, what type of reforestation – if any – might be used and where it should be carried out. Large-scale reforestation will usually mean trade-offs and compromises must be made. Some of these are discussed in the following chapter together with a consideration of a way to approach reforestation at a landscape scale.

## Conclusions

The direct economic consequences of increasing forest cover depend on the type of new forest established and its location. Large areas of contiguous plantation are usually more attractive to industrial growers but small individual plantings can be economically attractive if appropriate species are used. These plantings are likely to be even more rewarding if collective marketing arrangements can be made to link separate owners.

The ecological consequences of increasing forest area vary depending on the process or function considered. Reforestation usually leads to a reduction in run-off at least in small watersheds. It may also reduce dry season flows although this may be improved if reforestation occurs on severely degraded soils and improves surface infiltration rates.

Increasing the area reforested is also likely to reduce erosion and sedimentation although simple monocultural plantations without much litter or understorey will be less effective than more structurally complex new forests and those in which the soil surface is protected, especially on steeper land.

Biodiversity conservation is also likely to be improved by increased forest cover. It is usually difficult to specify thresholds that must be exceeded and much depends on the landscape context and history. Reforestation may homogenise some currently heterogeneous agricultural landscapes. But much depends on the biota of concern and habitat specialists will respond differently to habitat generalists.

Large-scale reforestation will remove carbon from the atmosphere but the storage capacity of the forest depends on how it is managed. The greatest storage will be found in forests that are protected and become self-sustaining. The capacity of most forest soils to store additional carbon appears to be modest, at least in the short term.

# 7 Creating multi-functional landscapes

## Choices and trade-offs

### Introduction

Opportunities for forest restoration in a particular landscape depend on how much degraded or under-used land is present and on the goods or ecosystem services needed by the people living in, or relying on, that landscape. The capacity of forest restoration to supply these goods and services will be determined by the type of reforestation undertaken as well as by its location and spatial extent. Choices concerning each of these three variables will be made by individual landholders and be determined by the way they perceive the market opportunities and the potential costs and benefits of reforestation on their land. But others are also likely to see themselves as being affected by such choices and may wish to be involved in the decision-making process. These stakeholders may include neighbours as well as non-residents such as downstream water users, conservation bodies or townspeople concerned about the goods and services supplied by the lands in question. In addition, governments are also likely to want to influence local land use decisions to ensure that regional or national goals are achieved.

In short, what may seem to a landowner to be a relatively simple decision about a particular area of land may be, in fact, a rather more complicated matter of how to modify a complex landscape mosaic in a way that satisfies a diverse range of interests. The problem is most formidable, of course, when large areas are being reforested. In many situations the rights of a landholder are paramount but the demands of other stakeholders and the need for collective action to overcome various environmental problems is beginning to intrude on these rights. This chapter explores some of the choices and trade-offs that are likely to confront landholders and how these options might be coordinated by those promoting reforestation on a large scale in order to create a multi-functional landscape.

## Some questions confronting those undertaking large-scale forest restoration

Following the discussion in Chapter 1 concerning the context in which future reforestation will be carried out, perhaps the primary question for those contemplating large scale forest restoration across a particular landscape is whether it should be used for food production or forest. But there are a variety of other questions likely to arise. Some concern choices confronting individual landholders while others concern those seeking to coordinate land use activities to achieve broader community or national goals. The following explores some of these beginning with the agriculture versus reforestation choice (see Table 7.1).

*Table 7.1* Some questions for those undertaking large-scale forest restoration

| *Key question* | *Response* |
|---|---|
| 1. Must we choose between either using cleared land for food production or for reforestation to produce goods or generate ecosystem services? | A choice must not always be made. There is considerable deforested land that is now unsuitable for agriculture. |
| 2. How should we decide when to rely on natural regeneration and when to undertake planting or seeding? | No intervention will be needed when tree regrowth is widespread, evenly distributed and is sufficient to generate useful goods or ecosystem services. Otherwise planting or seeding will be necessary. |
| 3. Should plantings seek to maximise timber production or create multi-purpose forests? | There will be locations where maximum timber production is the primary goal but many future planted forests are likely to have other objectives and need appropriate silvicultural systems to achieve these. |
| 4. Should Ecological Restoration be undertaken or should the creation of novel ecosystems be allowed? | Ecological Restoration may not always be possible because landholders are sometimes dependent on the new forests for their livelihoods. |
| 5. How should we balance improving forest production and maintaining water yields? | Options include careful choice of species used, locations planted and stand management. Choices depend on the relative importance of the other ecosystem services required. |
| 6. Is it possible to compensate or offset planned environmental damage in one location by restoring forest cover in another? | There is scope for using offsets but uncertainty over whether existing biodiversity can always be replaced. Performance bonds and offset ratios might be used to improve outcomes. |
| 7. How should reforestation programmes be designed appropriately to accommodate a changing climate? | Reforestation should include seed sources and species adapted to the types of climate like to develop in particular area, Reforestation should help maintain the populations of vulnerable species and assist them to migrate across the landscape. |

***1. Must we choose between either using cleared land for food production or for reforestation to produce goods or generate ecosystem services?***

Some have suggested the scale of the challenge to achieve food security may overwhelm concerns such as supplying various ecosystem services or conserving the world's remaining biodiversity (Cunningham *et al.* 2013). That is, the need to produce more food trumps all other options. On the other hand, it was argued in Chapter 1 that part of the reason for present day food malnutrition in various parts of the world is because of poverty and inequality rather than an inability to produce enough food. Further, it was suggested that food production could be substantially increased by simply helping more farmers to use existing knowledge (Foley *et al.* 2011).

But there are additional reasons for thinking the dilemma may be less difficult than it might seem. Chief of these is that much cleared land is now clearly marginal for agriculture and some of this might be reforested without significantly limiting overall food production. Such lands are likely to include those that have lost fertility, have erratic rainfalls or are too distant from transport systems and markets to be economically viable. Some will argue that the definition of 'marginal' is quite subjective but much farming receives agricultural subsidies of various kinds and the removal of these would help clarify where reforestation has a real opportunity cost and where, by contrast, it may be entirely beneficial.

Other lands where some reforestation might be undertaken with limited adverse consequences for food security are those where long-term agriculture is not sustainable and where current farming is causing environmental damage. These lands include steep mountain areas, places subject to wind erosion and some riverine areas. Again, these are areas where the national environmental benefits of reforestation are likely to outweigh any local agricultural opportunity costs. In some cases it may be necessary to compensate local landholders to facilitate the change but, in others, the benefits of reforestation may be self-evident and it can be achieved by technical assistance or capacity building.

Finally, there are areas where some reforestation could actually improve agricultural productivity and contribute to achieving food security. These might include grazing lands where trees can provide shelter for animals as well as areas where strategically located belts of trees would limit overland water flows and help reduce soil erosion. Many crops rely on pollination by native insect fauna also dependent on nearby forests (Garibaldi *et al.* 2011).

*Conclusion*

In many places it will be unnecessary to make this choice. Many landscapes have areas of degraded or marginal land unsuited to agriculture where reforestation will improve ecosystem functioning and the supply of ecosystem services. Many other agricultural landscapes are also likely to contain areas where some reforestation is not only possible but also desirable. In any case, reforested areas can be deforested again and replaced by crops if it becomes necessary to do so to boost food production.

### 2. *How should we decide when to rely on natural regeneration and when to undertake planting or seeding?*

Natural regeneration can be an attractive reforestation option, especially when a landholder wishes to reforest a large area. The dilemma lies in deciding the circumstances under which deliberate reforestation may still be the preferred option. It depends on the types of species being established, their spatial distribution and the time it takes for trees to become established and begin generating some form of benefit. Natural regeneration is most attractive when seedling regrowth is widespread and evenly distributed across the areas available for reforestation. Provided these trees are present in appropriate densities, generate appropriate goods and services and can be incorporated into farming systems or land use plans, there should be no reason to consider planting.

But natural regrowth can be unreliable. Trees may be patchy and unevenly distributed and may include unwanted weed species. It may not always occur in areas where tree cover is needed, and, even when it does, the rate at which colonisation occurs can be slow. This is especially likely where there are few areas of natural forest remaining or where seed dispersal agents are limited. It may also be difficult to protect from disturbances. These possible constraints mean that there can be a high degree of uncertainty associated with the development of regrowth. Dorrough *et al.* (2008) describe an approach for balancing risk and benefit in a farmland setting.

But perhaps the main question is whether natural regrowth will generate the goods and services required within an acceptable time period. If the primary objective is the provision of fuelwood then patchily distributed trees – even exotic species – may be quite acceptable and may be easily managed and incorporated into a farming system. On the other hand, if the primary purpose is to stabilise hillslopes and protect soils from erosion then this type of regrowth may be unacceptable because it takes too long and the outcome is too uncertain. Clearly, different stakeholders may hold contrasting views about these matters.

#### *Conclusion*

Natural regeneration is most attractive when tree regrowth is widespread, evenly distributed and the trees generate useful goods or ecosystem services. If this is not the case the choice of whether, or not, to undertake planting necessarily depends on the costs of doing so and the perceived benefits such planting may offer.

### *3. Should plantings seek to maximise timber production or create multi-purpose forests?*

Most of the forests established during the twentieth century were established as 'production forests' and the primary purpose was to generate goods such as sawlogs or pulpwood in as short a time as possible. Over time a variety of techniques were developed to maximise the productivity of these forests. These included using a relatively small number of species grown in plantation monocultures, growing these on short rotations and removing all competing vegetation. These silvicultural systems were mostly pioneered by state forestry agencies who then passed on the technology to corporate plantation owners and to smallholders. These industrial plantation systems became the standard method to use where ever reforestation was needed. The corporate owners developed the technology even further and, by the end of the twentieth century, some of the productivity levels being achieved would have astonished foresters from an earlier time.

But there are two sources of concern about this approach. One is that enhanced productivity has been achieved at the expense of resilience. The yields from each hectare may have increased but the systems are brittle and risky (Walker and Salt 2006). The risks are from ecological events such as windstorms, fires, pests, diseases or climate change. But they also arise from economic factors such as a lack of flexibility to withstand changes in markets or land prices, both of which can affect the financial viability of the plantation and its opportunity costs. The second concern about these types of forest is that management objectives are changing and there is a rising demand for not just timber but the kinds of ecosystem services that only forests can provide. This means the old certainties are being lost and those undertaking tree planting can no longer be sure that planted trees will necessarily be harvested nor which ecosystem services will be in demand in particular areas.

Both factors mean that intensively managed and single purpose plantations may not be as useful as was once the case. Corporate growers will continue to focus on productivity but many smallholders and state forestry agencies are likely to be concerned about risk minimisation and multi-purpose objectives. Some context is needed here. A global survey of plantation ownership by FAO found there were 27 million ha of plantations managed by corporate owners but 127.8 million ha, or nearly five times that area, were plantations established and being managed by state agencies and smallholders (Del Lungo *et al.* 2006). Some state-owned plantations and some smallholders would have necessarily used the industrial plantation model because they were growing industrial timbers. But many others are likely to be interested in other silvicultural systems able to supply a wider variety of goods and services. Rather than simply seeking maximum timber production, these forest owners are likely to prefer systems that provide more flexibility and increased resilience. (e.g. Pichancourt *et al.* 2014).

*Conclusion*

There will continue to be locations where intensively managed silvicultural systems will be the preferred choice. But these can no longer automatically be taken as the standard reforestation method in future. Instead, reforestation methods should evolve to encompass a variety of approaches to suit the range of ecological and economic conditions now present and the capacity to be modified if objectives change.

### *4. Should Ecological Restoration be undertaken or should the creation of novel ecosystems be allowed?*

Many believe the best way of overcoming degradation is to restore the original ecosystem using the process described earlier as Ecological Restoration (Chapter 4). This involves re-assembling the original biota and ecological processes to re-establish the historic ecosystem. The merits of the approach seem self-evident, especially for those concerned with restoring habitats of endangered species or conserving biodiversity in general. But critics argue that in many places, too many biotic and abiotic changes may have taken place for Ecological Restoration to be practical. They also point to the fact that the likely impacts of global climate change will make Ecological Restoration increasingly difficult. The pragmatic alternative they offer is to accept the fact that many ecosystems of the future will have to be 'novel ecosystems' that lack some of the original biota and that may even include exotic species (Choi 2007, Jackson and Hobbs 2009, Hobbs *et al.* 2013).

The concept of novel ecosystems has caused some debate. Some ecologists are concerned about the implications of accepting the concept because they think it will lead to further biodiversity losses and an acceptance of the inevitability of biological homogenisation (Clewell and Aronson 2013). Others worry that it signals a lowering of standards, especially for mining companies and others with a legal obligation to repair any environmental damage they may have caused. Interestingly, Katz (2012) levels a similar criticism against Ecological Restoration itself. He argues it allows those same mining companies and other developers of land the opportunity to over-exploit and damage land because they can claim any damage will be subsequently repaired (see discussion below on environmental offsets). Whether or not one accepts that novel ecosystems (or Ecological Restoration) will lead to a lowering of currents standards regulating the future activities of organisations such as mining companies, it is the case that large areas of degraded land have now accumulated because of past activities and decisions must now be reached on how to treat these.

Any discussion over novel ecosystems necessitates that the term be carefully defined. Hobbs *et al.* (2013) argue that novel ecosystems are self-organising systems with novel qualities differing from those that prevailed historically. They explicitly exclude from their definition ecosystems being intensively managed by humans. The requirement to be self-organising seems unnecessarily restrictive. New managed ecosystems have evolved in some cultural landscapes such as those of Western Europe and other parts of the world that are the result of human activity over extended periods of time. Many of these are now relatively stable and have their own predictable (and novel) combinations of plant and animal species. If abandoned these would not necessarily return to their historic state but could evolve over time to develop a new condition (and only then become a novel ecosystem under the definition of Hobbs *et al.* (2013)). Likewise, one might argue that landscapes containing planted forests managed using a clear-felling or some kind of selection system, can acquire over time a particular combination of biota and generate a predictable set of ecosystem services. These, too, might be viewed as being novel ecosystems. These are not self-organising because managers manipulate the forest to achieve successive harvests. On the other hand, many would be self-sustaining because, once established, they no longer need to be replanting or seeded. In what follows, novel ecosystems will be broadly defined to include all forms of multi-species forest, other than those involving Ecological Restoration, that are established with the intention of these being maintained over the longer term. This is obviously broader in scope but necessarily includes the condition covered by Hobbs *et al.* (2013).

The essence of the question above is whether Ecological Restoration is the only objective worth pursuing when dealing with cleared or degraded landscapes? There is a degree of unworldliness about the question. It may be worth debating if the areas are only one or two hectares but is not terribly realistic in situations where there are large areas occupied by poor rural communities. In practice there are probably three options available.

- *The preferred conservation option*: all of the original biota, including seed dispersers, still remain nearby in the landscape in remnant patches of undisturbed forest. Further, soils and other abiotic attributes of the environment are not too badly damaged or can be repaired. In this case there should be good prospects, at least in the immediate future, of restoring a successional trajectory leading to an ecosystem resembling that which was once historically present. In such sites Ecological Restoration represents a worthwhile objective although the means by which this is carried out will depend on circumstances. Where the areas are large it will probably be largely dependent on natural regeneration rather than interventions such as sowing seed or planting seedlings.
- *The pragmatic conservation option*: past biotic and abiotic changes across the landscape make it impossible (or too expensive) to undertake Ecological Restoration. In this case the objective should be to use species tolerant of the new environmental conditions – including exotic species where this is necessary – and assemble a multi-species forest

containing a diversity of functional types. Over time the aim should be to restore ecosystem functionality as far as this is possible and create uneven-aged, self-sustaining and resilient new forests. Ideally the numbers of species used should resemble the number thought to have been originally present but this may not be always possible especially when depending on natural regeneration. Options to improve species diversity may improve over time (an example of this approach is shown in Box 7.1). Where this kind of reforestation is being practiced to fulfil legal obligations, it should be possible to develop prescriptions ensuring monitoring is carried out and that appropriate standards are actually met.

- *The working landscape option* – landowners require new forests to produce goods for commercial purposes or household needs. The conservation of biodiversity and the provision of ecosystem services may also be important for these landowners but are, nevertheless, secondary considerations. In this case production forests are established on cleared land by natural regeneration (Chapter 3) or planting (Chapter 4) and managed as a new, long-term land uses. Periodic harvesting of forest products may take place but management ensures that regeneration always occurs either by replanting or through regeneration induced by the harvesting process. The extent to which ecosystem functioning is restored and various services are also delivered depends on the particular type of reforestation used and the landscape context. Changing ecological or economic circumstances may lead to management changes which, over time, also change the nature of the ecosystems and the services these forests provide. This change may be positive (e.g. when monocultural forests are deliberately enriched or colonised by species from nearby natural forests) or negative (e.g. when the forests are felled and replaced by some alternative land use such as an agricultural crop).

### *Conclusion*

Ecological Restoration should remain the preferred option where ever it is ecologically and economically possible. But there are likely to be many other situations where this type of restoration is simply not feasible and where the establishment of some kind of novel ecosystem involving multiple species is a worthwhile alternative. This is especially likely when reforestation is being implemented on a large scale. In such cases ecological functioning can be substantially improved using combinations of species chosen on the basis of their ecological tolerances and capacities to restore particular ecosystem functions, goods or services. The nature of this new ecosystem may evolve over time as ecological and economic circumstances change. In many cases the real question is not whether or not to accept the development of novel ecosystems but how to design and assemble these new forests.

**Box 7.1 Multi-species plantings on a large scale in Hunan, China**

In 2008 a severe ice storm degraded large areas of forest and plantations in southern China (Stone 2008, Shao *et al.* 2011). In some cases wildfires subsequently burned the damaged sites and converted the land to grassland, shrubs or bamboo thicket. There was evidence that plantations containing native species were less damaged by the storm than those with exotic species and that multi-species plantations suffered less than monocultures, especially at higher altitudes. Authorities in Hunan Province decided to reforest 28,000 ha of degraded area using species mixtures rather than the monocultures that had been more widely used beforehand (World Bank 2013). The objective was to create resilient new forests that would stabilise hillslopes and protect watersheds rather than to re-establish new production forests.

The obvious question was what type of forest should be established? The composition of the original forests at many sites in these mountain areas is not fully known and there are no areas that might act as reference sites. Nor are there many wildlife species remaining, including species able to act as seed dispersers. This meant that Ecological Restoration was not feasible. On the other hand, the type of forest once found at various elevations is broadly understood. It was decided to create a series of novel ecosystems using mixtures of up to five species at any one reforestation site and a total of around forty species, mostly of local origin, across the province as a whole. Further investigations will explore if these numbers can be expanded in future and particular attention will be given to the possibility of including endangered species in the plantings. The species being planted at each site are those thought to be tolerant of the particular soil and climatic conditions. The proportions of each species in the mixture are being left to local foresters to decide but no species is to have more than 70 per cent of the total tree density. Some additional plantings will also be done to enrich another 30,000 ha of residual plantations that were damaged by the storm but which were not entirely destroyed.

The approach represents a radical new form of reforestation in the province which has hitherto largely relied on simple plantation monocultures and it will take considerable organisation to achieve over such a large area. It is no small task to raise large numbers of seedlings of this variety of species to be planted out at the same time and across an extensive area. It is also being done across a variety of land ownerships including individual households, communal land, forest cooperatives and state-owned land. The new forests will be managed by the 26,000 households and communities who have rights to the land being used.

The future ecological trajectory of these new forests is unclear but the long-term objective in Hunan is to create self-sustaining forests. The initial plantings will necessarily be even-aged forests and, at some point, these will have to be converted into uneven-aged forests if they are to become self-sustaining. It is anticipated this will be done using some selective fellings to open plantation canopies to encourage seedling regeneration. Some of this timber may be removed for household use or sale but the primary purpose of any thinning will be to trigger seedling regeneration, not to produce timber. The fact that many people have temporarily moved off the land to work in urban areas and no longer rely on these lands for income has probably made this transition, from forests for goods to forests for services, easier.

***5. How should we balance improving forest production while maintaining water yields?***

There is a widespread public perception that reforestation in local watersheds enhances stream flows but the reverse is often the case because of the higher rates of evapo-transpiration following reforestation (but see the qualification discussed in Chapter 6 where reforestation can improve dry season water flow because it improves the infiltration capacity of degraded soils). This means there is a dilemma for those promoting reforestation, especially in drier landscapes: how to balance the advantages of reforestation against the consequent reduction in water yield?

There are a variety of ways water losses might be reduced. These include:

- limiting the spatial extent of plantings and having a number of small forest patches rather than a few large plantations (Van Dijk and Keenan 2007);
- using slow-growing native species with sparse crowns rather than using fast-growing exotic species (to reduce the rate of evapo-transpiration);
- planting trees at low densities and reducing tree density whenever monitoring (e.g. of depth to water table) shows it is necessary to reduce impacts on water resources;
- planting mostly upper topographic areas rather than lower parts of the landscape where water tables are shallow;
- growing trees on short rotations and, after harvesting, allowing soil water resources to be replenished before replanting trees in a second rotation.

Careful placement of trees may maximise the benefits (e.g. shelterbelts, soil protection) while minimising the disadvantages (e.g. reducing crop yields). An example of such careful placement in hilly land in China was given in Figure 5.3. It should be noted that the impacts of new forests on water resources may not necessarily be experienced by the actual forest owner but by neighbours or downstream water users.

*Conclusion*

Reforestation will increase evapo-transpiration but the significance of the change depends on climate, the status of local water resources and on the relative importance of the other functional benefits and ecosystem services that reforestation will provide.

***6. Is it possible to compensate or offset planned environmental damage in one location by restoring forest cover in another?***

Sometimes an economically attractive development will cause unavoidable damage to the natural ecosystems at that location. Might the loss of biodiversity, the functional benefits or various ecosystem services currently provided by the forest at that site be offset by restoring forest and generating ecologically equivalent gains at another site that has already been cleared or degraded? The idea of offsets has attracted considerable interest (see http://bbop.forest-trends.org/ Accessed 16 May 2014.). Offsets could be useful if (i) there are biodiversity gains at the restored site and, consequently, no net losses and (ii) the same type of biodiversity lost at the first (impacted) location is regained at the second site at which reforestation occurs.

Alternatively, an offset might be deemed to be acceptable if a functionally similar forest (or one able to provide a particular ecosystem service) is restored at the second site even though it is not necessarily one with an identical composition. In both cases it is assumed the restored site can be protected at least as well as the impacted site would otherwise have been if not for the approved disturbance.

The most obvious reforestation approach to use is Ecological Restoration in which the objective is to restore a successional trajectory likely to lead to the re-establishment of the historical ecosystem. Knowing that destruction at the first site is certain, is Ecological Restoration sufficiently reliable? The evidence is that it has had a record of both successes and failures (Suding 2011, Maron *et al.* 2012). Successes are most likely where ample natural regrowth is already present and restoration largely involves excluding invasive weeds or herbivore pests. Restoration is more difficult at sites having floristically complex vegetation communities, at highly degraded areas and at locations that have experienced nutrient enrichment. In such cases successional trajectories can be diverted. For example, a series of below-average rainfall years in the early years after restoration commenced may cause the loss of some key species and limit further successional development. Or failures may involve unexpected imbalances in the relative abundances of certain species that then alter successional pathways. Much depends on the resources invested in restoration and both of these problems might be overcome by careful monitoring and adaptive management over the recovery period. It may be the absence of this monitoring rather than any inherent flaw in the idea of Ecological Restoration itself that is one of the key causes of failure. Likewise, failure may have as much to do with the landscape context in which the offset site is located (e.g. too great a distance to seed sources or wildlife colonists in natural forests) as with anything happening at the site itself.

A key component of offset programmes is the notion of like-for-like or replacing what is lost at one site by its equivalent at the other site. But many ecological attributes can be difficult to quantify. In particular, it is difficult to use a single measure that truly captures all components of biodiversity (Purvis and Hector 2000). A common approach is to develop an index of some kind (e.g. Parkes *et al.* 2003, Lindenmayer and Likens 2009). But these types of assessments can be problematic if the forest being lost provides a special ecosystem service or has some unusual conservation or community value. In such cases there can be disagreements amongst stakeholders over the scores to give different elements of the index. For example, how to balance the views of those especially concerned about biodiversity with those more concerned about water quality? Should they always be treated equally? Other factors that make quantitative assessments difficult include:

- *Scale*: over what proportion of a site must success occur? Would poorer growth in a relatively small proportion of the larger, successfully restored, area necessarily constitute failure?
- *Complementarity*: what if the restored area matches the size of the area of intact forest being damaged but is only a small proportion of the total area of degraded land in the landscape that needs treatment? In such cases, success depends on the ways these external areas are managed and whether the offset site is able to complement these other landscape management goals. Will the restored area be complementary, will it trigger a broader recovery process or will it be irrelevant?
- *Location*: is the area being restored located in such a way as to maintain connectivity for wildlife or maintain the supply of particular ecosystem services (e.g. control erosion). Again, success may depend on how well the offset site is able to complement other landscape management activities.
- *Single large or several small areas of offset*: in most cases a single large area is preferable to several small areas but much depends on the landscape context and the availability of land for reforestation. It may be the case that several small restored areas are better for maintaining connectivity or conserving soils.
- *Spatial variability*: where large areas covering different site conditions are being restored there may be a need for different reforestation methods and, consequently, different assessment methods.

A particular difficulty in assessing the quality of the compensatory site lies in the time it can take before it is clear that a successional trajectory is on-track and that a successful outcome is likely. Some such attributes may only develop after several decades. An example is the tree hollows needed by certain wildlife which will only begin to form when trees are several decades old. Similarly, some trees may only begin to flower and produce seed after several decades. This time element means it may take some years before the ecosystem at the offset site is able to contribute to the functions operating at a landscape scale such as enabling wildlife affected by the new development to move across the area. But it also means there is a long period of (legal) uncertainty over whether the offset has actually compensated for the truly certain losses that will be incurred when the development proceeds.

There are several ways these problems might be addressed. One is to specify criteria for success that match the circumstances of the particular offset arrangement. These might cover the structure of the new forest (tree density, canopy height), the biodiversity represented (at a local as well as at a landscape scale) or some assessment of the functionality of the forest (e.g. extent to which surface erosion is controlled or water quality is maintained). It might also be useful to develop an indicator of the extent to which the new forest has become self-regenerating such as the extent to which canopy trees are regenerating. The criteria that might be used will clearly need to be agreed upon by experts, landholders and community members. Forms of monitoring and adaptive management that might be used for these purposes are discussed by Lindenmayer and Likens (2009).

A second approach would be to incorporate a degree of insurance into the programme by requiring that the offset area is several times greater than the area of the impacted site. This ratio is sometimes referred to as the offset ratio and the size of the ratio could be proportional to the importance of the asset being impacted by the development as well as the risk that the compensatory programme might not succeed or might be accidentally damaged by wildfire or some other disturbance. A larger ratio means that localised failure in one place has more chance of being compensated for elsewhere. Bull *et al.* (2013) quote a South African example where 30 ha of land must be restored for every hectare of critically endangered habitat that is destroyed.

A third way of dealing with uncertainty and the complications arising from the time it may take to achieve desired outcomes might be to require a performance bond. This is commonly used in many mining operations to ensure site rehabilitation is completed once mining ceases and before the legal responsibilities for managing the site are relinquished. The dilemma facing mining regulators is similar to that facing regulators of offset programmes: that it can be some time before a post-mining rehabilitation scheme can be said to be successful (and it may be a long time after the mine itself ceases operations). Any bond should be sufficient to cover the cost of restoration if a third party has to step in and take over because of the failure of the original parties to achieve an acceptable outcome. The size of the bond might be reduced over time as certain benchmarks are achieved.

## *Conclusion*

There is scope for offsetting the damage caused by certain land uses by reforesting degraded land at another site. It is likely that forest of a similar structure and function can be established although there is less certainty that all biodiversity can be restored. These concerns mean there should be some limit on the overall extent to which offsets are allowed in a particular region or landscape. Much depends on the conditions at the offset site and on the resources available. The use of offset ratios and performance bonds could help overcome some of the difficulties in achieving the desired outcomes.

### *7. How should reforestation programmes be designed appropriately to accommodate a changing climate?*

Recent projections of climate change by the Intergovernmental Panel on Climate Change (IPCC) suggest that over the next 100 years global temperature will warm by more than 1.5 degrees over that experienced between 1850–1900 and that there will be more frequent hot periods and less frequent cold periods over most land areas. The contrasts between wet and dry regions and wet and dry seasons will also increase (IPCC 2013). Another expected change is an increase in the frequency of extreme storm events. The effects of these changes on natural ecosystems are likely to be profound (Bellard *et al.* 2012). They are also likely to influence where agriculture is practiced and where people live. In the particular case of reforestation they probably mean that many of the species and methods of reforestation used at particular locations are unlikely to be appropriate in future.

Forests tree species will be affected by climate change in two ways. The more direct effects will come through physiological responses such as changes to the rates of photosynthesis, water use and in flowering patterns. The indirect effects will be felt through the impact of on important mutualists such as mycorrhizae, pollinators and seed dispersers as well as through changes in the distribution of pests and diseases. Climate change is also likely to affect abiotic variables such as fire frequencies. The overall consequence of these many changes on the distribution of species and the functioning of systems is difficult to predict (McMahon *et al.* 2011). Each species has a climatic envelope which encompasses the areas its populations can use. As the climate changes, this envelop shifts across the earth's surface and the distribution of species populations must move to stay within it. Species unable to track the changes will go extinct. Changes in the distribution of a number of species across altitude and latitude can be seen already and the empirical evidence shows many plant species have tracked climate change either poorly or not at all (Corlett and Westcott 2013). This may be because current distribution patterns are limited by factors other than climate and species can accommodate some climate change, at least in the short term. It is also likely that some species may benefit from climate change and even become invasive (Pawson *et al.* 2013). But, in human dominated landscapes, there are likely to be more losers amongst plant species than winners, especially over the longer term. The plant species most likely to be able to adapt to changes are the more readily dispersed species with shorter generation times such as those usually found in early successional stages.

An analysis of the potential effects of climate change on ecosystems was undertaken by Watson *et al.* (2013). This was based on the degree to which the original vegetation in an area was intact and the likely future climatic stability of that area. Four conditions could be recognised ranging from those where future climate changes might be modest to those where considerable change is likely (Table 7.2). The most vulnerable areas are those where ecosystems have been substantially modified and where significant future climate change is likely. They include areas of southern and south-eastern Asia, western and central Europe, eastern South America and southern Australia.

In order to improve the adaptive capacity of these natural ecosystems, reforestation in the more vulnerable areas should seek to increase the habitat areas of the most vulnerable species, thereby increasing their population sizes, and restore habitat connectivity across landscapes, thereby improving the capacity of endangered species to migrate to more appropriate locations. The most obvious question is what types of species and what types of plantings should be used? In most cases it will probably be appropriate to continue to use species currently found at these sites but seed might be collected from provenances of these species likely to be more representative of future climatic conditions, insofar as this is known (Booth and Williams 2012, Booth *et al.* 2012). Emphasis should also be placed on using species mixtures or fine-grained spatial mosaics of species

*Table 7.2* Effect of future climate changes on intact and modified ecosystems (Watson *et al.* 2013)

| | *High levels of natural vegetation intactness* | *Low levels of natural vegetation intactness* |
|---|---|---|
| High relative climatic stability | Low turnover of species within ecoregion due to climate change; continuation of ecological processes that sustain the adaptive capacity of species | High number of declining species but small turnover of species within ecoregion; impairment of ecological functioning which will affect the adaptive capacity of species |
| Low relative climatic stability | High turnover of species within and beyond the ecoregion as species track their climate niche; some species able to persist in the changing climate but the adaptive capacity of other species may be exceeded; chance of species extinction unless preventative action is taken | High number of declining species and turnover of species within and beyond the ecoregion as species track their climate niche; many ecological processes affected by climate change which alters the adaptive capacity of species; many species will be confined to refugia and others will face local extinction |

monocultures to enhance the resilience of the new forests (Pawson *et al.* 2013). Some consideration should also be given to including new species – possibly even exotic species – able to maintain key ecological processes, supply resource gaps introduced by climate-sponsored phenological shifts or to add functional redundancy in order to increase resilience (Lunt *et al.* 2013).

Reforestation involving multi-species plantings should also have a role to play in improving the capacity of agricultural systems to adapt to climate change. This might be by providing additional fodder and food reserves for stock; by limiting overland water and soils movement induced by storms; or by helping maintain populations of predators of insect pests of crops.

Some reforestation may be undertaken with the primary aim of storing carbon. The type of markets and the market rules will determine the most financially rewarding types of forests and where these should be located but, from an ecological perspective, the most appropriate type of reforestation will be one able to store large amounts of carbon for extended periods as well as generate other environmental benefits. These are most likely to be mixed-species plantings that are uneven-aged and self-sustaining. Depending on location they may have species able to tolerate occasional wildfires, droughts or storms and preferably recover unaided. Pichancourt *et al.* (2010) describe ways of establishing and managing new forests to achieve carbon storage and other outcomes.

### *Conclusion*

Future reforestation methods will need to include species and genotypes able to tolerate the new environmental conditions. But the precise nature of these conditions is unclear. Nor is it clear what kind of land will be available for reforestation since the competition between agriculture and forestry will probably change. Until these issues are clarified the most prudent option will be to enhance the diversity of species and functional types used at particular sites, as well as across the landscape as a whole, in order to improve resilience.

## Designing new multi-functional landscapes

There will always be places where land is managed for a single specific purpose such as a food crop or a pulpwood plantation. But in most landscapes there will usually be a variety of land uses being practiced and a diverse range of possible future land use objectives. A consensus is emerging from within the scientific community that this is desirable; not only should such diversity be favoured but landscape heterogeneity should be protected. This is despite the economic attractions of large and intensive agricultural and timber plantation schemes. Landscape diversity can be promoted by protecting residual forest patches and adding to these through reforestation

(Sekercioglu *et al.* 2007, Perfecto and Vandermeer 2010, Clough *et al.* 2011, Tscharntke *et al.* 2012, Dawson *et al.* 2013). These plantings will help conserve soils, water and regional biodiversity. They should also help diversify farm incomes (e.g. George *et al* 2012).

But this 'scientific' vision is not necessarily shared by everyone and many farmers are driven to make choices based on more immediate concerns; such people have a limited capacity to forgo production benefits now in return for the possibility of other benefits in the longer term. An example of this viewpoint is given in Box 7.2. The example is hardly an isolated one and the fact that different stakeholders may have contrasting views on issues like biodiversity conservation is well known (Pannell *et al.* 2001, McShane *et al.* 2011).

More generally, people's attitudes vary because they form different mental models about the environments in which they live. This gives them a conceptual tool with which to manage risk and evaluate opportunities. An example comes from Uruguay, where Vihervaara *et al.* (2012) interviewed members of the public, including town and country people as well as some technical staff, about the impacts of plantation establishment on a natural

**Box 7.2 Differences in perceptions of the value of biodiversity in the agricultural landscapes of Ghana**

People can have quite different views about the importance of biodiversity. This is illustrated by the contrasting attitudes about the need for forest restoration in the forest-savannah transition zone in southern Ghana (Awanyo 2007). The area is a global biodiversity hot spot but has been degraded by extensive clearing and many farmers continue to use a bush fallow system of agriculture. An international effort is being made to rehabilitate the region but local farmers are sceptical of the advantages of restoring tree diversity on their agricultural lands. Instead, they are happy to continue using a short (<3 year) fallow period dominated by the exotic weed *Chromolaena odorata* because they believe it has significant agronomic advantages (*C.odorata* can restore soil fertility more rapidly in this region than a tree fallow). They do not regard their lands as being degraded and neither do they think their livelihoods are being threatened by the current loss of biodiversity across the region. Not surprisingly most are unenthusiastic about adopting the longer tree-based fallows and weeding needed to encourage the growth of tree seedlings that are being recommended by scientists to promote biodiversity conservation.

There may be scope for compromise. Although *C.odorata* can restore soil fertility there is also evidence showing it competes with agricultural crop plants once the fallow period is over. It may be possible to develop longer tree-based fallows that help reduce *C.odorata* populations although these fallows are likely to have fewer tree species and lower tree densities than are recommended by the non-resident outsiders who are promoting biodiversity conservation.

grassland. They found there were significant differences between the groups in their knowledge of how plantation establishment affected the delivery of various ecosystem services. The differences were greatest concerning impacts on hydrological cycles and carbon storage with most laypeople being quite unaware of these impacts. Of course, there are equally likely to be differences within the 'technical' community concerning the nature of reforestation programmes (Paruelo 2012).

These differences in the mental models held by various stakeholders are likely to be quite common. They mean that even amongst those interested in reforestation (and not all landholders will be) there are likely to be differences of opinion over the merits of such things as intensively managed timber plantations and multi-functional plantings as well as over the need for, and necessary scale of, landscape heterogeneity. The different mental models will determine where in the landscape reforestation will be done, what type of reforestation will be carried out and how much will be undertaken. If 'land sharing' of the type discussed in Chapter 5 is to be achieved then some form of land use planning is clearly preferable to the *ad hoc* decision making by individual landholders that these different models are likely to generate.

### *Top-down and bottom-up planning*

Some of these differences in objectives or aspirations can be resolved by careful planning. One approach involves planning from a largely regional or national perspective. This might be caricatured as a 'top-down' approach. The advantage of this is that it enables planners to make judgements and trade-offs from a national perspective that optimise the goods and services generated by reforestation. Not all areas of a landscape are equally able to supply ecosystem services. Nor are those seeking these services evenly distributed across a region. This means that top-down planning can allow government planners to maximise the return from the public funds invested in reforestation. All the technical information is assembled and both the environmental and economic issues can be balanced in ways that meet government policy objectives. Planning can be assisted by various economic techniques to assess the economic merits of alternative land use plans including cost-benefit analyses, cost-effectiveness analyses, multi-criteria analyses and risk-benefit analyses (Mayers and Bass 2004).

A variety of computer-based tools have also been developed to assist this approach by exploring spatial options (e.g. Polasky *et al.* 2008, Nelson *et al.* 2009, Bateman *et al.* 2013). Some of these were originally used to identify locations to add to existing protected area networks but similar models are now being developed to identify preferred locations for reforestation (see Chapter 5). These types of models have both advantages and disadvantages. The advantages are that they can utilise all the technical information that is available about a landscape and the species that might be planted. The models then map the environmental conditions present in various areas across the landscape and determine which areas should be planted to generate particular functional outcomes. Some models enable users to pose 'what-if' questions such as what might be the consequences if the price of timber or a certain ecosystem service doubles, or is reduced, over time (Booth 2012). The apparent authoritativeness of these models makes them attractive to planners, especially those dealing with situations when all the land is owned by the government or where there is only a small number of landholders.

But models like these have some disadvantages. One is that they are heavily dependent on the availability of environmental data including the natural distribution patterns of species in undisturbed landscape as well as spatial information about climates and soils. But data such as these are commonly scarce in many developing counties. Another disadvantage is that it is difficult to assess where landowners might undertake reforestation because the opportunity costs of doing so will be known by a farmer but are unlikely to be known to anyone else. This means it may be possible to estimate the profitability of reforestation in a region but not on that particular farm. And, of course, the more landholders there are in a region, the greater the uncertainty about just how much, where and what type of reforestation might be possible across that region. This means that the real value of these types of spatial models is not that they can optimise outcomes but, rather, that they can help facilitate a discussion whereupon land use planning becomes a mixture of both top-down and bottom-up information flows. In this way the regional or national perspective is retained but local landholder's knowledge and views about the inevitable ecological and financial trade-offs are given more weight.

### *Participatory models*

A form of top-down and bottom-up planning is one becoming known as participatory modelling. These can be defined as modelling processes that explicitly collect and incorporate key stakeholder perspectives in order to develop shared understandings and identify mutually acceptable outcomes (Lynam *et al.* 2007, Leys and Vanclay 2011, Sinclair 2012). There are a wide variety of these. Some are conceptual models that describe the system from a stakeholder's perspective. Others are simulations that seek to model the system's behaviour and the way a project might develop over time. All of them share the fact that technical modellers link up with non-expert stakeholders to develop the models. They are most useful in complex situations where a high level of engagement is needed to resolve key issues but where there may be only a limited amount of data, some of which may only be held by local stakeholders. Most of those who have used this approach say the value lies more in the process than in the actual outcome arising from the models. Sandker *et al.* (2010) even refer to temporary 'throw-away' models which were primarily useful because they enabled a discussion about land use planning to take place that might otherwise have been difficult to arrange and not because of the actual outputs they generated. On the other hand, some models can lead to practical outcomes. Bourgoin and Castella (2011) as well as Pullar and Lamb (2012) describe participatory modelling that generated practical outcomes in the field.

## Forest Landscape Restoration

An approach to implementing large-scale forest restoration that takes account of these perspectives has evolved over time and is being referred to as Forest Landscape Restoration (Reitbergen-McCracken *et al.* 2007, Worboys *et al.* 2010, Lamb 2011, Sayer *et al.* 2013). The main steps are as follows:

(i) *Develop a landscape view of the problem:* gather background material about the landscape mosaic in which reforestation is to be undertaken (or, in the case of natural regrowth, managed) including maps of current land use activities, the distribution of plant and wildlife biota and the location of areas requiring reforestation to prevent erosion, protect watersheds or conserve biodiversity. Also identify the potential constraints that may influence the way in which any reforestation is undertaken (e.g. present and possible future environmental conditions, land tenure uncertainties or disputes, areas with rapidly escalating land prices). Likewise, identify possible opportunities that reforestation may offer such as markets for forest products and services.

*Identify stakeholders and forms of governance:* identify the landholders, residents and other stakeholders with a legitimate interest in land use across the area, especially those whose livelihoods may be affected by any reforestation. This means local as well as national bodies and cross-sectoral groups as well (e.g. agricultural departments as well as forestry or conservation departments). Care should be taken to ensure the body is representative of the community as a whole and is not dominated by members of a non-resident political elite. Develop institutional arrangements to guide how these representatives will meet, share knowledge, resolve disputes and make decisions. Some representatives may need assistance to improve their technical capacity to contribute. Some may need to be compensated for their time and the expenses they incurred in order to participate.

(ii) *Identify reforestation possibilities:* based on the background material and views of stakeholders, identify the problems that reforestation may be able to address and develop alternative reforestation scenarios to solve these problems. These scenarios might include where reforestation should be done as well as the type of reforestation (e.g.

monocultures, mixed-species plantings, natural regeneration). Various modelling tools may be useful at this stage. This process should identify the costs of each scenario (including the opportunity costs) as well as an estimate of the actual economic benefits different stakeholders might expect to receive. Stakeholder representatives must be able to discuss these options with their constituents.

(iii) *Choose preferred scenario and timetable:* consider land use options and decide upon a vision for the future landscape mosaic. This means reconciling national plans with the goals of local stakeholders in order to achieve multiple objectives. Decide on priorities, trade-offs, a timetable and a budget. Agreement amongst stakeholders is more likely where information is shared and they have the capacity to change their minds in future if they wish to do so. Nonetheless, disagreements may still occur and it may not be possible to get unanimous agreement; in such cases a pilot project may be useful. In other cases some kind of compensation will be necessary if the full project is actually implemented.

(iv) *Implement:* promote the plan and begin implementing reforestation according to the priorities identified; enhance the technical capacity of those responsible for implementation.

(v) *Monitor and adaptively manage:* monitor the ecological, social and economic changes that take place as the reforestation programme is implemented. Modify the scheme if necessary.

This process will need to be undertaken within a particular policy and institutional framework and the role of these will be discussed further in Chapter 8.

## Conclusions

Reforestation can help overcome land degradation and improve the supply of various goods and services but this will be at the expense of existing land uses. If it is accepted that reforestation should be undertaken there are a wide variety of choices concerning which type of reforestation to use, which location within the landscape to reforest and how large an area is reforested.

There are technical guidelines to assist in making such choices but, inevitably, there will be differences of opinion over each. One difference will be between those taking a national viewpoint and those with a more local perspective. In addition, there will be differences between sectors as diverse as the timber industry, agricultural producers, conservation bodies and water authorities. There will also be a wide difference of opinion amongst individual landholders over whether to engage in reforestation and, if so, how to do so.

Mixed-species plantings often offer the possibility of satisfying multiple objectives at a single site but it is usually easier to make trade-offs when working at a landscape scale.

The development of multi-functional landscapes requires a planning mechanism that can receive inputs from a variety of sources including landholders and other stakeholders as well as government agencies. Local participation in the decision-making process will increase the likelihood that any trade-offs made are politically acceptable and, consequently, that a reforestation programme will be successful.

# 8 Making it happen

## Policies and institutions

### Introduction

Large-scale forest restoration will be difficult to achieve notwithstanding the widespread acceptance that it is a desirable thing to do. The main problem does not lie in deciding which species to use or the types of plantations to establish. Rather, the task will be to develop policies to promote reforestation and to create institutions that will allow these policies to work, including under the socio-economic and environmental conditions of the future. This will be difficult because of the differences in the goals and time frames of individual land managers and the state. Some existing policies will be quite adequate and will continue to facilitate reforestation. But this will not always be true. Indeed, some current policies may be responsible for the continuing degradation of forests and land (Emerton 2012). But the advent of new challenges such as the need to ensure food security, overcome rural poverty, adapt to climate change or respond to rising levels of concern about environmental degradation are all likely to require new reforestation policy responses.

Policy tools can be described as the methods employed to form collective action to provide public goods. Broadly speaking, the tools available to implement policy include property rights, economic tools, information and contractual/regulatory tools (Mayers and Bass 2004). The choice of which of these are the most appropriate to use in reforestation depends on one's viewpoint and is likely to be the subject of considerable debate. For example, decisions will be needed on how much land should be reforested and how any reforestation within the landscape mosaic might impact on food production. Likewise, there will be debates over the rights of landholders, who might be responsible for undertaking any reforestation and the appropriate level of government funding to support reforestation. Many of the policies used to guide reforestation in the past are still relevant but it is likely that new policies will be needed if the future scale of reforestation is to increase.

The importance of having a an appropriate policy framework is illustrated by the changes occurring in places such as Niger and Nepal where massive amounts of reforestation have taken place simply as a consequence of a policy reversal that empowered landholders and enabled them to make their own decisions about where and how forests would be developed (Chapter 3). In each of these cases millions of trees were established without any cost to the respective governments but to the undoubted benefit of individual households and the state. In cases such as these the new policy reflected changing ecological and socio-economic circumstances. But policies should also be grounded in scientific reality and there are a number of common misconceptions about reforestation and the impacts of reforestation which can lead to inappropriate policy development if not corrected (Table 8.1).

*Table 8.1* Common misconceptions that can hinder reforestation policy development

| *Common viewpoint* | *Factual evidence* |
|---|---|
| Only professional foresters and ecologists can restore forests | Farmers and other smallholders are quite capable of re-establishing forests on cleared land if they believe it is in their interest to do so and if they have the necessary resources and technical knowledge. |
| Land degradation is largely a technical problem and amenable to technological solutions | Degradation is mostly solved using social and economic tools although appropriate technical knowledge is also needed. |
| Forests can only be established by planting seedlings or sowing seeds | Trees can sometimes regenerate naturally on seemingly deforested lands when these are protected from further disturbances. Their capacity to do so depends on the presence of residual seed or stumps, or on nearby forests from which seed can be dispersed. |
| New forest prevents floods | Reforestation can reduce flooding from moderate storms but the benefits are less clear when heavily degraded sites are reforested. Any impact is likely to diminish as watershed size increases due to a dilution effect and the size of rainfall fields relative to watershed size. |
| New forest makes rivers flow again | Reforestation usually causes a reduction in river flow. Dry season flows may increase at badly degraded sites if reforestation is able to improve the hydraulic conductivity of soils sufficient to overcome losses caused by evapo-transpiration. |
| All forests are equally effective at preventing erosion | Only structurally complex forests or those with dense understories and thick litter layers will limit erosion. |
| Timber plantations are unable to provide wildlife habitats | Most plantations, but especially older multi-species plantings, can provide habitats for some wildlife but they are unlikely to be suitable for most habitat specialists. |

Institutional arrangements are also important and affect the effectiveness of policy implementation. This is because policy changes have political consequences; some people may benefit from the changes but others may be disadvantaged. Thus an increase in forest cover is likely to be supported by members of the timber industry as well as by managers of water resources who require better protection of watersheds. Conservation groups concerned with diversifying the agricultural landscape surrounding protected areas are also likely to be supportive. On the other hand, some farming groups may be opposed to a loss of agricultural land, even when it is not especially productive land. This is because the change threatens what they see as the primacy of agriculture as a land use.

New institutional arrangements are also likely to generate considerable debate within government bureaucracies. In many cases forestry departments are seen as being of marginal political importance (especially as the area of natural forests shrink) and much less influential than, say, departments of agriculture or mining. Such departments often dominate discussions concerning land use policy and neither will necessarily be enthusiastic supporters of policies that threaten their influence. In short, any new reforestation policies will have to be developed within a broad and noisy policy framework where there are competing demands for land for food production, infrastructure development, watershed protection, biodiversity conservation and a myriad of other national economic projects. That means policy development must involve multiple sectors and government agencies and not just forestry departments. The institutions needed to reconcile these various interests must be representative of the various stakeholders as well as being accountable to society as a whole (Le *et al.* 2012).

This chapter reviews some of the main issues that should be addressed to develop an appropriate policy and institutional framework to encourage large-scale forest restoration.

National reforestation master plans are often developed but can be problematic when circumstances change and where there is a difference between the view from the capital and the political reality on the ground. New policies must reflect political realities within communities, existing patterns of land ownership and the fact that many land owners are poorly informed about the opportunities that reforestation can offer. In most cases policy must evolve over time through a process of debate, negotiation and analysis of the impact of past policies. Some might take the view that household activities are largely irrelevant in determining national economic outcomes and that it is the larger industrial-scale operations that are important. Previous chapters have provided compelling evidence that this is not the case when it comes to reforestation. Further, households are often the critical pathway through which economic well-being is enhanced within the community. This means that new policies should aim to foster reforestation activities that benefit smallholders as well the more traditional approach of facilitating reforestation by large industrial plantation companies.

The chapter has been arranged to emphasise the three sets of policies that might be used to encourage forest restoration. One set of policies are those concerned with simply providing enabling conditions to allow households and other land managers interested in reforestation to carry it out. A second set goes beyond this largely passive response and offer deliberate incentives to encourage landholders and other land managers to undertake reforestation in situations where they might not otherwise do so. Finally, there are regulatory policies needed to cover situations where certain areas must be reforested or certain types of reforestation must be employed to achieve national objectives. This may be because of the unacceptable level of damage at the site caused by past activities (e.g. mining) or because of the consequences that degradation is now having on neighbouring landholders and those downstream in the watershed. The chapter concludes by considering the institutional arrangements needed to implement these different types of policies.

## Enabling policies

Governments have funded much of the reforestation carried out in the twentieth century but most cannot afford to continue doing so. Or, rather, they cannot do so on the scale necessary to overcome the backlog of degraded land that has now accumulated. Many state-owned plantations have now been sold and ways of attracting private funds will be needed if further large-scale reforestation is to be implemented. Private landholders willingly commit funds to land use activities such as agricultural crops or grazing animals so that the policy task is to find an appropriate framework that can make reforestation an attractive alternative land use. Some possible policy directions are summarised in Table 8.2.

### *Remove disincentives*

The logical first step in enabling reforestation is to remove policies that currently act to discourage landholders from doing so. The most obvious of these is unregulated logging of natural forests. As long as this continues the supply of cheap but illegally-sourced timber entering the market will undercut producers of plantation-grown timbers. A second step is to remove perverse laws having unintended consequences. One of these is a law found in many states that was originally introduced to prevent illegal logging in natural forests. This requires those wishing to fell trees to obtain a harvesting permit from a government official. Other laws sometimes oblige landholders to sell logs of certain tree species grown on their land to the government itself at a price determined solely by the government. These types of regulations can be found in many countries throughout Asia, Africa and Latin America. It is doubtful that they have been especially effective in preventing the degradation of natural forests but they have proved to be a

*Table 8.2* Policy areas needed to enable reforestation by private landholders

| *Policy direction* | *Reason* |
|---|---|
| Remove disincentives | Some existing policies may actively discourage tree planting. |
| Establish land tenure and property rights | Legally secure land tenure and rights to harvest trees are necessary if landholders are to engage in a long-term activity like forestry. |
| Empower landholders or communities | Landholders having the right to make their own land use decisions are more likely to find appropriate ways of incorporating tree growing within the landscape although ways will need to be found to ensure an appropriate balance is found between individual rights and national requirements (e.g. for watershed protection or biodiversity conservation). |
| Popularise tree growing amongst landholders | Growing trees in plantations may be an unfamiliar land use option for many landholders and one they are unwilling to use unless they are encouraged and their capacity to do so is increased. |

major disincentive to landholders who might otherwise have been interested in planting trees. It is sometimes the case that landholders even remove seedlings of particular tree species that they find regenerating naturally on their properties to avoid the likelihood of being caught in a bureaucratic tangle (Vera Lex Engels: pers. comm.). An especially insidious type of disincentive can be fees, 'road taxes' and charges imposed by local officials without the knowledge of the central government. The impact of these disincentives can severely damage the profitability (and hence the attractiveness) of tree growing (Le *et al.* 2004). The collective removal of these types of policies is likely to mean landholders will be more interested in responding to price signals from the forest products market.

### *Establish land tenure and property rights*

A second critical policy likely to improve the attractiveness of reforestation is the granting of land tenure and property rights to those managing deforested land.

Tenure can take many forms including include rights to access land, rights to harvest any products grown on that land, rights to manage forests grown on the land or rights to alienate the land (thereby making the land owners landless). These rights determine who has access to land, the ways in which it can be used and those entitled to enjoy the benefits arising from this use. Some forms of tenure allow use of the land but not the rights to exclude others since this resides with the state. Some impose limitations on certain harvesting rights, especially of trees. Other forms of community tenure mean that a person has rights of access and usage but not rights to sell the land because this resides with the community or its leaders. It is not uncommon to see land over which a government asserts ownership, but does not manage, being degraded because it is now regarded by the community as being an open-access resource. In many cases privatising such land (especially restoring ownership to customary owners) can begin the process of overcoming degradation at no further cost to the wider community.

Secure tenure is critical for reforestation because it means landholders (or their children) are likely to benefit from any reforestation that they undertake (Lambini and Nguyen 2013). Zhang and Owiredo (2007) note that even within the same village in Ghana, farmers responded differently to tree-planting programmes depending on the type of tenure they had. With legal access to land, together with rights to unrestricted harvesting of any trees they grow (or that regenerate naturally), a landholder with full tenure can decide how they can balance tree growing with other activities they may wish to adopt across their land. Without tenure and harvest rights it is difficult to see why anyone other than a government agency would bother with a long-term activity like reforestation. Obidzinski and Dermawan (2010) argue that the right to transfer or inherit land is also crucial and the absence of this right has inhibited the extent of reforestation by smallholders in Indonesia. Secure tenure is re-enforced if the community accepts particular land ownership claims but this community acceptance should be accompanied by a central land registry where all such ownership arrangements are registered.

The situation can be more complex when land is owned by the community rather than by individuals. Community tenure is common in many traditional societies and reforestation can be effectively practiced in these communities, especially when natural regeneration takes place (Carter and Gilmour 1989). This is because there are trusted leaders, established social relationships and accepted ways of making decisions, enforcing rules and sharing benefits. However, traditional communal decision-making processes can be disrupted when members of the community move to urban areas. This can temporarily paralyse the ability of the community to resolve these land use issues (D. G. Gilmour: pers. comm.).

Some community members can find collective ownership of assets such as planted trees is burdensome and may wish to pass the ownership of the trees they planted to their children rather than to the community as a whole (i.e. a form of quasi-privatisation). The idea is often contentious. There are examples of this being done for some perennial agricultural crops (e.g. via leases or some form of occupancy registration) and similar arrangements could probably be developed for tree plantations. Something similar might even be done by people who are not members of the customary landholder group (Fingleton 2005, Curtin and Lea 2006).

The situation is different when land held by the state is privatised and allocated to households or industry. The allocation of land to households can be a difficult process because the quality of land varies so that land near roads or with good soils will be more valuable than more remote land with infertile soil. Care is needed to ensure the allocation process is equitable and that members of a political elite do not capture the better land at the expense of less well-connected members of society. There can also be problems in allocating poorer quality land or land in remote locations because nobody may be interested in acquiring it (unless, perhaps, it was formerly part of their customary land holdings). The problem may be exacerbated if future 'ownership' is conditional on the allocated land then being reforested. In such cases it may be difficult to ensure the new land owners actually use the land and fulfil the formal conditions (Box 8.1). Privatisation programmes can also be difficult if people regarding themselves as customary owners do not benefit and there are many examples of trees growing on land allocated to a plantation timber company being cleared and replaced with agricultural crops by those regarding themselves as the customary owners of that land (e.g. Obidzinski and Dermawan 2010).

An unusual example of reforestation being enabled because of the granting of conditional tenure is being tested in Indonesia. In this case the government has provided assistance to several NGOs to obtain a 'restoration concession' over 98,000 ha of heavily-logged tropical forest in Sumatra for the specific purpose of restoring natural forest across the area (Normile 2010). Funding will come from the NGO group but the reforestation was unlikely to have been carried out without the government's assistance in acquiring the concession.

### *Empower communities and improve governance*

A third enabling policy is one that empowers communities and improves local governance arrangements. By itself, the granting of land tenure will not necessarily result in large-scale reforestation. Tenure also needs to be accompanied by policies and political institutions that distribute political and economic power widely in a pluralistic manner. If a local community is not involved they will have no vested interest in any proposed land use changes and may even work to undermine a new policy (see Box 8.1). Rather than choices being dictated by governments, households need to be able to judge for themselves how to invest their labour and capital.

There are a variety of ways in which power can be devolved and the problems of doing so are well known (Hobley 1996, Sandberg 2007). In the case of reforestation the best way is likely to be by seeking a convergence of interest between the state and landholder. In this way the government's aim of increasing the national forest cover is linked to the landholder's aim of improving their own economic circumstances. Empowerment is often facilitated by the development of local community organisations that act to gather information about silvicultural techniques and markets and disseminate this information to members (Gilmour *et al.* 2004). These and other types of institutional arrangements for empowering communities are discussed further on the next page.

**Box 8.1 Local modifications to national land allocation programmes in Vietnam**

It can be difficult to ensure that national land allocation policies are actually implemented in practice. Experiences in north-western Vietnam illustrate the difficulties that can sometimes occur (Sikor 2001). In the 1990s Vietnam embarked on a programme of allocating state-owned land to farmers. One class of lease was for land granted to farmers to be used for cropping. A second type of lease was for land to be specifically used for reforestation to help meet national reforestation goals. Farmers were granted long-term usage rights but the state reserved for itself the right to allocate land to specific uses. In north-western Vietnam land allocated for reforestation was mostly on upper slopes of deforested areas and along the main national road network. Land to be used for agriculture was away from the immediate roadsides. Villagers were happy to accept the allocated land but did not accept the way in which the state expected them to use this land. They undertook reforestation in order to receive the financial assistance being offered for this purpose but the trees were not protected and were allowed to die. Cropping was carried out between these trees and eventually replaced them, especially in areas near roads. Despite these practices, regional tree cover did increase. This was not because of tree planting but because changing farming practices allowed certain areas to be abandoned. For example, farmers replaced cassava crops with corn and constructed terraces, switching from growing dryland rice to growing wet rice. These changes resulted in less land being needed for agriculture and enabled natural regrowth to take place on former swidden fields. Local officials were aware of the difference between the state-specified land use plans and what was actually happening on the ground but were reluctant to enforce these plans because of their close personal ties with villagers and their much weaker links with the central government. The experience illustrates how local communities are usually better placed to decide the forms of land use likely to be of most benefit to themselves.

### *Popularise tree planting as a valid land use option*

A fourth enabling policy is to popularise tree growing as a legitimate and profitable land use activity. Even if reforestation is seen to be in the national interest it may still be an unpopular activity in some farming communities because it is not viewed as a traditional activity. Indeed, some farmers may have spent a good part of their lives trying to remove trees to establish their farms. For example, Barlow and Cocklin (2003) describes the response by an Australian farming community where the market for traditional farm produce had declined and where private companies were buying land on the open market to grow plantation timber. The visual impact was dramatic and the plantations were seen as undermining the community and its values, notwithstanding the economic merits of timber plantations and the local employment it generated. This type of response is probably not uncommon but is not necessarily a universal view. A contrasting experience occurred in a pastoral area of Uruguay where many people welcomed the commencement of timber plantations because of the economic advantages these could bring even though the species being used were non-native eucalypts (Vihervaara *et al.* 2012).

Different places have different outcomes but the point remains that policies may be needed to persuade farmers that tree planting is in their interest and that they will receive some tangible benefit from doing so. Once reforestation begins other opportunities such as for private nurseries and tree planting companies can develop (see Figure 8.1).

*Figure 8.1* A privately owned tree nursery in Vietnam selling seedlings to farmers. The nursery is not subsidised and sells a variety of tree species.

## Incentives

With the exception of land allocation and granting of tenure, most of these enabling policies are likely to imposed minimal cost on government budgets, at least in comparison with the cost involved if the government itself carried out reforestation.

But in many situations, enabling policies of the type described above may not be enough. In these cases governments may need to commit funds in the form of incentives or subsidies to induce landholders to protect regrowth or plant trees. Bull *et al.* (2006) note there is some confusion over the use of these two terms so they will be used interchangeably in what follows. The circumstances under which incentives may be useful is when the net present value of tree growing is lower than alternative land uses but where the wider environmental and social benefits to the community exceed those of the alternatives. In this situation the purpose of the incentives or subsidy is to overcome the disadvantage suffered by the landholder (Keipi 1997, Enters *et al.* 2003).

The types of incentives used to make reforestation more attractive can be classified as either (i) indirect and non-financial or (ii) direct financial incentives. Some of these are outlined in Table 8.3.

*Table 8.3* Financial and non-financial incentives to encourage reforestation where it might not otherwise occur

| *Type* | *Format* |
|---|---|
| Indirect and non-financial assistance | Development of new roads to link plantations and markets<br>Seedlings (free or at subsidised rates)<br>Fertilisers (free or at subsidised rates)<br>Tools (free or at subsidised rates)<br>Improved access to off-farm employment<br>Provision of silvicultural knowledge<br>Provision of market and marketing knowledge<br>Provision of fair and equitable markets |
| Direct financial assistance | Direct payments to retire land from agriculture and protect natural regeneration<br>Grants to purchase and plant seedlings<br>Cheap loans with long periods before repayments fall due.<br>Loan guarantees<br>Taxation concessions for plantation investors<br>Direct grants to undertake reforestation (sometimes in specific areas)<br>Creation of new markets for ecosystem services |

***Indirect incentives***

There are a variety of indirect incentives that might induce landholders to undertake reforestation. These include the development of roads to help transport goods to markets as well as free (or cheap) seedlings or fertiliser. Another policy direction that can act as an indirect incentive for reforestation is to provide increased opportunities for off-farm employment meaning there will be less dependency on land to provide immediate cash income. This may make an activity like tree growing quite attractive because little on-site labour is needed once the trees are established.

But there are several other kinds of indirect incentives both of which concern providing landholders with more knowledge about tree growing. One is the provision of knowledge about the potential benefits of reforestation and technical advice on how it might be done. Many farmers may have grown a few trees around their farms but have no experience of growing them in plantations. Advice on the advantages of doing so and just how this might be done could lead some more innovate landholders to test whether they could benefit. Some care is necessary here. The silvicultural techniques suitable for large industrial plantations will not necessarily be those suitable for landholders only willing, or able, to establish small plantations. This means that specific research programmes may be needed to develop appropriate silvicultural systems that match the environmental and economic circumstances of these particular landholders.

A second related incentive is the provision of knowledge about how landholders might gain financially from tree growing. It is not uncommon to see reforestation projects that have encouraged (and perhaps even paid) landholders to plant trees on their land. Farmers go along with the project but never replicate it. Indeed, some even clear and replace the trees when the project ends. They have obviously judged that tree planting does not meet their needs and that there are better alternatives available. This suggests that landholders may need forms of assistance other than just technical assistance if they are to benefit financially from reforestation. This might include knowledge of the various markets available and the prices of various goods in these markets.

Once this is done the expectation of a financial benefit can be a powerful incentive. An especially interesting illustration of this was provided by Demurger and Yang (2006) who analysed the way farmers allocated land to agriculture or tree planting in 25 provinces across China between 1984 and 2001. They found the perceived relative profitability of trees was a key determinant of land use patterns during this period. There was also a view amongst landholders that tree growing offered a means of diversifying income and reducing risk. Similar views have been reported by Barrow and Mlenge (2003) and Haglund *et al.* (2011).

Other policy initiatives likely to act as indirect incentives for reforestation because they improve grower's income include:

- Removing taxes on the harvesting of trees;
- Ensuring that the prices of timber from government-owned plantations are sold on the open market at prices that reflect the real costs of production and not some discounted price;
- Controlling the activities of middlemen who can benefit small holders by helping them access market when they have only small amounts of timber but who can also limit the returns received by small producers;
- Minimising licences, fees and taxes, especially those imposed by local authorities when logs are carried through their districts (Le *et al.* 2004);
- Ensuring farmers acting as out-growers for large industrial timber companies receive fair prices (Mayers 2000, Mayers and Vermeulen 2002, Nawir *et al.* 2003);
- Assisting farmers to reach the standards needed to have their timber certified by relevant certification bodies and assisting them access these better markets;
- Improving road networks linking growers and markets since the costs of transporting goods from isolated plantations will be high;
- Assisting in accessing markets and in the development of new markets (see discussion below on markets for ecosystem services);
- Providing a stable policy environment without sudden changes in emphasis or direction.

Finally, an example of what can only be described as a decidedly indirect incentive comes from Gujarat State in India and is reported by Pant (1979:21):

> To lend impetus to free-planting a religious leader, Pandit Shamboo Maharaj, has offered to recite the 'Mahabharat Katha' free of charge if any farm owner plants 500 trees. The narration lasts for five or six days and Pandit Shamboo Maharaj is well known for his masterly presentation. His offer has prompted many a farmer to plant trees.

### *Direct incentives*

Sometimes more direct incentives are needed. The most obvious direct incentive is to provide cash payments. One approach might be to provide payments to encourage landholders to retire certain specified lands from agriculture and to protect these in order for natural regeneration to take place. Another could be to provide funds to landholders to purchase and plant seedlings. Other forms of direct financial assistance are cheap loans, loan guarantees or various types of taxation assistance (although this might only be attractive to larger companies since many smallholders may not pay tax).

The merits of incentives like these have been widely debated because of the large amounts of public money that can be involved and because they have sometimes lead to unwanted outcomes (Enters *et al.* 2003). For example, they have been sometimes been used to enable natural forest to be cleared and replaced by planted forests. In other cases they have stimulated reforestation but the plantations were subsequently abandoned once the incentive payments ceased. This was true of the tax concessions that lead to the managed investment schemes in Australia (Pacoe 2009). Rather than acting as an incentive for growers to participate in a worthwhile new land use, these payments were the sole reason for these investors' participation.

Incentive schemes are also open to abuse. Growers often need to obtain assistance to actually access money allocated through incentive schemes and there can be long period of grace between the awarding of loans and the time when they must be repaid. Both factors mean that many incentive schemes offer opportunities for rent-seeking and misuse by third parties (Obidzinski and Dermawan 2010).

These potential disadvantages may be lessened when the incentives are specifically directed to encourage reforestation at particular locations. Such might be the case where the objective is to quickly create a resource large enough to sustain a new timber industry. An example is the case described by Boulay and Tacconi (2012) designed to encourage farmers to act as out-growers for a new pulp mill in Thailand.

Likewise, it may be important to undertake reforestation at particular locations in order to control erosion or for conservation purposes. Some of the types of locations that might be involved were discussed in Chapter 5. In each case only landholders at these locations are eligible to receive these incentives. Further, reforestation at such locations usually requires that particular types of reforestation should be undertaken. In such cases it may be necessary for the incentive payment to be sufficient to cover the costs of the type of reforestation required as well as the added cost of undertaking reforestation at that site.

These forms of targeted incentives can be particularly useful when seeking to foster a coordinated programme of reforestation at a landscape scale, something referred to earlier as Forest Landscape Restoration. In this case the intent of the incentive is to ensure national goals (e.g. in watershed management or conservation) are not achieved at the expense of individual landholders. Financial incentives can obviously help facilitate this process but should only be offered after the development of land use plans following a process like that discussed in Chapter 7. Kleijn and Sutherland (2003) and Hanley *et al.* (2012) describe how cooperative reforestation agreements have been developed in parts of Europe for this type of purpose.

Direct incentives or subsidies have not been as widely used in forestry as in agriculture. Bull *et al.* (2006) quote data suggesting agriculture may receive 37 per cent of all global subsidies while forestry receives only 3.3 per cent. Nonetheless, incentives have been widely used to initiate many large national reforestation programmes as well as to ensure the reforestation of degraded landscapes (Cossalter and Pye-Smith 2003). A claimed advantage of incentives paid to timber companies is that these transform the investment climate because companies receiving the payment then go on to invest additional funds of their own to create a resource and provide employment and other economic benefits to the state. Without the incentive these companies may decide reforestation is financially unattractive and invest their funds elsewhere. Nonetheless, Enters *et al.* (2003) argue that incentives should be catalysts but not the principal drivers of change.

## Payments for ecosystem services

An additional incentive for forest restoration could come from the emerging new markets for ecosystem services. Perhaps the key services that new regrowth or plantation forests could supply are carbon storage, watershed protection and, possibly, the creation of habitats able to conserve biodiversity. Interest in such markets is likely to grow as areas of natural forest shrink (Kemkes *et al.* 2010, Wunder 2013). The prospect of forest owners being paid for a service most beneficiaries current receive for free could be a major new incentive for reforestation.

*Table 8.4* Potential difficulties in making payment schemes for owners of new plantations supplying ecosystem services act as an incentive for reforestation

| *Issue* | *Resolution* |
|---|---|
| Incompatibility of timber production and some services | Grow new plantation on long rotations or entirely forgo seeking to benefit from the timber market |
| Service delivery influenced by type of reforestation | Design the type of plantings according to the ecosystem service required; this will probably mean using mixed-species plantations |
| Service delivery depends on plantation age | Ensure that forests are protected long enough for services to be generated; consider the possibility of interim payments |
| Service delivery depends on landscape coordination | Develop a planning group to coordinate landscape reforestation and target key locations |
| Service delivery depends on plantation area | Develop a planning group to coordinate landscape reforestation |
| Verification that services are being delivered | Establish a body to carry out monitoring to verify that landholders are complying with contractual reforestation obligations |
| High transaction costs if many landholders are involved | Develop a coordinating body to administer |
| Land ownership contested | Clarify tenure |

However, there are several difficulties in to be overcome before a market is possible (Table 8.4). The first is that this new market may be at the expense of the timber market. In the case of carbon storage, for example, there are likely to be limits on the capacity of a grower to receive credits for storing carbon in biomass and then, after a few years, receive payment for felling that biomass. Formal market rules are still being developed but it is likely that trees will have to remain undisturbed for many decades (possibly 80 years) for their carbon storage capacity to qualify. This could, of course, still leave open the possibility of harvesting particularly high-value timber trees grown on long rotations. And there may be scope for large industrial growers with an expanding plantation estate to produce timber and be paid for storing carbon but much depends on how these market rules are developed.

A second issue concerns the type of forest. Forests are not equally able to supply various ecosystem services and different services may require different types of forest. But this dilemma may be more apparent than real. For example, rapid carbon sequestration is best achieved by plantation monocultures of fast-growing species but stable forests able to store carbon over long time periods will probably involve slower-growing species grown in mixtures. These kinds of mixed-species plantings will probably be useful for the delivery of other ecosystem services.

A third issue concerns timing. It may take some time for a new forest to reach the age where the required service is being delivered. Should the landholder be paid in the meantime? In most cases some payments are probably needed as a mark of good faith although this exposes the buyer to the risk that the plantations may fail (e.g. because of wildfires or storms) before any service is provided. Some form of insurance may be needed here. A final group of issues concerns resolving ownership disputes and coordinating the planning, establishment and management of new forests across large landscape areas. The problem will be greatest when there are multiple landholders involved. The best way of addressing these issues is through the mechanisms and institutions discussed in Chapter 7. Some of the institutional arrangements are discussed further below.

There are several policy challenges in using payments for ecosystem services as means of encouraging reforestation. The first is to help design and facilitate the development of the market and bring potential participants together in the absence of this market evolving by itself. The second is to develop a legal framework to ensure suppliers and buyers comply with their contractual obligations. The third is to help ensure that small landholders as well as large landholders are able to benefit from participating in the new market. All of these could be expensive undertakings.

By themselves, payments for ecosystem services may be insufficient to persuade a landholder to voluntarily change land uses and grow trees instead of an agricultural crop although they may be enough to tip the balance in marginal cases. Much depends on the actual prices paid and on the merits of competing alternative land uses. But part of the problem is also that the funds actually reaching growers will also be reduced by the transaction costs. These are the costs of verifying service delivery and managing payments to many landholders scattered across a large landscape. Kemkes *et al.* (2010) argue that transaction costs are likely to be lower when there is a single buyer, or a monopsony, (e.g. a water authority requiring watershed protection) since there is no need for coordination amongst forest owners although there will still be a need to identify the relative share each owner should receive of any payment based on the extent to which their particular plantations contribute to supplying the service. On the other hand, where a single landholder is the sole supplier (i.e. a monopoly) buyers will have no alternative but to purchase the desired ecosystem service from this single supplier. In the case of an essential service such as water this may lead to calls for a more regulatory or prescriptive approach to control the supply and pricing of the service. Having multiple buyers for a bundle of services might be a way of increasing political support for reforestation payment schemes as well as insuring against the possibility of one buyer being unable to sustain the market (Kemkes *et al.* 2010).

Markets for the ecosystem services provided by forest are still evolving and, to date, there has been limited experience of them being used to induce reforestation where it would not otherwise have occurred. One of the few studies of the effectiveness of payment in increasing forest cover was a study carried out in Costa Rica showing that a payment scheme was associated with an increase in forest cover across the country from 11 per cent to 17 per cent in only eight years (Arriagada *et al.* 2012). This is encouraging and suggests payment schemes have considerable potential to attract landholders. The topic is likely to be one that continues to engage policy makers in future.

## Obtaining value for money from incentive schemes

A key issue in any kind of incentive scheme is how a funding body can obtain value for money. If funds are to be used to foster reforestation – especially if public funds are involved – then how can it be ensured that they are being used to maximum effect? Bull *et al.* (2006) argue that Brazil's taxation incentive programme for industrial plantations covered 100 per cent (or more) of the actual planting costs but that this was poorly used and only 4 million ha of the 6 million ha thus planted were commercially viable. There are also risks that those allocated funds in the form of grants or subsidies for particular purposes may use them for unintended purposes (e.g. funds provided to buy fertilisers for trees will be used to fertilise crops). Likewise, subsidies in the form of free seedlings may be planted but not cared for if the landowner has no real long-term commitment to reforestation.

Some argue that the best kinds of incentives and subsidies are not those that improve private interests but, rather, those that generate public benefits (Birch *et al.* 2010). Governments often justify financial subsidies to large timber companies on the basis that they will make reforestation more attractive and therefore help increase rural employment: there is a public benefit as well as a private benefit. But, in the longer term, a better type of assistance might be that which provides a broader range of assistance to a wider range of beneficiaries, including small landholders as well as large companies. For example, assistance to develop new silvicultural systems could be a good way to maximise the return on public money invested because these new forms of reforestation could provide a specific outcome or generate a wider variety of ecosystem services. The new techniques could then be applied by a wider range of landholders.

Financial analyses of the merits of alternative timber production schemes often make comparisons based on metrics such as net present value. It can be more difficult to evaluate the benefits arising from alternative reforestation approaches when the purpose is to provide an ecosystem service. The most appropriate way of doing so is to compare the environmental outcome with the project intervention versus the outcome without the intervention. This is much better than simply comparing the benefit before and after the intervention because the system may have been changing anyway. Thereafter, the changes might be assessed as either the biophysical changes that take place or as the change in the value of ecosystems services generated by reforestation.

But, as discussed above in the context of payments made by recipients of ecosystem services to the providers of these services, it can take time for services to become evident. Moreover, there can be considerable transaction costs associated with the need to deal with multiple landholders which further complicate making cost-benefit analyses. Some of the issues involved in evaluating the benefits of incentive payments for environmental projects are discussed by Pannell *et al.* (2013).

Reforestation costs depend on local circumstances and it is difficult for funding agencies to know the actual costs of reforestation in a particular location let alone the opportunity costs for those landholders. Each landholder is likely to be in a far better position to assess their actual costs than an external government agency. The result is that funding bodies may offer a subsidy that is too little to entice a landholder to plant trees or, alternatively, to provide more funding than is necessary. One way of dealing with this problem is to carry out a reverse auction or a competitive tender (Box 8.2).

**Box 8.2 Reverse auctions**

Reverse auctions (or conservation auctions) differ from normal auctions in that sellers compete to obtain business from buyers. They do so by offering increasingly lower bids (i.e. rather than buyers competing with each other to win business from a seller by offering increasingly higher bids). In the case of reverse auctions for reforestation, incentive funding is usually limited and landholders are invited by a funding body to compete to obtain support by submitting sealed bids in which they outline what reforestation will be done and how much it will cost. Minimum standards can be specified that describe the species to be used or the survival rates that must be achieved. The quality of the bid is then judged on the basis of value for money (e.g. the area reforested or on the provision of a particular ecosystem service) and the bidder offering the best outcome per unit of funding wins the auction (Stoneham *et al.* 2003, Connor *et al.* 2008). The advantage of the approach is that the responsibility of assessing costs is borne by landholders who are in a better position to do this than, say, a government agent. The disadvantage of the approach is that it might be too complicated for many landholders to bother (Grammatikopoulou *et al.* 2012). However, a successful trial was organised amongst farmers in Sichuan Province in China after they had been briefed on the way the auction was to work. This found that outcomes could be 15 per cent more effective than a fixed price scheme (Wang *et al.* 2012). The authors concluded that, although reverse auctions have been largely used in developed countries so far, there is no reason why they might not be used more widely in future.

The winners of normal reverse auctions can be in isolated farms spread across a landscape. This can mean there is no aggregation of newly forested areas or any inter-connectedness between plantings and natural forest areas even though this is often needed to obtain an optimal functional outcome. Ways of modifying the auction process to encourage connectivity might include using an agglomeration bonus or using multiple bidding rounds with incentives for cooperative behaviour. Some of the methodologies and policy implications arising are described by Windle and Rolfe (2008) and Reeson *et al.* (2011).

## Regulatory policies

The third set of policies needed to facilitate large-scale forest restoration concern situations where, despite enabling policies and various incentive schemes, appropriate forms of reforestation are still not being carried out. This means some kind of regulatory approach might be needed. An obvious example is where mining companies are obliged to rehabilitate land after mining is complete. Other examples include where severe erosion at a particular location is polluting water supplies or where habitat restoration is necessary at a key location within a landscape to conserve an endangered wildlife species. In such cases the costs of coercion can be high but the benefits can also be significant provided the required standards are specified. In each of these cases the obligations must be technically feasible, fair and apply to specifically defined areas. It may be appropriate for penalties to be imposed if the regulation is not complied with or the standards are not met.

In the particular case of mine site rehabilitation there are several possible approaches. Miners are commonly granted a licence to mine and, increasingly, they are legally required to rehabilitate the site once mining is complete. Some jurisdictions require miners to restore the former ecosystems after mining ceases (using undisturbed reference sites as benchmarks) while others only require that a diverse and self-sustaining new ecosystem be established although not necessarily one that replicates the original ecosystem. In each case, policies must define the standards to be reached and the criteria that must be met before legal liability for rehabilitating the site can be relinquished. Simple requirements to replant certain species are rarely sufficient. Instead those responsible should be required to provide evidence that the new community is self-sustaining and no longer needs external support (e.g. weed control, fertilisers, fire exclusion or protection from herbivores). Some regulators may also require evidence that wildlife as well as plant species have recolonised the site. Others may specify that slopes are stable and there is no erosion, or that there are no leachates from the former mine reaching rivers. These criteria mean it may be some years after mining is completed before legal liability is relinquished. One way of ensuring these standards are reached is to require miners to lodge a performance bond at the commencement of mining.

A second type of regulatory policy is that where landholders are obliged to undertake reforestation of certain lands even though they still retain ownership of the land. An example is the Sloping Land Conversion Program in China (Bennett 2008, Kelly and Huo 2013). In this case land steeper than 25 degrees is being reforested thereby displacing all previous land uses. Landholders have been granted compensation for the change (initially in cash and food for up to eight years) and villages and township governments have been made responsible for implementing the actual work although, again, there is some regional variation in the compensation provided and the way the work has been carried out. In some cases the acceptability of the regulations may have been helped by the fact that household income from these steeper lands was modest and many landholders have left to work in urban areas with the actual reforestation being carried out by cooperatives of which they are members.

Another example of normal ownership rights being over-ridden and a landholder being obliged to reforest their land occurs in Brazil where landowners are required to reforest 20 per cent or more of their land depending on the size of their holdings and their location (Economist 2012). State legislation in São Paulo also specifies the minimum reforestation standards that must be met. These include requiring that the restored forest must contain at least 80 native woody species (established by planting or via natural regeneration) within a defined time period (Aronson *et al.* 2011). The numbers of different types of species and the minimum number of individuals of each species are also specified. The law has caused considerable debate, including amongst those who would see themselves as conservationists but who believe the law is unnecessarily prescriptive. These critics would prefer the over-arching goals to be specified but not the precise steps that must be implemented to achieve these goals (Rodrigues *et al.* 2011).

A final example of the regulatory approach occurs when a cleared area is restored to offset the loss of biodiversity when another area is damaged (discussed in Chapter 7). State regulation is needed to specify the standards that must be met to ensure that the biodiversity losses in one area are truly compensated by gains in another. As discussed earlier, Ecological Restoration is an inexact science and success is not guaranteed even though a genuine effort to succeed might have been made. In these cases the policy dilemma lies in knowing how to monitor whether the regulatory obligations are being met and, if they are not, resolving how to deal with non-compliance.

## Institutions for implementing reforestation policies

Policies developed to encourage reforestation must be accompanied by arrangements to ensure these policies are implemented. These arrangements for undertaking collective action can be described as institutions. Ostrom and Cox (2010) defined institutions as codes of behaviour that can facilitate collective action and mediate self-interest. Expanding the scale of reforestation is likely to need a variety of institutional arrangements to implement reforestation policies and, where there are national targets, to drive the process to ensure these targets are met.

Perhaps the key role of institutions is to provide coordination across a multitude of policies and laws in order to limit policy inflation and prevent a collapse in the capacity of the community to do anything (Mayers and Bass 2004). Institutions coordinating and driving reforestation programmes must deal with questions such as why reforestation should be undertaken, where it will be carried out, how it will be funded and who will do it. In addition, they will have to work with a variety of other sectors (e.g. agriculture, water, conservation) that may have conflicting policy mandates and be operating on different spatial and temporal scales. Ideally, institutional arrangements will be such that they involve representatives of all the main stakeholders, are transparent and accountable and are able to adapt to unexpected changes. Reforestation is often carried out to generate an improved supply of ecosystem services and many of these can only be supplied when landscapes rather than just sites are reforested. This means institutions will need to develop ways of collectively managing landscapes while maintaining the ownership mosaic.

Some desirable attributes of institutions relevant to large-scale reforestation programmes, including those likely to involve small landholders, are shown in Table 8.5.

### *Scale*

There are several key issues affecting the development and operation of reforestation institutions. The first of these concerns the scale at which different institutions will operate. Some institutions must clearly operate at a national scale because of the need for strategic national planning of land use, budgeting and resource distribution. And certain issues including the design of trading schemes to arrange payments for ecosystem services like carbon storage are likely to require sophisticated forms of accounting that are probably best done by central governments. But policies developed in a national capital may not always be appropriate in some regional settings and there is widespread agreement that more localised institutions are also needed to implement land use changes such as reforestation.

Provincial or district level institutions are likely to be aware of the attitudes of landholders towards reforestation (whether it be through natural regrowth or planting) and be able to assess what enabling policies are needed or whether different types of incentives will be required. They are also likely to be in a better position than national bodies to manage and assess the operational effectiveness of national funding programmes,

*Table 8.5* Desirable attributes for institutions involved in large-scale reforestation

| *Attribute* | *Justification* |
|---|---|
| Involve representatives of all main stakeholders | To ensure that all perspectives are heard and that equitable outcomes are achieved |
| Involve representatives of related sectors | To ensure that reforestation activities are integrated with other sectors such as agriculture, conservation and water management |
| Are transparent and accountable | To ensure the broader community is aware of the issues being considered and trade-offs being made; to enhance institutional legitimacy |
| Work at a landscape scale | To manage ecological processes that largely operate at these broader scales |
| Are able to monitor feedback and practice adaptive management | Because unexpected events can occur and it is important to learn about any changes in the nature of the social-ecological system from these |
| Take a long-term perspective | Because many landholders must deal with short-term problems and may be unaware of historical influences or the need for long-term planning |

including incentive schemes. Local institutions will be especially aware of tenure issues and the location of any local land disputes. They may also be in a position to help resolve many of these (Elmqvist *et al.* 2007, Lebel and Daniel 2009). Local institutions are also more likely to be effective at minimising 'land grabs' of the type discussed in Chapter 1.

Because of this diversity of tasks a wide variety of institutions are likely to be needed at the local level. Some of these will be concerned with government activities but others will be largely based on community activities. The latter might include grower's organisations concerned with managing the newly established forests, marketing organisations concerned with selling forest products or groups concerned with controlling wildfires. Some of these could be built onto pre-existing local institutions such as farmer groups and NGOs may have a particular role to play in helping these existing groups broaden their activities. Buchy and Hoverman (2000) suggest community participation in decision making can stimulate a learning process within the community by increasing awareness of collective responsibilities and this should be seen by professional agencies as a major asset rather than as a threat. This is because it reduces reliance on a top-down flow of information. Indeed, van Oosten (2013) views such a learning process as crucial for reforestation at the landscape scale and believes that research is needed to: 'understand how processes of landscape learning are linked to larger policy networks, to what extent these could be formalized, how such more-or-less formalized arrangements would look, and which are the key factors and preconditions for their success.'

Ostrom (2010) referred to this arrangement of having multiple bodies or institutions operating at different scales as a polycentric system of governance and argued it was especially useful for managing environmental problems. The role of polycentric systems of governance in building trust and social capital is discussed further by Biggs *et al.* (2012). Each individual institution will have a defined geographic area of interest across which it has the authority (and resources) to initiate and coordinate reforestation activities.

If institutions are to be polycentric and if local institutions are to be effective then ways must be found to devolve power and decision-making authority from national bodies to the local level (or from existing local institutions who might see the new body as being a competitor for political influence). An example of a typology of participatory decision making showing the potential stages in this process of devolution is outlined in Box 8.3 and others are discussed by Ross *et al.* (2002). Many traditional forestry departments find it difficult to believe that untrained local communities will be able to manage reforestation more successfully than themselves and are reluctant to move far down this 'ladder' of responsibility (it should be added that the reluctance to relinquish power to community groups is something that affects many government bureaucracies other than just forestry departments). Moreover, national bodies may be more interested in rapid decision making than in social justice. The result is that so-called local institutions supposed to facilitate reforestation at a local level can be composed of hand-picked, compliant members or a political elite only concerned with their own self-interest while real decision-making power remains with central government agencies.

**Box 8.3 Typology of participation showing the ways governments might involve people in the management of rural development projects such as reforestation**

*Manipulative participation*

Participation is simply a pretence. There are 'people's' representatives on official boards but these are unelected and have no power.

*Passive participation*

Representatives are told what has been decided, sometimes after it has already happened. It involves unilateral announcements by decision-makers paying minimal attention to people's responses. The information being shared belongs only to external professionals.

*Participation by consultation*

People participate by being consulted or by answering questions. External agents define the problems and information gathering processes. This form of consultation does not concede any share in decision making, and professionals are under no obligation to consider people's views.

*Participation for material incentives*

People participate by contributing resources, for example labour, in return for food, cash or other material incentives. Farmers may provide the fields and labour, but are involved in neither experimentation nor the process of learning.

*Functional participation*

Participation is seen by external agencies as a means to achieve project goals and to reduce costs. People may participate by forming groups to meet predetermined objectives related to the project. Such involvement may be interactive and involve shared decision making, but tends to arise only after major decisions have already been made by external agents.

***Interactive participation***

People participate in joint analysis, development of action plans and in the formation or strengthening of local institutions. Participation is seen as a right, not just the means to achieve project goals. The process seeks multiple perspectives and makes use of systemic and structured learning processes. Groups have a stake in maintaining structures or practices because they are in control of local decisions and determine how available resources are used.

***Self-mobilisation***

People participate by taking the initiative, independently of external institutions, to change systems. They develop contacts with external institutions for the resources and technical advice they need but retain control over how these are used. Self-mobilisation can spread if governments and NGOs provide an enabling framework of support. Such self-initiated mobilisation may or may not challenge existing distributions of wealth and power.

Source: Bass *et al.* 1995.

Local institutions may need some assistance to increase their capacity to contribute to undertake activities such as reforestation in ways that contribute to national goals as well as improving local livelihoods. It is important that this assistance is provided since local institutions are the places where many of the conflicts between local aspirations and the larger 'big-picture' vision of the national institutions must be resolved. But stability in both the policy settings and in staffing arrangements is crucial. Siregar *et al.* (2012) give an example of how a complex reforestation scheme in Indonesia involving multiple stakeholders took some time to negotiate and plan but was effectively terminated not long after it eventually started because the high government staff turnover meant the local institutions could not be sustained.

***Representativeness***

A second difficulty to be faced in the development of appropriate institutions is the identity of those being represented. Ideally it should involve representatives of the full range of local interests (e.g. large and small landholders, men and women, industry, NGOs, different ethnic groups and local government organisations) as well as representatives from key national or regional agencies. It may be difficult to identify appropriate representatives in places where customarily owned land once controlled by clan leaders is being privatised. All representatives would have an equal voice in the group's decision making and care may be needed to avoid dominance by a local political elite (Shrestha and McManus 2008). The obvious advantage of this is that landholder representatives can identify preferred types and locations for reforestation while the government representatives are able to provide technical and market information to the group and perhaps assistance in accessing funding at the national level or from external sources. They could also ensure the community was aware of national land use and reforestation goals so that local activities supported or were at least consistent with these. Institutions are more likely to be regarded as legitimate and be able to resolve conflicts if their members are truly representative of the community and its interests.

Sometimes it may be preferable to build on an existing local body but in others it may be necessary to create an entirely new institution. This can be difficult. A review of a number of integrated landscape planning initiatives in Africa by Milder *et al.* (2014) noted that it was sometimes hard to involve the private business sector in multi-stakeholder processes because they regarded these as being too time consuming. The absence of the private sector (or the absence of other politically powerful stakeholders) increases the risk that the institution will be by-passed and that these groups will seek to achieve their objectives by other means. Likewise, it may be difficult to persuade small farmers to give up their time to travel and attend meetings with people they perceive as their social superiors.

### *Monitoring and feedback*

A key task of most institutions is that they develop arrangements to monitor the effects of the policies being implemented to ensure they are generating the desired outcomes and to take account of any ecological, social or economic changes that may be occurring. Are incentive schemes effective in ensuring that degraded or eroding lands are being reforested? Are the costs and benefits of reforestation being shared? Is reforestation impacting adversely on food production? Monitoring can show whether the policy settings adopted are functioning as intended (including, in particular, those involving financial incentives) or whether these policies need some modifications. Monitoring, and perhaps enforcement, will be particularly important to ensure that legally required reforestation is indeed being done and that the specified standards are being met.

## Institutional case studies

The following four case studies cover several different institutional arrangements. In each case the overall policy objective was to increase (or, in one case, to maintain) forest cover. They differ in the way this was approached and in the degree of participation by various stakeholders.

**Case study 1: Village forestry councils in Tamil Nadu, India**

Many poor people in India rely on forest products for a significant part of the livelihood. But the fragmented pattern of land ownership means it is difficult for the state to reforest large areas. Reforestation policies have focussed instead on finding methods by which communities undertake reforestation in ways that suit local circumstances (Sreedharan and Matta 2006). In the state of Tamil Nadu much of the administration of local reforestation since 1997 is being undertaken by Village Forest Councils (VFCs). These link the State Forestry Department and village communities in a programme to enable small-scale reforestation across the state. Apart from emphasising the devolution of rights to local communities the initiative seeks to generate economic and social benefits (especially for the rural poor and women) and ensure that any new forests are managed sustainably. The programme began by the VFC identifying areas most in need of reforestation. Funds were then allocated to the VFCs to undertake reforestation using appropriate species chosen by the community. Though still in an early stage, the new forests have begun to provide fuel, fodder, green manure and non-timber forest products (NTFPs). The VFCs regulate access to the new forests and any harvesting operations. They are also responsible for resolving intra-village conflicts and protecting planted areas from goats, cattle, wildlife and other infringements. A key responsibility of the VFCs is to ensure that benefits are shared equitably. This is done by distributing these products amongst members and selling any surplus. Funds are then distributed amongst members except for 25 per cent which is placed in a village development fund. The scheme has also encouraged the development of a number of small but complementary self-help projects via revolving micro-credit funding. One of the reasons for these is to help limit illegal forest harvesting. Women are a particular focus of these since they were especially dependent in the past on forest resources such as fuel and other sources of income. The initiative involves over 1,300 VFCs representing 466,000 people and it covers 480,000 ha of new forests. It is hard to imagine any other institutional arrangements being able to facilitate and manage a reforestation programme of this scale and intricacy.

### Case study 2: Opportunities for participatory planning in a landscape dominated by privately owned planted forests

Finnish forest policy is based on the fact that most forest land in Finland is privately owned and much of this is contained in relatively small holdings; around 20 per cent of Finns own forests, either alone, or as a family member. This present day situation has arisen after a long period of time during which considerable changes in the way the forests are managed have taken place (Palo and Lehto 2012). The forests are managed on long rotations and are mostly replanted after harvesting. There are several levels of forest planning (Pappila and Pölönen 2012). These include the national level (reflected in Finland's National Forest Programme), the regional level (reflected in regional forest programmes), the landscape or local level (described in regional forest plans at the forest owner level and depicted in private voluntary forest management plans) and, finally, the operational level, seen in the forest use declarations submitted before cutting. The key institutions are those involved in planning at the national and regional level.

In recent years there has been increasing amounts of discussion concerning the need for forests to supply ecosystem services to the public as well as timber to the forest-based industry (Pappila and Pölönen 2012). This has led to discussions about how there might be greater public participation in the planning process. In theory, the public are able to express their opinions about the ways private forest management is practiced but this is unlikely to have much impact on the practices of individual forest owners. In fact there are currently no institutional arrangements that allow the public to be engaged in discussions about local level planning and management (for example, concerning clear felling near particular housing areas). Pappila and Pölönen (2012) argue the institutional arrangements of the Finnish forest planning system and the localised scale of operations makes it difficult for other members of the community who are not landholders (but who might consider themselves to be stakeholders) to participate in the management process.

## Case study 3: Large-scale reforestation in Brazil coordinated by an NGO

An interesting and complex large-scale reforestation project is now underway in Brazil and an unusual set of institutional arrangements have developed to coordinate this (Brancalion *et al.* 2013, Melo *et al.* 2013). The project area is in the Atlantic Forest region of Brazil. This has been extensively cleared and less than 12 per cent of the original forest remains with most fragments now being less than 50 ha. There is widespread agreement that some reforestation should take place and there are national and state laws requiring that this happen. Nonetheless, until recently, the institutional arrangements necessary to coordinate such a reforestation effort have been lacking. However, in the last few years, a coalition of NGOs, research institutions, private companies and government agencies has come together to form the Atlantic Forest Restoration Pact (Calmon *et al.* 2011). The group has more than 160 member bodies and aims to restore high diversity tropical forests to 15 million ha of degraded lands in the region by 2050. The group is governed by the Coordinating Board composed of 20 members representing the various categories of stakeholder and is governed by the Executive Secretariat. That is, government agencies are involved but do not drive the process.

Besides the legal policy framework in which it operates, the group is aware that many landholders in the area have only modest incomes so that reforestation must improve economic opportunities as well as generating ecological benefits. It is tackling this by having working groups covering reforestation techniques, information sharing, marketing, policy development and fund-raising. It operates by seeking to promote biodiversity conservation, job generation and income opportunities through the restoration supply chain (e.g. seed collection and nurseries), providing ecosystem services to the wider community and by establishing serious incentives for landowners to comply with the Forest Act. The group has developed monitoring protocols to track outcomes (Melo *et al.* 2013). It has also developed a website to communicate with the public (Accessed 16 May 2014. Available at www.pactomataatlantica.org.br/index.aspx?lang=en).

Some 80 projects covering 60,000 ha are underway (Melo *et al.* 2013) but these institutional arrangements and the group's future activities are dependent on the willingness of landholders to comply with the law and on the law being maintained in an unmodified state.

## Case study 4: Reforesting degraded farmland in Ghana

One aim of the national forest policy in Ghana is to increase forest cover, especially on degraded former forest land. Patterns of farming across the landscape make this difficult for the national Forestry Commission to achieve even on forest reserve land (Blay 2012). A recent development assistance project explored the use of a new local institutional arrangement to facilitate reforestation within and outside forest reserves. The project did this by creating a novel community-based institution which involved community heads (including chiefs and elders), district assembly officials, local community-based forestry organisations and farmer groups. This organised awareness-raising workshops and reforestation training sessions. Discussions also considered how benefits arising from reforestation would be shared and the responsibilities of different project participants. It was agreed that farmers would be the owners of the forests products together with the legal owners of the land (i.e. the traditional authorities through whom they have access to land), the Forestry Commission and communities living adjacent to the plantations. Each would share in the benefits according to their contribution. Thus the farmers were to do most of the planting and maintenance, the Forestry Commission was to arrange seedlings, tools and training, the landowners were to provide the land while adjacent villagers were responsible for protecting the new plantations from disturbances. A steering committee was elected with representatives from each stakeholder group. This was responsible for identifying the projects' goals and priorities while farmers identified the tree species to be used. Reforestation was carried out using a modified *taungya* system. Land was allocated to farmers on which trees were planted but crops could be grown with the trees until canopy cover occurred. Thereafter, project funding allowed assistance in other employment activities such as bee keeping.

The scheme was strongly supported by the chiefs and the broader community because it restored forests and provided some economic benefits. Some also valued the access to land provided by the project while farmers were especially pleased with the financial and non-financial incentives provided to them. The project initially involved ten communities and was able to restore 250 ha of forest within only a four year period.

Blay *et al.* (2008) believe the institutional arrangements developed by the project have been unique in Ghana and helped enhance people's commitment to the scheme.

Subsequent reviews of the way the modified *taungya* system is practiced in Ghana conclude it has been responsible for a major boost in reforestation across the country and that the institutional arrangement remains basically suitable although some fine-tuning concerning benefit sharing and social safeguards would be useful (Kalame *et al.* 2011, Ros-Tonen *et al.* 2013). Blay (2012) specifically noted the need for better institutional arrangements to keep track of the areas planted by individual farmers to ensure benefits are shared equitably and to clarify who is responsible for tree management once the cropping phase is completed and farmers move away.

## Conclusion

Policies to encourage forest restoration fall into three broad groups. These are policies that (i) remove impediments and enable reforestation when landholders wish to practice it, (ii) policies that actively encourage reforestation through the use of incentives or subsidies and (iii) policies for dealing with situations where reforestation is deemed necessary by either the state or the community.

The primary enabling policy is the granting of tenure and the right of landholders to benefit from any reforestation carried out on their land including the right to sell forest products. Other enabling policies involve the sharing of technical knowledge, assistance for forest growers to access markets and the removal of disincentives such as discriminatory taxes and charges. In many situations the implementation of such enabling policies may be sufficient for market forces to then attract landholders to undertake reforestation.

If market forces are not sufficient to encourage forest restoration then incentives may be needed to engage some landholders. Incentives can be indirect and non-financial or involve various forms of monetary assistance.

Incentives in the form of direct financial assistance may be especially useful if priority is being given to reforest certain locations within a landscape and if certain types of reforestation are needed. This is likely to be the case where the intention is to undertake Forest Landscape Restoration to create multi-functional landscapes.

A system of payments for the provision of the ecosystem services provided by new forests could be a powerful new form of incentive attracting landholders to reforest some of their land. However, there is still only limited experience in the development of these schemes for reforestation purposes and a number of problems remain to be clarified.

Overall, the best kinds of incentives, from a public policy perspective, are those that generate a social benefit rather than a private gain.

In some situations forest restoration cannot be left to market forces or to decisions made by individual landowners. Policies will be needed to define situations where reforestation is obligatory and here certain types of reforestation must be undertaken for the wider public benefit.

Policies may need to be revised from time to time to reflect changed circumstances or unexpected outcomes (but frequent policy changes or an unstable policy environment can be counter-productive).

A variety of institutions will be needed to administer policies and drive reforestation programmes. The best institutional arrangements are polycentric and involve complementary national and local institutions. They should also be representative of a variety of actors and sectors and be transparent in their operations.

# 9 Final discussion

## Introduction

The question raised in Chapter 1 was how forest restoration might be carried out on a large scale in ways that could improve both functional outcomes and human livelihoods? It was suggested that, despite a rising demand for food caused by rising populations, there was still likely to be a large area of degraded land and marginal farmland that might be reforested without adversely affecting future food security. Moreover, reforestation of this land could improve the sustainability of agriculture on other land and generate considerable benefits for landholders. Even so, the task remains a difficult one. Not all landowners are necessarily interested in undertaking reforestation, including those with degraded or marginal land. Nor is it clear just what type of reforestation might be most suitable. And finally, any large-scale reforestation project is likely to require a coordinated programme of activities involving landholders and a variety of other organisations and sectors (Le *et al.* 2012). This will be difficult to arrange since landholders and some of these other organisations may have contrary perspectives and aspirations.

However, there are good reasons for being optimistic. One is that a considerable amount of experience has now been acquired about how to reforest different types of land, including degraded lands. The benefits would be immense even if only some of this new knowledge could be applied. A second reason for optimism is that a wider variety of landholders are now beginning to consider reforestation on part of their land. Rather than being seen as unusual, reforestation is becoming seen as a worthwhile use of agricultural land in some circumstances. A third reason for optimism is because the whole 'idea' of reforestation is beginning to capture the imagination of governments and the public (Lovejoy and Ashton 2011). Evidence ranges from the numbers of landowners now growing trees on their land, to the variety of NGOs involved in forest restoration and the increasing numbers of national and international reforestation schemes now underway.

It was not many years ago when talk of large-scale reforestation schemes simply meant the establishment of big pulpwood plantations. Now it might mean schemes to protect watersheds in China, the development of transcontinental wildlife corridors across Central America or projects like the Great Green Wall for the Sahara and the Sahel Initiative (Chapter 5). And, some groups are planning not just reforestation but the 'rewilding' of large areas of land to recreate habitats for wildlife once common before humans began transforming these landscapes several thousand years ago (Chapter 3). Timber production is still an important reason for reforestation but there is a rapidly rising variety of other objectives and silvicultural approaches and many of these are being applied across extensive areas. Once they capture people's imagination, ideas such as these become powerful agents of change and supersede the idea of reforestation being driven largely by markets. FAO statistics now show that net forest losses are being reversed in more than 70 countries (FAO 2010). All this is occurring despite global financial crises, concerns over food security, widespread poverty and great uncertainty about the spatial impacts of future changes in global climates.

## Some issues for the future

Most of the main issues needing to be resolved if forest restoration is to take place on a large scale have been discussed in previous chapters. But there are several that deserve further mention.

### *Is it useful to set spatial targets?*

A common motivational technique is to set targets or goals against which progress can be measured. The technique is used in raising money for community infrastructure such as school buildings and it has also been used to promote national and international reforestation programmes. For example, the Five Million Hectare Reforestation Program in Vietnam (McElwee 2009), the Bonn Challenge to promote international forest restoration (IUCN 2012) and the One Billion Tree Program launched in Africa by UNEP in 2006 (UNEP 2011) all used the idea of a target to motivate the public. These types of targets can capture people's imagination and lead to substantial gains in forest cover. In the case of the One Billion Tree Program, the scheme was said to have easily reached its goal and exceeded 12 billion trees by 2011. It also appears to have prompted reforestation in a large number of countries far beyond its starting point in East Africa (UNEP 2011).

A particular advantage of large global goals like these is that they can help re-enforce national programmes. They also help force planners to reconcile the needs of agriculture and forest conservation. Yet there are reasons for caution.

- Sometimes the race to reach a spatial target can be at the expense of quality; trees are planted but subsequently die because there are no funds to maintain them. (If they are replanted they may be counted a second time!)
- In the race to achieve the target and spend the allocated funds on time, the seedlings used are simply those most readily available and may be quite inappropriate for the types of sites being planted (Cao *et al.* 2011, Xu 2011).
- Broad spatial targets may not tackle some of the areas most in need of reforestation unless this is specified. Indeed, in the rush to achieve the target area, some of the more difficult or costly areas may be deliberately excluded.

Targets have been valuable devices to focus public attention on reforestation but targets or 'grand visions' can have adverse consequences if they are highly prescriptive and the vision is not widely shared. This means that any target setting should follow a sustained period of information sharing and public education during which landholders and the wider community are provided with some context and feedback is sought. In this way the need for reforestation can be explained and the benefits of reforestation more widely understood. Progress towards the target should also be carefully monitored (and the results publicised) to ensure the underlying objective is being achieved. Funds should be set aside and available for corrective action if this is needed.

### *Who should be responsible for implementing large-scale forest restoration in future?*

Most land use decisions are made by individual landholders although they can be influenced by the policy framework in which they live. In that sense, nobody is 'in charge'. But perhaps the most impressive examples of large-scale reforestation have been when governments have taken the lead in developing reforestation techniques and implementing these on a large-scale. Examples include projects in China, Vietnam and South Korea although there have also been examples of government sponsored plantation failures such as the 'Green Dam' of Algeria (Chapter 2).

Some governments may be willing to play this coordination role in future but many are likely to prefer doing so in partnership with others including large timber companies and small farmers in order to share costs. Corporate plantation owners are highly visible and their previous role in reforesting large areas is well known and the collective importance of smaller landholders in reforesting large areas is becoming recognised (Del Lungo *et al.* 2006). But, in recent years, perhaps the most innovative contributions to thinking about reforestation have come from NGOs (e.g. ideas about Ecological Restoration, ecological corridors and the rewilding movement).

There are clear advantages in involving all of these stakeholders because each has something special to contribute. Any future policy framework should ensure this is the case. The way in which this might be done when reforestation is being implemented on a particular landscape was discussed in Chapter 7 while the institutional arrangements outlined in Chapter 8 suggest an overall framework that might be used.

### *How critical is devolution and governance?*

Foresters and ecologists interested in restoring forests often assume that technical issues are the main constraint. They are undoubtedly important but rarely as important as some of the social and political constraints. Landholders are unlikely to participate in reforestation unless they believe they will benefit in some way. Those believing they will be disadvantaged may actively seek to destroy new forests by lighting fires or not controlling grazing stock. Devolution and forms of governance lie at the heart of these issues because they involve the allocation of land, the rights of landholders and the overall process of policy development and decision making by governments.

Much large-scale reforestation is concerned with creating a public good but it can be at the expense of certain individuals. Governments and industry can form a powerful reforestation duopoly but unless care is taken they can dominate and perhaps even take over markets for ecosystem services including those such as the future carbon markets. In so-doing, smallholders and customary landowners can be pushed aside, supposedly for the sake of efficiency and the public good. On the other hand, small landholders often lack the financial or technical resources needed to establish the types of multi-species forests needed to generate ecosystem services. Nor do they have the organisational capacity to implement reforestation on a large scale.

The best way of balancing these public and private interests is to ensure landholders and other stakeholders have a role in policy development and institutional decision making and that this is not seen as being solely the prerogative of government (Le *et al.* 2012). In particular, private landholders should be part of the land use planning process so that local knowledge can be used to fine-tune decisions being made in the national interest and so that any decisions made will be politically acceptable to the community living in a particular area (Chapter 8). Recent experience of so-called land grabs (discussed in Chapter 1) show what can happen if local communities are ignored. By contrast, the stunning changes in Niger and Nepal discussed in Chapter 3 are wonderful examples of what can happen when local communities are empowered to undertake reforestation. Recent experiences in Brazil where a large and complex reforestation programme is underway are likely to provide lessons for those undertaking large-scale reforestation projects in other parts of the world (Melo *et al.* 2013).

***Can future forest restoration simultaneously generate conservation benefits and improve livelihoods?***

Many large forest restoration projects seek to generate conservation benefits and, at the same time, improve livelihoods. Those living in poverty are often seen as a key target group. But is this realistic? Can forest restoration simultaneously improve conservation and livelihoods (measured as increased income and assets or reduced exposure to various risks)? The significance of the question arises from the fact that many communities living in degraded landscapes are at the poorer end of the socio-economic spectrum (i.e. low cash income, fewer economic alternatives, greater exposure to risk). Should the primary goal be to overcome degradation or, instead, deal with the economic situation?

Improving conservation and livelihoods outcomes are not necessarily complementary objectives and might be in conflict in some situations. On the other hand, certain types of reforestation may lead to improvements in each. There appear to be four situations:

*Forest restoration improves conservation outcomes but at the expense of poorer people*

In this case, degraded land is reforested, perhaps even using Ecologically Restoration, but it is carried out by displacing people who regard themselves as customary landowners. Their ownership claims are not recognised and no compensation is offered. Those undertaking the reforestation might argue that the national benefit is being served (e.g. watersheds and conservation areas are being protected) and that this outweighs any disadvantage the local community might suffer.

*Forest restoration damages conservation values and fails to improve livelihoods*

Extensive industrial timber plantations are established on land having certain conservation values. This may be because the land has species-rich natural regrowth or because it includes small patches of residual natural forest. All these natural forests are cleared and replaced by extensive plantation monocultures. Again, the claims of customary land owners are disregarded and they fail to even benefit from employment by the plantation owners.

*Forest restoration damages conservation outcomes but improves livelihoods*

Industrial timber plantations are established on land once having certain conservation values. In this case, the land being cleared is owned by both timber companies and smallholders whose land rights are recognised. The latter are able to form partnerships with timber companies and benefit by selling timber grown in small plantations on their land. Livelihoods are improved by the income from selling timber trees and from the social infrastructure (roads, schools, health clinics, etc.) provided by the company to the community.

*Forest restoration improves conservation outcomes and improves the livelihoods of poorer people*

A variety of different types of reforestation are carried out on cleared former agricultural land. Much of the land is owned by smallholders and these people are able to benefit from the sale of forest products, from payments they receive for certain ecosystem services or from protecting natural regrowth. Some may also benefit from ancillary activities such as nurseries or tree-planting businesses. Reforestation increases landscape heterogeneity and the capacity for wildlife to cross the landscape while the commercial benefits of the new forests reduce pressure (e.g. hunting, logging) on residual patches of remaining natural forest. Not all the community benefits are financial. Community members also benefit by being empowered to participate in decision-making processes such as land use planning and by the improved relationships with government agencies brought about by their participation in the reforestation programme.

In short, the possibilities range from lose-lose to win-win (or, perhaps more realistically, lose-less and win-more). Making choices and trade-offs is difficult (McShane *et al.* 2011). But it is easier if it is being done at a landscape scale rather than at a particular site and it should not be beyond the wit of policy makers to move away from damaging conservation values and not improving people's livelihoods towards outcomes where there are conservation gains and poverty is reduced.

Forest restoration is not necessarily the best way of improving conservation outcomes in every situation. It obviously also depends on the type of reforestation carried out and its landscape context. For example, some plantations will not be attractive to certain wildlife needing specialised habitats and there may be places where certain wildlife species prefer non-forest habitats such as grasslands. Likewise, not all poorer people will benefit from reforestation. Some may find the benefits from reforestation are too slow to arrive while those with small landholdings may find the opportunity costs simply too great for them to plant any trees other than perhaps a few fruit trees. In such cases other, more specifically-targeted, ways to overcome poverty will have to be developed.

## *How to mobilise funding for large-scale forest restoration*

It is clear that commercial tree plantations can often be very profitable. Likewise, analyses have shown that the ecosystem services provided by reforestation can also be cost effective (Bullock *et al.* 2011, De Groot *et al.*

2013). Is the implication of this that future reforestation can be safely left for the market to resolve? In most situations the answer is probably no. While the market is potentially a powerful mechanism for promoting reforestation there is limited evidence that it has been always able to do so, especially in the locations where reforestation is most needed. Further, market signals are often delayed. For example, there was no sign of market forces enticing private enterprise into major reforestation projects in the early years of the twentieth century when the modern reforestation era began. In that case it was government forestry agencies who initiated plantation establishment and it was only after a long lag period that market signals eventually persuaded private enterprise to become involved.

Market opportunities may attract investment by landholders, including small landholders, once the necessary policy and institutional arrangements are in place (Program on Forests 2011). Ways in which government incentives and subsidies (including the allocation of land) might also encourage reforestation were discussed in Chapter 8, but how else might funds be mobilised? There are several ways in which this might be done:

- Funds are provided by landholders who are legally obliged to undertake reforestation in certain areas. Examples include post-mining rehabilitation or compensatory biodiversity plantings to offset losses caused by environmental damage in certain areas. An unusual example of obligatory reforestation applies in Brazil where landholders are required to maintain a certain proportion of the landscape under forest cover depending on the location and size of their landholdings (Economist 2012).
- Funds are raised from levees imposed on industrial users of natural forests for the specific purpose of assisting reforestation on their former logging leases. Such a levee applies in Indonesia and raised US$ 5.8 billion in the 20 years following 1989 (Barr *et al.* 2009).
- Funds are generated through joint ventures between landholders and governments that encourage private investment in situations where it might not otherwise occur. In these situations a government agency provides conditional finance to initiate a project in the expectation that the landholder provides further funding to complete the project. The financial profits arising at the end of the rotation are shared. The attraction of such a scheme is that risks for smallholders can be reduced and the marketing power of individual small forest owners is enhanced while, from a government's perspective, reforestation on private land is encouraged at a moderate cost. Similar schemes can operate between smallholders and larger timber corporations (e.g. Obidzinski and Dermawan 2010).
- A government supplies funds to address specific problems (e.g. watershed protection) on non-government land. These funds might be conditional grants or loans to landholders and may specify the location

where reforestation is to occur and the type of reforestation to be used to achieve particular outcomes. The funds might be provided directly or through conservation auctions (Chapter 7).

- Funds are provided by international development agencies or NGOs to address specific environmental problems or foster rural development. Examples of potentially large funding sources include the Global Environmental Fund (Accessed 16 May 2014. Available at www.globalenvironmentalfund.com) and the World Bank's BioCarbon Fund (Accessed 16 May 2014. Available at www.wbcarbonfinance.org) but there are a variety of other sources varying in size and in geographic focus. A potential new funding vehicle to support smallholder reforestation activities is the Landscape Fund which seeks to diversify risks and markets and link financial investors with credible and sustainable investment opportunities. A group of researchers, led by the Center for International Forestry Research (CIFOR), are currently testing a pilot project (Wollenberg 2014).
- Funds are generated by new markets that reward growers able to establish forests supplying particular ecosystem services. The most obvious example is the market for carbon storage (Alexander *et al.* 2011). Funds generated by the carbon markets or for other services such as watershed protection or biodiversity may be sufficient to overcome opportunity costs and make reforestation in certain locations attractive to landholders (Accessed 16 May 2014. Available at www.ecosystemmarketplace.com).

There may be a variety of other small-scale funding sources available to community-based reforestation programmes. Durst *et al.* (2005) give examples of some of these in a series of projects undertaken in the Asia-Pacific region.

The fact there are a variety of potential funding sources has several consequences. One is that decisions are often made across a landscape in a piecemeal way without reference to other investments or interventions. It follows from this that there is likely to be considerable scope for strategic partnerships between different organisations having a common interest in some form of reforestation. This is not just to leverage more funds to promote reforestation (e.g. greater cooperation between groups engaged in agroforestry, rural development, water management or biodiversity conservation) but also to leverage better economic and functional outcomes across the landscape from individual investment decisions.

The variety of funding sources also raises the question of whether these will collectively support the apparent trend towards reforestation being undertaken to supply ecosystem services as well as timber or whether they will largely support just production forestry and so reverse this trend? Much will depend on the way the REDD+ negotiations play out and the development of a formal market for carbon stored in forests. A market that could be accessed by smallholders as well as larger organisations and that would fund reforestation to store carbon by requiring trees to remain intact for, say, 80 years or more could transform both the way reforestation is undertaken in future as well as its scale.

### *What are future research priorities?*

Expanding the scale of reforestation will necessitate further research in what might be thought of as traditional research areas such as how to raise seedlings in nurseries and how to match species with the sites available for planting. New forms of reforestation such as natural regeneration, direct seeding and Ecological Restoration also raise a variety of intriguing research questions. But there is a more fundamental series of problems that also need to be part of any future research agenda and they concern the resilience of these new forest ecosystems.

Forest restoration is being undertaken in an environment where a variety of changes – ecological, social and economic – are underway. Underlying all of these is the worrying prospect of future climate changes. All these changes will affect whether the new forests survive and whether they will produce the benefits expected. Some forests are more likely to tolerate and withstand these future changes better than others. For example a plantation monoculture is likely to be more sensitive to environmental changes or insect pests than a multi-species forest containing a variety of species having different traits and tolerances. Likewise a monoculture producing a single product for a single market will be more sensitive to changes in market prices than a forest able to produce a variety of goods and services. The capacity of a forest to tolerate disturbances and change is referred to as its resilience and the topic is one that has attracted considerable attention in recent years (Gunderson and Holling 2002, Walker and Salt 2006, Biggs *et al.* 2012).

In the case of newly established forests there are four ways in which reliance might be enhanced (Table 9.1). These involve ecological, economic, social and institutional elements. In each of these four elements resilience is improved when diversity is increased. In each case scale is also important and resilience acts at local as well as large scales.

Ecological resilience is improved by moving away from monocultures and establishing multi-species plantings involving a variety of functional types (i.e. tolerant of a variety of environmental conditions). This diversity can be achieved at both a site and a landscape level. It is obviously enhanced by retaining areas of original forest where ever this is possible. Economic resilience is also favoured by seeking ways of diversifying market opportunities at local as well as regional or larger scales. Social resilience is concerned with making communities more self-sufficient and less dependent on external sources of advice or information. This means there is more emphasis on growers being part of the experimentation process to increase the knowledge base and share this knowledge. Likewise, greater

*Table 9.1* Building resilience in large-scale forest restoration programmes

| *Forms of resilience* | *Developed by* |
|---|---|
| Ecological | Fostering the use of multi-species plantings<br>Enhancing diversity at landscape level as well as at a site level<br>Conserving patches of residual natural forest and regrowth forest<br>Planting a diversity of species<br>Including a variety of functional types<br>Fostering connectivity between forests across the landscape |
| Economic | Encouraging the development of multiple markets for a wide variety of goods and services<br>Seeking markets at various scales (local, regional, international)<br>Providing information on markets and marketing<br>Promoting new enterprises that diversify economic opportunities |
| Social | Broadening participation by relevant stakeholders in the reforestation process and encouraging feedback between each<br>Fostering local innovation and experimentation by growers<br>Promoting networks to foster collaboration and share knowledge |
| Institutional | Promoting polycentric governance systems at different scales to develop policies and institutions for collective action<br>Establishing policies to create, store and share information<br>Developing economic policies to encourage prescribed forms of reforestation (e.g. via taxation, incentives, compensation)<br>Developing appropriate legal and regulatory frameworks<br>Establishing monitoring systems to assess feedback on ecological, economic and social resilience |

social resilience means developing networks and building trust. Growers are then able to access and share information about market opportunities and marketing so they are more self-sufficient and can adapt more readily to unexpected events. Finally, improved institutional resilience means having a variety of inter-connected institutional arrangements for policy development, information sharing and decision making in contrast to having an entirely top-down structure. From this resilience perspective, Forest Landscape Restoration is not the creation of a series of new forest ecosystems but, rather, the development of new socio-ecological systems.

Ways of enhancing these various forms of resilience are now being developed but the need to do this comprehensively by creating new socio-ecological systems rather than simply new plantations opens up a new research agenda. For example, just how much resilience is needed? And are there trade-offs to be made between the various forms of resilience such that, say, increased social resilience will compensate for a reduction in ecological resilience?

Many plantations will continue to be established to produce various kinds of forest products, including speciality timbers, but the need to optimise timber productivity is decreasing while the need to create forests able to generate various ecosystem services is rising. Doing this while also tackling the four aspects of resilience outlined in Table 9.1 will have to be done with the assistance and collaboration of specialists knowledgeable about rural sociology and economics and not just by ecologists or silviculturists. It has always been difficult to get collaboration between, say, conservation biologists, hydrologists and foresters. Each group has their own research priorities, their own jargon, organise their own journals and go to their own conferences. This means it will be even harder to develop research programmes that also include anthropologists, sociologists, development specialists and economists. And yet that is what will be needed if robust forms of reforestation are to be implemented on a large scale in ways that are adapted to future environmental conditions.

## The consequences of scale

What other changes are likely to occur as the areas of new forest increase? Economies of scale are a well-known phenomenon in commercial plantation development but scale matters in other ways as well. For example, biologists have been intrigued for some time by changes in organismal structures and physiology as biota increase in size. Many biological traits can be approximated by mass raised to power of ¼. It is believed that this reflects network constraints controlling energy and resource distribution through an organism's body (West and Brown 2005). In recent years attention has turned to investigating whether there are scaling rules which apply to landscapes. It now appears that landscapes often exhibit fractal-like properties, especially stream networks. For example, Zhang *et al.* (2011) found forest expansion in parts of China also exhibited fractal growth properties. These properties may also affect the abundance, distribution and diversity of biota found within landscapes (Ritchie and Olff 1999, Zhang *et al.* 2011).

Studies of scaling rules in cities have found outcomes concerning social developments and innovation that may also be relevant to large-scale reforestation. These studies have concluded that, not unlike biological systems, many properties of cities are power-law functions of population size and are related to the use of urban material infrastructure (Bettencourt *et al.* 2007). In particular, increasing city size was associated with greater levels of innovation and wealth creation but, conversely, there were economies of scale associated with providing infrastructure and service delivery. Similar changes might be expected as the number of people actually engaged in reforestation increases. In this case, the level of expertise is likely to grow as knowledge is shared through social networks. Likewise, the costs of reforestation should fall as infrastructure is developed.

The importance of reforestation scale for such factors as biodiversity conservation, hydrology and economics was discussed in Chapter 6. The existence of similar types of scaling relationships in a diverse range of physical, biological and social systems is intriguing and the results of further investigation could have implications for the way in which large-scale reforestation programmes are designed (Zhang *et al.* 2011).

Lastly, might extensive reforestation provoke a negative reaction? Some traditional farming districts are already sceptical about the spread of industrial reforestation being carried out at the expense of crops and pastures (Schirmer and Bull 2014). The impact of ever more extensive areas of forest could increase this hostility. Much depends on how it is done. Large numbers of small plantings on individual landholdings are unlikely to be controversial. More extensive plantings on degraded or poorer quality lands are also unlikely to be contentious if a diverse range of reforestation methods other than just industrial-style plantation monocultures are used. These are likely to be even more acceptable if reforestation leads to new economic activities such as reforestation businesses or eco-tourism that provide benefits to local communities.

## What happens next?

There appears to be a significant amount of reforestation now underway through natural regeneration and planting despite competition from the agricultural sector and other potentially inhibitory factors discussed in Chapter 1. But it is difficult to predict just how much will take place over the longer term. Some are now referring to the present era as the Anthropocene and define this as the period when the impact of humans on the biosphere rivals that of geophysical processes (Steffen *et al.* 2011). They point to a series of worrying trends in atmospheric composition, ocean acidification, water use, pollution and biodiversity losses that are beginning to push the global system processes towards critical thresholds from which recovery may be difficult. Deforestation has been responsible for some of these changes and is the reason why reforestation is now being promoted as a means of reversing the changes. What might happen as this process unfolds?

### *Global climate change*

Perhaps the most worrying of the changes underway in the Anthropocene is global climate change. A number of studies have explored the possible consequences of various future climate scenarios on the world's natural vegetation. Not surprisingly the results have varied depending on the models used and assumptions made (Sitch *et al.* 2008, Ostberg *et al.* 2013). The overall conclusion is that global warming is likely to have profound impacts on terrestrial ecosystems. This will be difficult to manage and that no part of the world will be immune from changes unless warming is kept below a two degree rise above pre-industrial levels. But even then, the changes will be substantial.

There is considerable international activity seeking to find ways of minimising these impacts. In the context of forest ecosystems, this activity aims to prevent further releases of carbon into the atmosphere by limiting deforestation and to remove some atmospheric carbon by increasing reforestation. There have been many studies exploring the capacity of forests to sequester and store carbon (Pan *et al.* 2011, Richter and Houghton 2011, Houghton 2013). The beneficial effect this might actually have on global climates is less easily studied but some intriguing evidence comes from the medieval period when a reduction in atmospheric carbon dioxide was accompanied by a decline in global temperatures. Ruddiman (2007) and Dull *et al.* (2010) note this occurred between 1500 and 1750 at a time when diseases introduced by European settlers colonising the Americas were causing the deaths of up to 95 per cent of the pre-Columbian population. These deaths caused a decline in agricultural activity and allowed substantial forest regrowth. Similar diseases depopulated parts of Europe in the mid-1300s when the Black Death wiped out a third of the population. Again the decline in agricultural activity allowed natural forest regrowth to flourish. Ruddiman (2007) and Dull *et al.* (2010) argue this regrowth sequestered and stored large amounts of atmospheric carbon and was at least partially responsible for the global temperature decline between 1200 and 1700. The effect lasted until human populations in both continents recovered and began clearing forests once more. Of course it is difficult to know in retrospect the precise cause of the medieval cooling period. However the experience suggests what might happen if net reforestation was able to occur on a sufficiently large scale.

Most prudent forecasters only explore possible changes over relatively short periods and commonly less than 100 years. But two bolder forecasts have attempted to describe vegetation patterns over the next 300 years. One of these explored the impact of changing climate on natural ecological communities and found that, averaged across a number of models, natural plant communities on 49 per cent of the earth's surface would be affected by the end of the twenty-first century and 37 per cent of the world's ecosystems would undergo biome-scale changes over this period. By the end of the twenty-third century, plant communities on about 55 per cent of the earth's surface would be affected (Bergengren *et al.* 2011). These suggest significant changes in the lifetimes of many of the trees being planted now with many changes likely to make environmental conditions quite unsuitable for some of these species.

The second forecast built upon this and explored what might happen to global forest cover over the next 300 years if large-scale reforestation continues (Blaser and Gregersen 2013). In contrast to dynamic global vegetation models that rely on the responses of certain plant functional types to predict future trends, they based their forecast on the assumption that humans are economically responsive to the evolving changes and will react accordingly. Further, they assumed political systems will be capable of organising collaborative global systems of governance to manage the changes that take place. While recognising that vegetation patterns in particular locations will change they suggested forest cover will increase from about 3,800 million ha at present to 5,000 million ha by 2313 provided the increase in mean global temperature does not exceed four degrees (Table 9.2). To put this in context, there were probably around 2 billion ha of forest 14,000 years ago at the end of the last glacial period and about 9 billion ha of forest 7,000–9,000 years ago in the mid-Holocene when conditions had improved and agriculture was first beginning to spread.

They considered four types of forest (Table 9.2). The first type is the existing areas of contiguous primary or old-growth forests. These they expect to shrink because of climate change. The second is what they term the forest landscape mosaic which includes small patches of residual forest and natural regrowth scattered across an otherwise cleared agricultural landscape. These, too, will shrink. The third category includes well managed natural and semi-natural forests as well as plantation forests. They estimate the total area of these will triple over the time period to 3,000 million ha with many being used for intensive production. Of this total area, about 2,000 million ha will be planted forest. Finally, they identified areas of agroforestry and urban forestry that are also likely to increase in area from <100 million ha to >500 million ha.

They reasoned that all natural forests will pass to the protected area network and that degraded lands will become too valuable to be left unused. Further, they forecast that food needs will be met by more intensive forms of agriculture and that the competition between forestry and agriculture for land will not be as great as some fear.

Who knows if Blaser and Gregersen (2013) will be right? Three hundred year time periods may be unremarkable in terms of tree growth but they are beyond the time frame of most resource managers who are usually concerned with more immediate issues. Nonetheless, these types of forecasts are timely because they remind us that times are changing and that the old

*Table 9.2* Forms of forest cover likely to be present in 300 years' time (Blaser and Gregersen 2013)

| *Type of forest* | *Present* | *In 2313* |
|---|---|---|
| Primary or natural forest | <800 | <500 |
| Forest landscape mosaic* | >1,900 | >1,000 |
| Planted and semi-natural forest | *c.* 1,000 | >3,000 |
| Agroforests and urban areas | <100 | >500 |
| Total | 3,800 | 5,000 |

certainties are slipping away. The environments in which trees will mature are likely to be quite different to those in which they are now being planted. This is something no forester has had to face in the past. Likewise, the future economic and political environments are also likely to be different in ways that are not yet clear.

These concerns do not alter the need to undertake reforestation on a large scale in the immediate future. Indeed, the need to help conserve biodiversity, prevent erosion and store carbon as well as provide a host of other ecosystem services demands that we do. But it does mean that more sophisticated methods of restoring forests will be needed. In addition, those undertaking reforestation, for whatever immediate purposes, must be aware that these purposes may have to change and therefore that the types of reforestation may have to change as well. In future the task of assessing feedback and adapting to change will have to be at the core of reforestation methodologies.

Finally, it may be fitting to end with an anecdote concerning the French Marshal Lyautey (1854–1934), well known in his time as a colonial administrator. The Marshall asked his gardener to plant a favoured tree. The gardener reminded the Marshal that it was a very slow-growing tree and would take 100 years to mature. Exasperated, the Marshall responded "In that case there is no time to lose, plant it this afternoon".

# References


Abernethy, B., and I. D. Rutherfurd. 1998. Where along a river's length will vegetation most effectively stabilise stream banks? *Geomorphology* 23:55–75.

Adjers, G., S. Hadengganan, J. Kuusipalo, K. Nuryanto, and L. Vesa. 1995. Enrichment planting of dipterocarps in logged-over secondary forests – effect of width, direction and maintenance method of planting line on selected *Shorea* species. *Forest Ecology and Management* 73:259–270.

Aerts, R., and O. Honnay. 2011. Forest restoration, biodiversity and ecosystem functioning. *BMC Ecology* 11:1–10.

Agestam, E., M. Karlsson, and U. Nilsson. 2006. Mixed forests as a part of sustainable forestry in southern Sweden. *Journal of Sustainable Forestry* 21:101–117.

Aide, T. M., and H. R. Grau. 2004. Globalization, migration, and Latin American ecosystems. *Science* 305:1915–1916.

Aide, T. M., J. K. Zimmerman, M. Rosario, and H. Marcano. 1996. Forest recovery in abandoned cattle pastures along an elevational gradient in northeastern Puerto Rico. *Biotropica* 28:537–548.

Alcantara, C., T. Kuemmerle, A. V. Prishchepov, and V. C. Radeloff. 2012. Mapping abandoned agriculture with multi-temporal MODIS satellite data. *Remote Sensing of Environment* 124:334–347.

Aldhous, J. R. 1997. British forestry: 70 years of achievement. *Forestry* 70:283–291.

Alexander, S., C. R. Nelson, J. Aronson, D. Lamb, A. Cliquet, K. L. Erwin, C. M. Finlayson, R. S. De Groot, J. A. Harris, and E. S. Higgs. 2011. Opportunities and challenges for ecological restoration within REDD+. *Restoration Ecology* 19:683–689.

Allen, W. 2001. *Green Phoenix: Restoring the Tropical Forests of Guanacaste, Costa Rica.* Oxford University Press, Oxford, UK.

Alvarez, N. L., and L. Naughton-Treves. 2003. Linking national agrarian policy to deforestation in the Peruvian Amazon: a case study of Tambopata, 1986–1997. *Ambio* 32:269–274.

Améztegui, A., L. Brotons, and L. Coll. 2010. Land use changes as major drivers of mountain pine (*Pinus uncinata* Ram.) expansion in the Pyrenees. *Global Ecology and Biogeography* 19:632–641.

Ammer, C., E. Bickel, and C. Kolling. 2008. Converting Norway spruce stands with beech – a review of arguments and techniques. *Austrian Journal of Forest Science* 125:3–26.

Andersson, E., S. Brogaard, and L. Olsson. 2011. The political ecology of land degradation. *Annual Review of Environment and Resources* 36:295–319.

Andréassian, V. 2004. Waters and forests: from historical controversy to scientific debate. *Journal of Hydrology* 291:1–27.

Angassa, A., and G. Oba. 2008. Herder perceptions on impacts of range enclosures, crop farming, fire ban and bush encroachment on the rangelands of Borana, southern Ethiopia. *Human Ecology* 36:201–215.

Angelsen, A., and D. Kaimowitz. 2001. *Agricultural Technologies and Tropical Deforestation.* CAB Publishing, Wallingford, UK.

Anon. 1999. *Integrated Report on The Multi-Storied Forest Management Project in Malaysia (1991–1999).* Forestry Department Peninsular Malaysia, Perak State Forestry Department, Japan International Cooperation Agency, Kuala Lumpur.

Anon. 2001. *Plantations for Australia: An Industry-Government Initiative for Plantation Forestry in Australia: the 2020 Vision.* Canberra.

Aronson, J., P. H. Brancalion, G. Durigan, R. R. Rodrigues, V. L. Engel, M. Tabarelli, J. Torezan, S. Gandolfi, A. de Melo, and P. Y. Kageyama. 2011. What role should government regulation play in ecological restoration? Ongoing debate in Sao Paulo State, Brazil. *Restoration Ecology* 19:690–695.

Arriagada, R. A., P. J. Ferraro, E. O. Sills, S. K. Pattanayak, and S. Cordero-Sancho. 2012. Do payments for environmental services affect forest cover? A farm-level evaluation from Costa Rica. *Land Economics* 88:382–399.

Arroyo-Rodriguez, V., E. Pineda, F. Escobar, and J. Benitez-Malvido. 2009. Value of small patches in the conservation of plant-species diversity in highly fragmented rainforest. *Conservation Biology* 23:729–739.

Audet, P., A. Gravina, V. Glenn, P. McKenna, H. Vickers, M. Gillespie, and D. Mulligan. 2013. Structural development of vegetation on rehabilitated North Stradbroke Island: Above/belowground feedback may facilitate alternative ecological outcomes. *Ecological Processes* 2:20.

Ausubel, J. H., I. K. Wernick, and P. E. Waggoner. 2013. Peak farmland and the prospect for land sparing. *Population and Development Review* 38:221–242.

Awanyo, L. 2007. A Janus-faced biodiversity change and the partiality of ecological knowledge in a world biodiversity hotspot in Ghana: implications for biodiversity rehabilitation. *Geoforum* 38:739–751.

Bacha, C. J. C. 2003. The determinants of reforestation in Brazil. *Applied Economics* 35:631–639.

——2006. The evolution of reforestation in Brazil. *Oxford Development Studies* 34:243–263.

Bae, J. S., R. W. Joo, and Y. S. Kim. 2012. Forest transition in South Korea: reality, path and drivers. *Land Use Policy* 29:198–207.

Baguette, M., S. Blanchet, D. Legrand, V. M. Stevens, and C. Turlure. 2013. Individual dispersal, landscape connectivity and ecological networks. *Biological Reviews* 88:310–326.

Bai, Z. G., D. L. Dent, L. Olsson, and M. E. Schaepman. 2008. Proxy global assessment of land degradation. *Soil Use and Management* 24:223–234.

Balvanera, P., A. B. Pfisterer, N. Buchman, J.-H. He, T. Nakashizuka, D. Raffaelli, and B. Schmid. 2006. Quantifying the evidence for biodiversity effects on ecosystem functioning and services. *Ecology Letters* 9:1146–1156.

Barlow, K., and C. Cocklin. 2003. Reconstructing rurality and community: plantation forestry in Victoria, Australia. *Journal of Rural Studies* 19:503–519.

Barr, C., A. Dermawan, H. Purnomo, and H. Komarudin. 2009. *Financial governance and Indonesia's Reforestation Fund during the Soeharto and post-Soeharto periods, 1989–2009.* Center for International Forestry Research, Bogor.

Barrow, E., and W. Mlenge. 2003. Trees as key to pastoralist risk management in semi-arid landscapes in Shinyanga, Tanzania and Turkana, Kenya. International Conference on Rural Livelihoods, Forests and Biodiversity, Bonn, Germany.

Bass, S., B. Dalal-Clayton, and J. Pretty. 1995. *Participation in Strategies for Sustainable Development.* International Institute for Environment and Development, London.

Bateman, I. J., A. R. Harwood, G. M. Mace, R. T. Watson, D. J. Abson, B. Andrews, A. Binner, A. Crowe, B. H. Day, and S. Dugdale. 2013. Bringing ecosystem services into economic decision-making: land use in the United Kingdom. *Science* 341:45–50.

Bauhus, J., P. Van der Meer, and M. Kanninen. 2010. *Ecosystem Goods and Services from Planted Forests.* Earthscan, London.

Beck, H., L. Bruijnzeel, A. Van Dijk, T. McVicar, F. Scatena, and J. Schellekens. 2013. The impact of forest regeneration on streamflow in 12 meso-scale humid tropical catchments. *Hydrology & Earth System Sciences Discussions* 10: 3045–3102.

Bellard, C., C. Bertelsmeier, P. Leadley, W. Thuiller, and F. Courchamp. 2012. Impacts of climate change on the future of biodiversity. *Ecology Letters* 15:365–377.

Benayas, J. M. R., J. M. Bullock, and A. C. Newton. 2008. Creating woodland islets to reconcile ecological restoration, conservation, and agricultural land use. *Frontiers in Ecology and the Environment* 6:329–336.

Benayas, J. M. R., A. C. Newton, A. Diaz, and J. M. Bullock. 2009. Enhancing of biodiversity and ecological services by ecological restoration: a meta-analysis. *Science* 325:1121–1124.

Benayas, J. R., A. Martins, J. M. Nicolau, and J. J. Schulz. 2007. Abandonment of agricultural land: an overview of drivers and consequences. *CAB Reviews: Perspectives in Agriculture, Veterinary Science, Nutrition and Natural Resources* 2:1–14.

Bengtsson, J., P. Angelstam, T. Elmqvist, U. Emanuelsson, C. Folke, M. Ihse, F. Moberg, and M. Nystrom. 2003. Reserves, resilience and dynamic landscapes. *Ambio* 32:389–396.

Benitez, P. C., I. McCallum, M. Obersteiner, and Y. Yamagata. 2007. Global potential for carbon sequestration: geographical distribution, country risk and policy implications. *Ecological Economics* 60:572–583.

Bennett, M. T. 2008. China's sloping land conversion program: institutional innovation or business as usual? *Ecological Economics* 65:699–711.

Bergengren, J. C., D. E. Waliser, and Y. L. Yung. 2011. Ecological sensitivity: a biospheric view of climate change. *Climatic Change* 107:433–457.

Berqvist, G. 1999. Wood volume yield and stand structure in Norway spruce understorey depending on birch shelterwood density. *Forest Ecology and Management.* 122:221–229.

Berthrong, S. T., G. Pineiro, E. G. Jobbágy, and R. B. Jackson. 2012. Soil C and N changes with afforestation of grasslands across gradients of precipitation and plantation age. *Ecological Applications* 22:76–86.

Bettencourt, L. M., J. Lobo, D. Helbing, C. Kühnert, and G. B. West. 2007. Growth, innovation, scaling, and the pace of life in cities. *Proceedings of the National Academy of Sciences* 104:7301–7306.

Bhagwat, S. A., K. J. Willis, H. J. B. Birks, and R. J. Whittaker. 2008. Agroforestry: a refuge for tropical biodiversity? *Trends in Ecology & Evolution* 23:261–267.

Biancalani, R., F. Nachtergaele, M. Petri, and S. Bunning. 2011. *Land Degradation Assessment in Drylands: Methodology and Results.* United Nations Food and Agriculture Organization, Rome.

Biggs, R., M. Schlüter, D. Biggs, E. L. Bohensky, S. BurnSilver, G. Cundill, V. Dakos, T. M. Daw, L. S. Evans, and K. Kotschy. 2012. Toward principles for enhancing the resilience of ecosystem services. *Annual Review of Environment and Resources* 37:421–448.

Birch, J. C., A. C. Newton, C. A. Aquino, E. Cantarello, C. EcheverrÃa, T. Kitzberger, I. Schiappacasse, and N. T. Garavito. 2010. Cost-effectiveness of dryland forest restoration evaluated by spatial analysis of ecosystem services. *Proceedings of the National Academy of Sciences* 107:21925–21930.

Blaikie, P., and H. Brookfield. 1987. *Land Degradation and Society.* Methuen, London.

Blaser, J., and H. Gregersen. 2013. Forests in the next 300 years. *Unasylva* 64:240.

Blay, D. 2012. Reforestation of deforested and degraded land in Africa. Pages 267–319 *in* J. A. Stanturf, P. Madsen, and D. Lamb, editors. *A Goal-Oriented Approach to Forest Landscape Restoration.* Springer, Dordrecht, Netherlands.

Blay, D., M. Appiah, L. Damnyag, F. K. Dwomoh, O. Luukkanen, and A. Pappinen. 2008. Involving local farmers in rehabilitation of degraded tropical forests: some lessons from Ghana. *Environment, Development and Sustainability* 10:503–518.

Boedhihartono, A. K., and J. Sayer. 2012. Forest landscape restoration: restoring what and for whom? Pages 309–323 *in* J. A. Stanturf, D. Lamb, and P. Madsen, editors. *Forest Landscape Restoration: Integrating Natural and Social Sciences.* Springer, Dordrecht, Netherlands.

Bond, W. J., and J. J. Midgley. 2001. Ecology of sprouting in woody plants: the persistence niche. *Trends in Ecology & Evolution* 16:45–51.

Bonell, M., B. Purandara, B. Venkatesh, J. Krishnaswamy, H. Acharya, U. Singh, R. Jayakumar, and N. Chappell. 2010. The impact of forest use and reforestation on soil hydraulic conductivity in the Western Ghats of India: Implications for surface and sub-surface hydrology. *Journal of Hydrology* 391:47–62.

Booth, T. H. 2012. Forest landscape restoration in Australia's Murray-Darling Basin. Pages 355–371 *in* J. A. Stanturf, P. Madsen, and D. Lamb, editors. *A Goal-Orientated Approach to Forest Landscape Restoration.* Springer, Dordrecht, Netherlands.

Booth, T. H., and K. J. Williams. 2012. Developing biodiverse plantings suitable for changing climatic conditions 1: Underpinning scientific methods. *Ecological Management & Restoration* 13:267–273.

Booth, T. H., K. J. Williams, and L. Belbin. 2012. Developing biodiverse plantings suitable for changing climatic conditions 2: using the Atlas of Living Australia. *Ecological Management & Restoration* 13:274–281.

Boserup, E. 1993. *The Conditions of Agricultural Growth: the Economics of Agrarian Change under Population Pressure.* Earthscan, London.

Böttcher, H., and M. Lindner. 2010. Managing forest plantations for carbon sequestration today and in the future. Pages 43–76 *in* J. Bauhus, P. van der Meer, and M. Kanninen, editors. *Ecosystem Goods and Services from Plantation Forests.* Earthscan, London.

Boulay, A., and L. Tacconi. 2012. The drivers of contract eucalypt farming in Thailand. *International Forestry Review* 14:1–12.

Bourgoin, J., and J.-C. Castella. 2011. Plup fiction: landscape simulation for participatory land use planning in Northern Lao PDR. *Mountain Research and Development* 31:78–88.

Bradby, K. 2013. Gondwana Link: 1000 kilometres of hope. Pages 25–35 *in* J. Fitzsimmons, I. Pulsford, and J. Westcott, editors. *Linking Australia's Landscapes: Lessons and Opportunities from Large-scale Conservation Networks.* CSIRO Publishing, Collingwood, Australia.

Brancalion, P. H., R. A. Viani, M. Calmon, H. Carrascosa, and R. R. Rodrigues. 2013. How to organize a large-scale ecological restoration program? The framework developed by the Atlantic Forest Restoration Pact in Brazil. *Journal of Sustainable Forestry* 32:728–744.

Brockerhoff, E. G., H. Jactel, J. A. Parrotta, and S. F. Ferraz. 2012. Role of eucalypt and other planted forests in biodiversity conservation and the provision of biodiversity-related ecosystem services. *Forest Ecology and Management* 301: 43–50.

Brockway, D., K. Outcalt, D. Tomczak, and E. Johnson. 2005. Restoring longleaf pine forest ecosystems in the southern USA. Pages 501–519 *in* J. A. Stanturf and P. Madsen, editors. Restoration of Boreal and Temperate Forests. CRC Press, Boca Raton, USA.

Brooks, T. M., S. L. Pimm, and N. J. Collar. 1997. Deforestation predicts the number of threatened birds in insular southeast Asia. *Conservation Biology* 11:382–394.

Brown, A. E., L. Zhang, T. A. McMahon, A. W. Western, and R. A. Vertessy. 2005. A review of paired catchment studies for determining changes in water yield resulting from alterations in vegetation. *Journal of Hydrology* 310:28–61.

Bruijnzeel, L. A. 2004. Hydrological functions of tropical forests: not seeing the soil for the trees? *Agriculture, Ecosystems and Environment* 104:185–228.

Bruijnzeel, L. A., M. Bonell, D. A. Gilmour, and D. Lamb. 2005. Conclusion: Forest, water and people in the humid tropics: an emerging view. Pages 906–925 *in* M. Bonell and L. A. Bruijnzeel, editors. *Forests, Water and People in the Humid Tropics.* Cambridge University Press and UNESCO, Cambridge, UK.

Bruijnzeel, L. A., M. Mulligan, and F. N. Scatena. 2011. Hydrometeorology of tropical montane cloud forests: emerging patterns. *Hydrological Processes* 25:465–498.

Buchy, M., and S. Hoverman. 2000. Understanding public participation in forest planning: a review. *Forest Policy and Economics* 1:15–25.

Bull, G. Q., M. Bazett, O. Schwab, S. Nilsson, A. White, and S. Maginnis. 2006. Industrial forest plantation subsidies: Impacts and implications. *Forest Policy and Economics* 9:13–31.

Bull, J. W., K. B. Suttle, A. Gordon, N. J. Singh, and E. Milner-Gulland. 2013. Biodiversity offsets in theory and practice. *Oryx* 47:369–380.

Bullock, J. M., J. Aronson, A. C. Newton, R. F. Pywell, and J. M. Rey-Benayas. 2011. Restoration of ecosystem services and biodiversity: conflicts and opportunities. *Trends in Ecology & Evolution* 26:541–549.

Cacho, O., R. Hean, K. Ginoga, R. Wise, D. Djaenudin, M. Lugina, Y. Wulan, Subarudi, B. Lusiana, M. van Noordwijk, and N. Khasanah. 2008. *Economic Potential of Land-use Change and Forestry for Carbon Sequestration and Poverty Reduction.* Australian Center for International Agricultural Research, Canberra.

Calmon, M., P. H. S. Brancalion, A. Paese, J. Aronson, P. Castro, S. C. da Silva, and R. R. Rodrigues. 2011. Emerging threats and opportunities for large-scale ecological restoration in the Atlantic Forest of Brazil. *Restoration Ecology* 19:154–158.

Calvo-Alvarado, J., B. McLennan, A. Sanchez-Azofeifa, and T. Garvin. 2009. Deforestation and forest restoration in Guanacaste, Costa Rica: Putting conservation policies in context. *Forest Ecology and Management* 258:931–940.

Campbell, J. E., D. B. Lobell, R. C. Genova, and C. B. Field. 2008. The global potential of bioenergy on abandoned agriculture lands. *Environmental Science & Technology* 42:5791–5794.

Campoe, O., J. L. Stape, and J. Mendes. 2010. Can intensive management accelerate the restoration of Brazil's Atlantic forests? *Forest Ecology and Management* 259:1808–1814.

Campos-Filho, E. M., J. N. Da Costa, O. L. De Sousa, and R. G. Junqueira. 2013. Mechanized direct-seeding of native forests in Xingu, Central Brazil. *Journal of Sustainable Forestry* 32:702–727.

Cao, S. X. 2008. Why large-scale afforestation efforts in China have failed to solve the desertification problem. *Environmental Science & Technology* 42:1826–1831.

Cao, S., L. Chen, and Z. Liu. 2009. An investigation of Chinese attitudes toward the environment: case study using the Grain for Green project. *Ambio*: 38:55–64.

Cao, S., L. Chen, D. Shankman, C. Wang, X. Wang, and H. Zhang. 2011a. Excessive reliance on afforestation in China's arid and semi-arid regions: Lessons in ecological restoration. Earth-Science Reviews 104:240–245.

Cao, S., G. Sun, Z. Zhang, L. Chen, Q. Feng, B. Fu, S. McNulty, D. Shankman, J. Tang, Y. Wang, and X. Wei. 2011b. Greening China naturally. *Ambio* 40:828–831.

Cardinale, B. J., J. E. Duffy, A. Gonzalez, D. U. Hooper, C. Perrings, P. Venail, A. Narwani, G. M. Mace, D. Tilman, and D. A. Wardle. 2012. Biodiversity loss and its impact on humanity. *Nature* 486:59–67.

Cardinale, B. J., J. P. Wright, M. W. Cadotte, I. T. Carroll, A. Hector, D. S. Srivastava, M. Loreau, and J. J. Weis. 2007. Impacts of plant diversity on biomass production increase through time because of species complementarity. *Proceedings of the National Academy of Sciences* 104:18123–18128.

Carle, J., and P. Holmgren. 2003. *Definitions Related to Planted Forests.* FAO, Forestry Department, Rome.

——2008. Wood from planted forests: a global outlook 2005–2030. *Forest Products Journal* 58:6–18.

Carle, J., J. B. Ball, and A. Del Lungo. 2009. The global thematic study of planted forests. Pages 33–46 *in* J. Evans, editor. *Planted Forests: Uses, Impacts and Sustainability.* CABI, Wallingford, UK.

Caro, T. 2007. The Pleistocene re-wilding gambit. *Trends in Ecology & Evolution* 22:281–283.

Carr, D., N. Bonney, D. Huxtable, and J. Bartle. 2009. *Improving Direct Seeding for Woody Crops in Temperate Australia – a Review.* Rural Industries Research and Development Corporation, Joint Venture Agroforestry Program, 09/047, Canberra.

Carrere, R., and L. Lohman. 1996. *Pulping the South: Industrial Tree Plantations and the World Paper Economy.* Zed Books, London.

Carron, L. 1990. A history of plantation policy in Australia. Pages 11–24 *in* J. Dargavel and N. Semple, editors. *Prospects for Australian Forest Plantations.* Center for Resource and Environmental Studies, Australian National University, Canberra.

Carter, A. S., and D. A. Gilmour. 1989. Increase in tree cover on private farm land in Central Nepal. *Mountain Research and Development* 9:381–391.

Chazdon, R. 2014. *Second Growth: The Promise of Tropical Forest Regeneration in an Age of Deforestation.* University of Chicago Press, Chicago, USA.

Chazdon, R. L., C. A. Peres, D. Dent, D. Sheil, A. E. Lugo, D. Lamb, N. Stork, and S. E. Miller. 2009. The potential for species conservation in tropical secondary forests. *Conservation Biology* 23:1406–1417.

Chirino, E., A. Bonet, J. Bellot, and J. Sánchez. 2006. Effects of 30-year-old Aleppo pine plantations on runoff, soil erosion, and plant diversity in a semi-arid landscape in south eastern Spain. *Catena* 65:19–29.

Choi, Y. D. 2007. Restoration ecology to the future: a call for new paradigm. *Restoration Ecology* 15:351–353.

Chokkalingam, U., Z. Zhou, C. Wang, and T. Toma. 2006. *Learning Lessons from China's Forest Rehabilitation Efforts.* Center for International Forestry Research, Bogor, Indonesia.

Chomitz, K. 2007. *At Loggerheads? Agricultural Expansion, Poverty Reduction and Environment in the Tropical Forest.* World Bank, Washington.

Clapp, R. 1995. Creating competitive advantage: forest policy as industrial policy in Chile. *Economic Geography* 71:273–296.

Clapp, R. A. 1998. Regions of refuge and the agrarian question: Peasant agriculture and plantation forestry in Chilean Araucania. *World Development* 26:571–589.

——2001. Tree farming and forest conservation in Chile: Do replacement forests leave any originals behind? *Society & Natural Resources* 14:341–356.

Clarke, W. C., and R. Thaman. 1993. *Agroforestry in the Pacific Islands: Systems for Sustainability.* United Nations University Press, Tokyo, New York, Paris.

Clewell, A. F., and J. Aronson. 2013. *Ecological Restoration: Principles, Values and Structure of an Emerging Profession.* Island Press, Washington.

Clough, Y., J. Barkmann, J. Juhrbandt, M. Kessler, T. C. Wanger, A. Anshary, D. Buchori, D. Cicuzza, K. Darras, D. D. Putra, S. Erasmi, R. Pitopang, C. Schmidt, C. H. Schulze, D. Seidel, I. Steffan-Dewenter, K. Stenchly, S. Vidal, M. Weist, A. C. Wielgoss, and T. Tscharntke. 2011. Combining high biodiversity with high yields in tropical agroforests. *Proceedings of the National Academy of Sciences* 108:8311–8316.

Cohen, A. L., B. M. P. Singhakumara, and P. M. S. Ashton. 1995. Releasing rain forest succession: a case study in the *Dicranopterus linearis* fernlands of Sri Lanka. *Restoration Ecology* 3:261–270.

Cole, R. J., K. D. Holl, and R. A. Zahawi. 2010. Seed rain under tree islands planted to restore degraded lands in a tropical agricultural landscape. *Ecological Applications* 20:1255–1269.

Cole, R. J., K. D. Holl, C. L. Keene, and R. A. Zahawi. 2011. Direct seeding of late-successional trees to restore tropical montane forest. *Forest Ecology and Management* 261:1590–1597.

Coll, B. 2011. *UNEP Billion Tree Campaign Reaches 12 Billion Milestone.* United Nations Environmental Program, Nairobi.

Collier, P. 2008. The Politics of hunger: how illusion and greed fan the food crisis. *Foreign Affairs* 87:67–79.

Conacher, A., and J. Conacher. 1995. *Rural Land Degradation in Australia.* Oxford University Press Australia, South Melbourne, Australia.

Connell, J. H., and R. O. Slatyer. 1977. Mechanisms of succession in natural communities and their role in community stability and organization. *American Naturalist* 111:1119–1144.

Connor, J. D., J. R. Ward, and B. Bryan. 2008. Exploring the cost effectiveness of land conservation auctions and payment policies. *Australian Journal of Agricultural and Resource Economics* 52:303–319.

Convention on Biological Diversity. 2010. *Tenth meeting of the Conference of the Parties to the Convention on Biological Diversity; Decision X/2: Strategic Plan for Biodiversity 2011–2020.* 18–29 October, Nagoya, Japan.

——2012. *Eleventh meeting of the Conference of the Parties to the Convention on Biological Diversity; Decision XI/16: Ecosystem restoration.* 8–19 October 2012, Hyderabad, India.

Conservation International. 2012. *Gabarone Declaration, Summit for Sustainability in Africa, 24–25 May 2012.* Conservation International, Gabarone, Botswana.

Cooper, R. 1984. *Resource Scarcity and the Hmong Response.* Singapore University Press, Singapore.

Corlett, R. 2009. Seed dispersal distances and plant migration potential in tropical East Asia. *Biotropica* 41:592–598.

Corlett, R. T., and D. A. Westcott. 2013. Will plant movements keep up with climate change? *Trends in Ecology & Evolution* 28:482–488.

Cornia, G. A. 1985. Farm size, land yields and the agricultural production function: An analysis for fifteen developing countries. *World Development* 13:513–534.

Cossalter, C., and C. Pye-Smith. 2003. *Fast Wood.* Center for International Forestry Research, Bogor, Indonesia.

Cramb, R. 1988. Shifting cultivation and resource degradation in Sarawak: perceptions and policies. *Review of Indonesian and Malaysian Affairs* 22:115–149.

Cramer, V. A., and R. J. Hobbs. 2007. *Old Fields: Dynamics and Restoration of Abandoned Farmland.* Island Press, Washington.

Cribb, J. 2010. *The Coming Famine: the Global Food Crisis and What We Can Do to Avoid It.* CSIRO Publishing, Melbourne, Australia.

Cunningham, S. A., S. J. Attwood, K. S. Bawa, T. G. Benton, L. M. Broadhurst, R. K. Didham, S. McIntyre, I. Perfecto, M. J. Samways, and T. Tscharntke. 2013. To close the yield-gap while saving biodiversity will require multiple locally relevant strategies. *Agriculture, Ecosystems & Environment* 173:20–27.

Cuong, C. V., D. Lamb, and M. Hockings. 2013. Simple plantations have the potential to enhance biodiversity in degraded areas of Tam Dao National Park, Vietnam. *Natural Areas Journal* 33:139–147.

Curtin, T., and D. Lea. 2006. Land titling and socio-economic development in the Pacific. *Pacific Economic Bulletin* 21:151–180.

Cyranoski, D. 2007. Biodiversity: Logging: the new conservation. *Nature* 446:608–610.

Dalton, G. 1993. *Direct Seeding of Trees and Shrubs. A Manual for Australian Conditions.* Department of Primary Industries, Adelaide, Australia.

Daniels, A. E., K. Bagstad, V. Esposito, A. Moulaert, and C. M. Rodriguez. 2010. Understanding the impacts of Costa Rica's PES: Are we asking the right questions? *Ecological Economics* 69:2116–2126.

Dargavel, J., and A. Semple, editors. 1990. *Prospects for Australian Forest Plantations.* Center for Resources and Environmental Management, Canberra.

Dauber, J., C. Brown, A. L. Fernando, J. Finnan, E. Krasuska, J. Ponitka, D. Styles, D. Thran, K. J. Van Groenigen, and M. Weih. 2012. Bioenergy from 'surplus' land: environmental and socio-economic implications. *BioRisk* 7:5–50.

Dawson, I. K., M. R. Guariguata, J. Loo, J. C. Weber, A. Lengkeek, D. Bush, J. Cornelius, L. Guarino, R. Kindt, and C. Orwa. 2013. What is the relevance of smallholders' agroforestry systems for conserving tropical tree species and genetic diversity in circa situm, in situ and ex situ settings? A review. *Biodiversity and Conservation* 22:301–324.

De Groot, R. S., J. Blignaut, S. Van Der Ploeg, J. Aronson, T. Elmqvist, and J. Farley. 2013. Benefits of investing in ecosystem restoration. *Conservation Biology* 27:1286–1293.

De Jong, W., D. D. Sam., and T. V. Hung. 2006. *Forest Rehabilitation in Vietnam: History, Realities and Future.* Center for International Forestry Research, Bogor, Indonesia.

De Schutter, O. 2011a. Food crises: five priorities for the G20. *The Guardian*; 16 June 2011, London.

——2011b. How not to think of land-grabbing: three critiques of large-scale investments in farmland. *The Journal of Peasant Studies* 38:249–279.

Deininger, K. 2011. Challenges posed by the new wave of farmland investment. *The Journal of Peasant Studies* 38:217–247.

Del Lungo, A., J. Ball, and J. Carle. 2006. *Global Planted Forests Thematic Study. Results and Analysis.* Food and Agricultural Organisation of the UN, Rome.

Démurger, S., and H. Wan. 2012. Payments for ecological restoration and internal migration in China: the sloping land conversion program in Ningxia. *IZA Journal of Migration* 1:1–22.

Demurger, S., and W. Y. Yang. 2006. Economic changes and afforestation incentives in rural China. *Environment and Development Economics* 11:629–649.

den Herder, M., J. Kouki, and V. Ruusila. 2009. The effects of timber harvest, forest fire, and herbivores on regeneration of deciduous trees in boreal pine-dominated forests. *Canadian Journal of Forest Research* 39:712–722.

Deng, L., Z.-P. Shangguan, and R. Li. 2012. Effects of the Grain-for-Green program on soil erosion in China. *International Journal of Sediment Research* 27:120–127.

Dent, D., Z. Bai, M. Schaepman, and L. Olsson. 2009. Letter to the editor. *Soil Use and Management* 25:93–97.

Diamond, J. 1987. Reflection on goals and on the relationship between theory and practice. Pages 329–336 *in* W. R. Jordan, M. Gilpin, and D. Aber, editors. *Restoration Ecology: A Synthetic Approach to Ecological Research.* Cambridge University Press, Cambridge, UK.

Dodet, M., and C. Collet. 2012. When should exotic forest plantation tree species be considered as an invasive threat and how should we treat them? *Biological Invasions* 14:1765–1788.

Donlan, J. 2005. Re-wilding North America. *Nature* 436:913–914.

Dorrough, J., P. A. Vesk, and J. Moll. 2008. Integrating ecological uncertainty and farm-scale economics when planning restoration. *Journal of Applied Ecology* 45:288–295.

Douglas, I., and J. Guyot. 2005. Erosion and sediment yield in the humid tropics. Pages 407–421 *in* M. Bonell and L. A. Bruijnzeel, editors. *Forests, Water and People in the Humid Tropics.* Cambridge University Press, Cambridge, UK.

Doust, S. J. 2004. Seed and Seedling Ecology in the Early Stages of Rainforest Restoration. Unpublished PhD Thesis, University of Queensland, Brisbane, Australia.

Doust, S. J., P. D. Erskine, and D. Lamb. 2006. Direct seeding to restore rainforest species: Microsite effects on the early establishment and growth of rainforest tree seedlings on degraded land in the wet tropics of Australia. *Forest Ecology and Management* 234:333–343.

——2008. Restoring rainforest species by direct seeding: Tree seedling establishment and growth performance on degraded land in the wet tropics of Australia. *Forest Ecology and Management* 256:1178–1188.

Drielsma, M., S. Ferrier, G. Howling, G. Manion, S. Taylor, and J. Love. 2014. The biodiversity forecasting toolkit: Answering the 'how much', 'what', and 'where' of planning for biodiversity persistence. *Ecological Modelling* 274:80–91.

Dugan, P. 2000. Assisted natural regeneration: methods, results and issues relevant to sustained particiaption by communities *in* S. Elliott, J. Kerby, D. Blakesley, K. Hardwick, K. Woods, and V. Anusarnsunthorn, editors. *Forest Restoration for Wildlife Conservation.* University of Chiang Mai, Chaing Mai, Thailand.

Dull, R. A., R. J. Nevle, W. I. Woods, D. K. Bird, S. Avnery, and W. M. Denevan. 2010. The Columbian encounter and the Little Ice Age: Abrupt land use change, fire, and greenhouse forcing. *Annals of the Association of American Geographers* 100:755–771.

Durst, P., C. Brown, H. Tacio, and M. Ishikawa, editors. 2005. *In Search of Excellence: Exemplary Forest Management in Asia and the Pacific.* Food and Agriculture Organisation of the United Nations and Regional Community Training Center for Asia and the Pacific, Bangkok.

Dwyer, J. M., R. J. Fensham, D. W. Butler, and Y. M. Buckley. 2009. Carbon for conservation: Assessing the potential for win–win investment in an extensive Australian regrowth ecosystem. *Agriculture, Ecosystems & Environment* 134:1–7.

Echeverria, C., D. Coomes, J. Salas, J. M. a. Rey-Benayas, A. Lara, and A. Newton. 2006. Rapid deforestation and fragmentation of Chilean Temperate Forests. *Biological Conservation* 130:481–494.

*Economist, The.* 2008. Cereal offenders: curbing food exports to feed hungry mouths is a recipe for trouble. *The Economist,* 27 March 2008.

——2011. When others are grabbing their land: evidence is piling up against acquisition of farmland in poor countries. *The Economist,* 5 May 2011.

——2012. Environmental law in Brazil: compromise or deadlock? *The Economist,* 31 May 2012.

——2013. More money, more forest planting. *The Economist* 408, Issue 8853.

Eden, S. 2009. The work of environmental governance networks: traceability, credibility and certification by the Forest Stewardship Council. *Geoforum* 40:383–394.

El Kateb, H., H. Zhang, P. Zhang, and R. Mosandl. 2013. Soil erosion and surface runoff on different vegetation covers and slope gradients: A field experiment in Southern Shaanxi Province, China. *Catena* 105:1–10.

Elliott, S., D. Blakesley, and K. Hardwick. 2013. *Restoring Tropical Forests: A Practical Guide.* Royal Botanic Gardens, Kew, UK.

Ellison, D., M. N. Futter, and K. Bishop. 2012. On the forest cover–water yield debate: from demand-to supply-side thinking. *Global Change Biology* 18:806–820.

Elmqvist, T., M. Pyykonen, M. Tengo, F. Rakotondrasoa, E. Rabakonandrianina, and C. Radimilahy. 2007. Patterns of loss and regeneration of tropical dry forest in Madagascar: the social institutional context. *PLoS ONE* 2:e402.

El-Swaify, S. A. 1997. Factors affecting soil erosion hazards and conservation needs for tropical steeplands. *Soil Technology* 11:3–16.

Emerton, L. 2012. *Rethinking Economics, Markets and Incentives: Using Economic Tools at the Landscape Level.* International Union for the Conservation of Nature, Gland, Switzerland.

Engel, V. L., and J. A. Parrotta. 2001. An evaluation of direct seeding for reforestation of degraded lands in central Sao Paulo State, Brazil. *Forest Ecology and Management* 152:169–181.

Enters, T., P. Durst, and C. Brown. 2003. What does it take? The role of incentives in forest plantation development in the Asia-Pacific region. *Unasylva* 54:11–18.

Erskine, P., D. Lamb, and M. Bristow. 2005. *Reforestation in the Tropics and Subtropics of Australia using Rainforest Tree Species Rural Industries Research and Development Corporation*, Canberra. Available at https://rirdc.infoservices.com.au/items/05-087 [Accessed 20 June 2014].

——2006. Tree species and ecosystem function: can tropical multi-species plantations generate greater productivity? *Forest Ecology and Management* 233:205–210.

Erskine, P. D., C. P. Catterall, D. Lamb, and J. Kanowski. 2007. Patterns and processes of old-field reforestation in Australian rain forest landscapes. Pages 119–144 *in* V. A. Cramer and R. J. Hobbs, editors. *Old Fields: Dynamics and Restoration of Abandoned Farmland.* Island Press, Washington.

Estes, J. A., J. Terborgh, J. S. Brashares, M. E. Power, J. Berger, W. J. Bond, S. R. Carpenter, T. E. Essington, R. D. Holt, and J. B. Jackson. 2011. Trophic downgrading of planet earth. *Science* 333:301–306.

Evans, J. 2009. *Planted Forests: Uses, Impacts and Sustainability.* CAB International and FAO, Rome.

Evans, J., and J. W. Turnbull. 2004. *Plantation Forestry in the Tropics: The Role, Silviculture and Use of Planted Forests for Industrial, Social, Environmental and Agroforestry Purposes.* Oxford University Press, Oxford, UK.

Fairhead, J., and M. Leach. 1995. False forest history, complicit social analysis: rethinking some West African environmental narratives. *World Development* 23:1023–1035.

Fairhead, J., and M. Leach. 1996. *Misreading the African Landscape: Society and Ecology in a Forest-Savanna Mosaic.* Cambridge University Press, Cambridge, UK.

FAO. 2002. *Land Tenure and Rural Development.* Food and Agriculture Organization of the United Nations, Rome.

——2010. *Global Forest Resources Assessment 2010: Main Report.* Food and Agriculture Organization of the United Nations, Rome.

——2011. *State of the World's Forests.* Food and Agriculture Organization of the United Nations, Rome.

——2012. *The State of Food Insecurity in the World.* Food and Agriculture Organization of the United Nations, Rome.

FAO, DFSC, and IPGRI. 2001. *Forest Genetic Resources Conservation and Management. Vol. 2: In Managed Natural Forests and Protected Areas (in situ).* International Plant Genetic Resources Institute, Rome.

Fearnside, P. M. 1998. Plantation forestry in Brazil: projections to 2050. *Biomass and Bioenergy* 15:437–450.

Felton, A., M. Lindbladh, J. Brunet, and O. Fritz. 2010. Replacing coniferous monocultures with mixed-species production stands: An assessment of the potential benefits for forest biodiversity in northern Europe. *Forest Ecology and Management* 260:939–947.

Fensham, R., R. Fairfax, and S. Archer. 2005. Rainfall, land use and woody vegetation cover change in semi-arid Australian savanna. *Journal of Ecology* 93:596–606.

Fenske, J. 2011. Land tenure and investment incentives: Evidence from West Africa. *Journal of Development Economics* 95:137–156.
Fernandez, R., S. Rangel, J. A. Stanturf, C. Arseneau, and P. Nantel. 2002. *Forest Plantations in North America.* XXI Session of the North America Forestry Commission, Plantations Technical Group, USA.
Fernández-Moya, J., A. Alvarado, W. Forsythe, and M. Marchamalo-Sacristán. 2013. Effect of teak (*Tectona grandis*) plantations on hydraulic conductivity and porosity of alfisols in Costa Rica. *Journal of Tropical Forest Science* 25: 259–267.
Finegan, B., and D. Delgado. 2000. Structural and floristic heterogeneity in a 30-year-old Costa Rican rain forest restored on pasture through natural secondary succession. *Restoration Ecology* 8:380–393.
Fingleton, J. 2005. *Privatising Land in the Pacific: A Defence of Customary Tenures.* The Australia Institute, Canberra.
Firn, J., P. E. Erskine, and D. Lamb. 2007. Woody species diversity influences productivity and soil nutrient availability in tropical plantations. *Oecologia* 154:521–533.
Fischer, J., B. Brosi, G. C. Daily, P. R. Ehrlich, R. Goldman, J. Goldstein, D. B. Lindenmayer, A. D. Manning, H. A. Mooney, and L. Pejchar. 2008. Should agricultural policies encourage land sparing or wildlife-friendly farming? *Frontiers in Ecology and the Environment* 6:380–385.
Fisher, R. J., S. Maginnis, W. Jackson, E. Barrow, and S. Jeanrenaud. 2008. *Linking Conservation and Poverty Reduction: Landscapes, People and Power.* Earthscan, London.
Fitzsimons, J., I. Pulsford, and G. Westcott, editors. 2013. *Linking Australia's Landscapes: Lessons and Opportunities from Large-Scale Conservation Networks.* CSIRO Publishing, Collingwood, Australia.
Florence, R. G. 2004. *Ecology and Silviculture of Eucalypt Forests.* CSIRO Publishing, Canberra.
Florence, R., B. L. Turner, R. N. Byron, K. Groves, and S. Davey. 1990. Problems in formulating and Australian plantation policy. Pages 25–38 *in* J. Dargavel and N. Semple, editors. *Prospects for Australian Forest Plantations.* Center for Resource and Environmental Studies, Canberra.
Foley, J. A., N. Ramankutty, K. A. Brauman, E. S. Cassidy, J. S. Gerber, M. Johnston, N. D. Mueller, C. O'Connell, D. K. Ray, P. C. West, C. Balzer, E. M. Bennett, S. R. Carpenter, J. Hill, C. Monfreda, S. Polasky, J. Rockstrom, J. Sheehan, S. Siebert, D. Tilman, and D. P. M. Zaks. 2011. Solutions for a cultivated planet. *Nature* 478:337–342.
Forbes, K., and J. Broadbent. 2013. *Forests and Landslides: The Role of Trees and Forests in the Prevention of Landslides and Rehabilitation of Landslide-Affected Areas in Asia.* Food and Agriculture Organization of the United Nations, Bangkok.
Forrester, D. I., and R. G. B. Smith. 2012. Faster growth of *Eucalyptus grandis* and *Eucalyptus pilularis* in mixed-species stands than monocultures. *Forest Ecology and Management* 286:81–86.
Forrester, D. I., J. Bauhus, A. L. Cowie, and J. K. Vanclay. 2006. Mixed-species plantations of *Eucalyptus* with nitrogen-fixing trees: A review. *Forest Ecology and Management* 233:211–230.
Forrester, D. I., S. Theiveyanathan, J. J. Collopy, and N. E. Marcar. 2010. Enhanced water use efficiency in a mixed *Eucalyptus globulus* and *Acacia mearnsii* plantation. *Forest Ecology and Management* 259:1761–1770.

Fukushima, M., M. Kanzaki, M. Hara, T. Ohkubo, P. Preechapanya, and C. Choocharoen. 2008. Secondary forest succession after the cessation of swidden cultivation in the montane forest area in Northern Thailand. *Forest Ecology and Management* 255:1994–2006.

Fyfe, R. 2007. The importance of local scale openness within regions dominated by closed woodland. *Journal of Quaternary Science* 22:571–578.

Gamfeldt, L., T. Snäll, R. Bagchi, M. Jonsson, L. Gustafsson, P. Kjellander, M. C. Ruiz-Jaen, M. Fröberg, J. Stendahl, and C. D. Philipson. 2013. Higher levels of multiple ecosystem services are found in forests with more tree species. *Nature Communications* 4:1340.

Gardiner, E., and J. Oliver. 2005. Restoration of bottomland hardwood forest in the lower Mississippi alluvial valley, USA. Pages 235–251 *in* J. A. Stanturf and P. Madsen, editors. *Restoration of Boreal and Temperate Forests*. CRC Press, Boca Raton, USA.

Garibaldi, L. A., M. A. Aizen, A. M. Klein, S. A. Cunningham, and L. D. Harder. 2011. Global growth and stability of agricultural yield decrease with pollinator dependence. *Proceedings of the National Academy of Sciences* 108:5909–5914.

Geddes, W. R. 1976. *Migrants of the Mountains: The Cultural Ecology of the Blue Miao (Hmong Njua) of Thailand*. Clarendon Press, Oxford, UK.

Gellrich, M., P. Baur, B. Koch, and N. E. Zimmermann. 2007. Agricultural land abandonment and natural forest re-growth in the Swiss mountains: A spatially explicit economic analysis. *Agriculture, Ecosystems & Environment* 118:93–108.

George, S., R. Harper, R. Hobbs, and M. Tibbett. 2012. A sustainable agricultural landscape for Australia: A review of interlacing carbon sequestration, biodiversity and salinity management in agroforestry systems. *Agriculture, Ecosystems & Environment* 163:28–36.

Gerber, J.-F. 2011. Conflicts over industrial tree plantations in the South: who, how and why? *Global Environmental Change* 21:165–176.

German, L., G. Shoneveld, and E. Mwangi. 2011. *Contemporary Processes of Large-Scale Land Acquisition by Investors: Case Studies from Sub-Saharan Africa*. Center for International Forestry Research, Bogor, Indonesia.

Gerrand, A., R. J. Keenan, P. Kanowski, and R. Stanton. 2003. Australian forest plantations: an overview of industry, environmental and community issues and benefits. *Australian Forestry* 66:1–8.

Ghimire, C. P., L. A. Bruijnzeel, M. Bonell, N. Coles, M. W. Lubczynski, and D. A. Gilmour. 2013. The effects of sustained forest use on hillslope soil hydraulic conductivity in the Middle Mountains of Central Nepal. *Ecohydrology* 7: 478–495.

Giambelluca, T. W. 2002. Hydrology of altered tropical forest. *Hydrological Processes* 16:1665–1669.

Gilbert-Norton, L., R. Wilson, J. R. Stevens, and K. H. Beard. 2010. A meta-analytic review of corridor effectiveness. *Conservation Biology* 24:660–668.

Gilmour, D. 1990. Resource availability and indigenous forest management systems in Nepal. *Society & Natural Resources* 3:145–158.

——2003. Retrospective and prospective view of community forestry in Nepal. *Journal of Forest and Livelihood* 2:1–5.

Gilmour, D. A., and R. J. Fisher. 1991. *Villagers, Forests and Foresters*. Sahayogi Press, Kathmandu.

Gilmour, D. A., and M. Nurse. 1991. Farmer initiatives in increasing tree cover in central Nepal. *Mountain Research and Development* 11:329–337.

Gilmour, D. A., G. C. King, G. C. Applegate, and B. Mohns. 1990. Silviculture of plantation forest in central Nepal to maximise community benefit. *Forest Ecology and Management* 32:173–186.

Gilmour, D., Y. Malla, and M. Nurse. 2004. *Linkages between Community Forestry and Poverty*. Regional Community Forestry Training Center for Asia and the Pacific (RECOFTC), Bangkok.

Golub, A., T. Hertel, H. L. Lee, S. Rose, and B. Sohngen. 2009. The opportunity cost of land use and the global potential for greenhouse gas mitigation in agriculture and forestry. *Resource and Energy Economics* 31:299–319.

Goosem, S., and N. Tucker. 1995. *Repairing the Rainforest: Theory and Practice of Rainforest Re-establishment in North Queensland's Wet Tropics*. Wet Tropics Management Authority, Cairns, Australia.

Grainger, A. 1988. Estimating areas of degraded tropical lands requiring replenishment of forest cover. *International Tree Crops Journal* 5:31–61.

Grammatikopoulou, I., A. Iho, and E. Pouta. 2012. Willingness of farmers to participate in agri-environmental auctions in Finland. *Food Economics* 9:215–230.

Greenwood, W., and J. Buttle. 2014. Effects of reforestation on near surface saturated hydraulic conductivity in a managed forest landscape, southern Ontario, Canada. *Ecohydrology* 7:45–55.

Gregersen, H., H. El Lakany, L. Bailey, and A. White. 2011. *The Greener Side of REDD+*. Rights and Resources Initiative, Washington.

Griess, V. C., and T. Knoke. 2011. Growth performance, windthrow, and insects: meta-analyses of parameters influencing performance of mixed-species stands in boreal and northern temperate biomes. *Canadian Journal of Forest Research-Revue Canadienne De Recherche Forestiere* 41:1141–1159.

——2013. Bioeconomic modeling of mixed Norway spruce—European beech stands: economic consequences of considering ecological effects. *European Journal of Forest Research* 132(3):511–522.

Griscom, H. P., and M. S. Ashton. 2011. Restoration of dry tropical forests in Central America: A review of pattern and process. *Forest Ecology and Management* 261:1564–1579.

Grove, R. 1995. *Green Imperialism: Colonial Expansion, Tropical Island Edens and the Origins of Environmentalism 1600–1860*. Cambridge University Press, Cambridge, UK.

Grundy, D. 2005. From plantation developer to steward of the nation's forests. Pages 246–265 *in* M. Garforth and J. Mayers, editors. *Plantations, Privatization, Poverty and Power: Changing Ownership and Management of State Forests*. Earthscan, London.

Gunderson, L. H., and C. S. Holling, editors. 2002. *Panarchy : Understanding Transformations in Human and Natural Systems*. Island Press, Washington, DC.

Haggar, J., R. Rheingans, P. Arroyo, and B. Alvarado. 2003. Benefits and costs of intercropping reforestation in the Atlantic lowlands of Costa Rica. *New Forests* 25:41–48.

Haglund, E., J. Ndjeunga, L. Snook, and D. Pasternak. 2011. Dry land tree management for improved household livelihoods: Farmer managed natural regeneration in Niger. *Journal of Environmental Management* 92:1696–1705.

Hall, R. L., and I. R. Calder. 1993. Drop size modification by forest canopies: measurements using a disdrometer. *Journal of Geophysical Research: Atmospheres (1984–2012)* 98:18465–18470.

Hanley, N., S. Banerjee, G. D. Lennox, and P. R. Armsworth. 2012. How should we incentivize private landowners to 'produce' more biodiversity? *Oxford Review of Economic Policy* 28:93–113.

Hardwick, K., J. Healy, S. Elliott, N. C. Garwood, and V. Anusarnsunthorn. 1997. Understanding and assisting natural regeneration in northern Thailand. *Forest Ecology and Management* 99:203–214.

Hardwick, K., J. R. Healey, S. Elliott, and D. Blakesley. 2004. Research needs for restoring seasonal tropical forests in Thailand: accelerated natural regeneration. *New Forests* 27:285–302.

Harper, R., S. Sochacki, K. Smettem, and N. Robinson. 2009. Bioenergy Feedstock Potential from Short-Rotation Woody Crops in a Dryland Environment. *Energy & Fuels* 24:225–231.

Harper, R., K. Smettem, P. Townsend, J. Bartle, and J. McGrath. 2012. Broad-scale restoration of landscape function with timber, carbon and eater investment. Pages 275–292 *in* J. A. Stanturf, D. Lamb, and P. Madsen, editors. *Forest Landscape Restoration: Integrating Natural and Social Sciences.* Springer, Dordrecht, Netherlands.

Harper, R. J., S. J. Sochacki, K. R. J. Smettem, and N. Robinson. 2013. Managing water in agricultural landscapes with short-rotation biomass plantations. Global Change Biology, Bioenergy: 10.1111/gcbb.12090.

Hassler, S. K., B. Zimmermann, M. van Breugel, J. S. Hall, and H. Elsenbeer. 2011. Recovery of saturated hydraulic conductivity under secondary succession on former pasture in the humid tropics. *Forest Ecology and Management* 261:1634–1642.

Hatton, T., and R. George. 2001. The role of afforestation in managing dryland salinity. Pages 28–35 *in* E. K. S. Nambiar and A. Brown, editors. *Plantations, Farm Forestry and Water.* Rural Industries Research and Development Corporation, Canberra.

Hecht, S. B., and S. S. Saatchi. 2007. Globalization and forest resurgence: changes in forest cover in El Salvador. *Bioscience* 57:663–672.

Hertel, T. W. 2011. The global supply and demand for agricultural land in 2050: A perfect storm in the making? *American Journal of Agricultural Economics* 93:259–275.

Hevia, J. N., J. C. Araújo, and J. M. Manso. 2014. Assessment of 80 years of ancient badlands restoration in Saldaña, Spain. *Earth Surface Processes and Landforms.* doi:10.1002/esp.3541.

Hillis, W. 1978. Wood quality and utilisation. Pages 259–289 *in* W. E. Hillis and A. Brown, editors. *Eucalypts for Wood Production.* Commonwealth Scientific and Industrial Research Organisation, Canberra.

Hobbs, R. J., E. Higgs, and J. A. Harris. 2009. Novel ecosystems: implications for conservation and restoration. *Trends in Ecology and Evolution.* 24:599–605.

Hobbs, R. J., E. Higgs, and C. Hall. 2013. Defining novel ecosystems. Pages 58–60 *in* R. J. Hobbs, E. Higgs, and C. Hall, editors. *Novel Ecosystems: Intervening in the New Ecological World Order.* John Wiley and Sons, Oxford, UK.

Hobley, M. 1996. *Participatory Forestry: the Process of Change in India and Nepal.* Overseas Development Institute, London.

Hodgson, B., and P. McGhee. 1992. Development of aerial seeding for the regeneration of Tasmanian eucalypt forests. *Tasforests* 4:77–85.

Hodgson, J. A., C. D. Thomas, B. A. Wintle, and A. Moilanen. 2009. Climate change, connectivity and conservation decision making: back to basics. *Journal of Applied Ecology* 46:964–969.

Holden, C. 1986. Regrowing a dry tropical forest: Dry tropical forest has all but vanished; now a Pennsylvania biologist is raising money to restore one from forest remnants in a corner of Costa Rica. *Science* 234:809.

Holl, K. D., and T. M. Aide. 2011. When and where to actively restore ecosystems? *Forest Ecology and Management* 261:1558–1563.

Holl, K. D., R. A. Zahawi, R. J. Cole, R. Ostertag, and S. Cordell. 2011. Planting seedlings in tree islands versus plantations as a large-scale tropical forest restoration strategy. *Restoration Ecology* 19:470–479.

Holmgren, P., E. J. Masakha, and H. Sjoholm. 1994. Not all African land is being degraded – a recent survey of trees on farms in Kenya reveals increasing forest resources. *Ambio* 23:390–395.

Hooper, D. U., F. S. Chapin, J. J. Ewel, A. Hector, P. Inchausti, S. Lavorel, J. H. Lawton, D. M. Lodge, M. Loreau, S. Naeem, B. Schmid, H. Setala, A. J. Symstad, J. Vandermeer, and D. A. Wardle. 2005. Effects of biodiversity on ecosystem functioning: A consensus of current knowledge. *Ecological Monographs* 75:3–35.

Hopper, S. D., and P. Gioia. 2004. The southwest Australian floristic region: evolution and conservation of a global hot spot of biodiversity. *Annual Review of Ecology, Evolution, and Systematics* 35:623–650.

Houghton, R. 2013. The contribution of land use and land-use change to the carbon cycle. Pages 52–76 *in* D. Brown, D. Robinson, F. N., and B. Reed, editors. *Land Use and the Carbon Cycle: Advances in Integrated Science, Management, and Policy.* Cambridge University Press, Cambridge, UK.

Houghton, R. A. 1990. The future role of tropical forests in affecting the carbon dioxide concentration of the atmosphere. *Ambio* 19:204–209.

Hubble, T. C. T., B. B. Docker, and I. D. Rutherfurd. 2010. The role of riparian trees in maintaining riverbank stability: A review of Australian experience and practice. *Ecological Engineering* 36:292–304.

Hunt, C. 2009. *Carbon Sinks and Climate Change: Forests in the Fight Against Global Warming.* Edward Elgar, Cheltenham, UK.

Hynynen, J., J. Repola, and K. Mielikainen. 2011. The effects of species mixture on the growth and yield of mid-rotation mixed stands of Scots pine and silver birch. *Forest Ecology and Management* 262:1174–1183.

Ilstedt, U., A. Malmer, E. Verbeeten, and D. Murdiyarso. 2007. The effect of afforestation on water infiltration in the tropics: a systematic review and meta-analysis. *Forest Ecology and Management* 251:45–51.

Indufor. 2012. *Strategic review on the Future of Forest Plantations.* Indufor Oy, Helsinki.

IPCC. 2013. Summary for policymakers *in* T. F. Stocker, D. Qin, G. Plattner, M. Tignor, S. Allen, J. Boschung, A. Nauels, Y. Xia, V. Bex, and P. Midgley, editors. *Climate Change 2013: The Physical Science Basis. Contribution of Working Group I to the Fifth Assessment Report of the Intergovernmental Panel on Climate Change.* Cambridge University Press, United Kingdom and New York.

Isbell, F., V. Calcagno, A. Hector, J. Connolly, W. S. Harpole, P. B. Reich, M. Scherer-Lorenzen, B. Schmid, D. Tilman, and J. van Ruijven. 2011. High plant diversity is needed to maintain ecosystem services. *Nature* 477:199–202.

Isutsa, D. K., and I. A. Merwin. 2000. *Malus* germplasm varies in resistance or tolerance to apple replant disease in a mixture of New York orchard soils. *HortScience* 35:262–268.

ITTO. 1993. ITTO guidelines for the establishment and sustainable management of planted tropical forests. *International Tropical Timbers Organisation*, Yokohama, Japan.

IUCN. 2012. *Bonn Challenge and Landscape Restoration*. International Union for the Conservation of Nature, Gland, Switzerland.

Jackson, R. B., E. G. Jobbágy, R. Avissar, S. B. Roy, D. J. Barrett, C. W. Cook, K. A. Farley, D. C. le Maitre, B. A. McCarl, and B. C. Murray. 2005. Trading water for carbon with biological carbon sequestration. *Science* 310:1944–1947.

Jackson, S. T., and R. J. Hobbs. 2009. Ecological restoration in the light of ecological history. *Science* 325:567–569.

Jacobs, M. 1955. *Growth Habits of the Eucalypts*. Forestry and Timber Bureau, Canberra.

Jactel, H., and E. G. Brockerhoff. 2007. Tree diversity reduces herbivory by forest insects. *Ecology Letters* 10:835–848.

Jandl, R., M. Lindner, L. Vesterdal, B. Bauwens, R. Baritz, F. Hagedorn, D. W. Johnson, K. Minkkinen, and K. A. Byrne. 2007. How strongly can forest management influence soil carbon sequestration? *Geoderma* 137:253–268.

Janzen, D. 1986. *Guanacaste National Park: Tropical Ecological and Cultural Restoration*. Editorial Universidad Estatal a Distancia, San José.

Janzen, D. 1988. Management of habitat fragments in a tropical dry forest: growth. *Annals of the Missouri Botanical Garden* 75:105–116.

Janzen, D. 2002. Tropical dry forest: Area de Conservacion Guanacaste, northwest Costa Rica. Pages 559–583 *in* M. Perrow and A. Davy, editors. *Handbook of Ecological Restoration*. Cambridge University Press, Cambridge, UK.

Jiang, B., and S. Liang. 2013. Improved vegetation greenness increases summer atmospheric water vapor over Northern China. *Journal of Geophysical Research: Atmospheres* 118:8129–8139.

Jiang, G., X. Han, and J. Wu. 2006. Restoration and Management of the Inner Mongolia Grassland Require a Sustainable Strategy. *Ambio* 35:269–270.

Johnson, D. L., S. H. Ambrose, T. J. Bassett, M. L. Bowen, D. E. Crummey, J. S. Isaacson, D. N. Johnson, P. Lamb, M. Saul, and A. E. Winter-Nelson. 1997. Meanings of Environmental Terms. *Journal of Environmental Quality* 26:581–589.

Jonson, J. 2010. Ecological restoration of cleared agricultural land in Gondwana Link: lifting the bar at 'Peniup'. *Ecological Management & Restoration* 11:16–26.

Kaimowitz, D., and A. Angelsen. 1998. *Economic Models of Tropical Deforestation: A Review*. Center for International Forestry Research, Bogor, Indonesia.

Kalame, F. B., R. Aidoo, J. Nkem, O. C. Ajayie, M. Kanninen, O. Luukkanen, and M. Idinoba. 2011. Modified taungya system in Ghana: a win-win practice for forestry and adaptation to climate change? *Environmental Science & Policy* 14:519–530.

Kamanzi, S. 2011. *Statement by Rwanda Minister for Environment and Lands at Global Launch of International Year of Forests*. 9th Session, United Nations Forum on Forests, New York.

Kanowski, J., C. P. Catterall, H. Procter, T. Reis, N. Tucker, and G. Wardell-Johnson. 2005. Biodiversity values of timber plantations and restoration plantings for rainforest fauna in tropical and subtropical Australia. Pages 183–205 *in* P. D. Erskine, D. Lamb, and M. Bristow, editors. *Reforestation in the Tropics and Subtropics of Australia using Rainforest Tree species*. Rural Industries Research and Development Corporation, Canberra. Available at: https://rirdc.infoservices.com.au/items/05-087 [Accessed 20 June 2014].

Kassa, H., M. Bekele, and B. Campbell. 2011. Reading the landscape past: explaining the lack of on-farm tree planting in Ethiopia. *Environment and History* 17:461–479.

Kato. 1994. The emergence of abandoned paddy fields in Negera Sembilan, Malaysia. *Southeast Asian Studies* 32:145–172.

Katz, E. 2012. Further adventures in the case against restoration. *Environmental Ethics* 34:67–97.

Kauppi, P. E., J. H. Ausubel, J. Fang, A. S. Mather, R. A. Sedjo, and P. E. Waggoner. 2006. Returning forests analyzed with the forest identity. *Proceedings of the National Academy of Sciences* 103:17574–17579.

Kaur, R., W. L. Gonzales, L. D. Llambi, P. J. Soriano, R. M. Callaway, M. E. Rout, and T. J. Gallaher. 2012. Community impacts of *Prosopis juliflora* invasion: biogeographic and congeneric comparisons. *PLoS ONE* 7:e44966.

Keenan, R., D. Lamb, and G. Sexton. 1995. Experience with mixed species rainforest tree plantations in North Queensland. *Commonwealth Forestry Review* 74:315–321.

Keenan, R., D. Lamb, O. Woldring, T. Irvine, and R. Jensen. 1997. Restoration of plant diversity beneath tropical tree plantations in northern Australia. *Forest Ecology and Management* 99:117–132.

Keenan, R., D. Lamb, J. Parrotta, and J. Kikkawa. 1999. Ecosystem management in tropical timber plantations: satisfying economic, conservation, and social objectives. *Journal of Sustainable Forestry* 9:117–134.

Keenleyside, C., and G. M. Tucker. 2010. *Farmland Abandonment in the EU: An Assessment of Trends and Prospects.* Report prepared for WWF. Institute for European Environmental Policy, London.

Keenleyside, K., N. Dudley, S. Cairns, C. Hall, and S. Stolton, editors. 2012. *Ecological Restoration for Protected Areas: Principles, Guidelines and Best Practices.* International Union for the Conservation of Nature, Gland, Switzerland.

Keeves, A. 1966. Some evidence of loss of productivity with successive rotations of *Pinus radiata* in the south east of South Australia. *Australian Forestry* 30:51–63.

Keipi, K. 1997. Financing forest plantations in Latin America: government incentives. *Unasylva* 48(188):50–56.

Kelly, P., and X. Huo. 2013. Land retirement and nonfarm labor market participation: an analysis of China's Sloping Land Conversion Program. *World Development* 48:156–169.

Kelty, M. J. 2006. The role of species mixtures in plantation forestry. *Forest Ecology and Management* 233:195–204.

Kemkes, R. J., J. Farley, and C. J. Koliba. 2010. Determining when payments are an effective policy approach to ecosystem service provision. *Ecological Economics* 69:2069–2074.

Kisdi, É. 2002. Dispersal: risk spreading versus local adaptation. *The American Naturalist* 159:579–596.

Kleijn, D., and W. J. Sutherland. 2003. How effective are European agri environment schemes in conserving and promoting biodiversity? *Journal of Applied Ecology* 40:947–969.

Knight, A., P. Beale, and G. Dalton. 1998. Direct seeding of native trees and shrubs in low rainfall areas and on non-wetting sands in South Australia. *Agroforestry Systems* 39:225–239.

Knoke, T. 2008. Mixed forests and finance – methodological approaches. *Ecological Economics* 65:590–601.

Knoke, T., C. Ammer, B. Stimm, and R. Mosandl. 2008. Admixing broadleaved to coniferous tree species: a review on yield, ecological stability and economics. *European Journal of Forest Research* 127:89–101.

Koca, D., B. Smith, and M. T. Sykes. 2006. Modelling regional climate change effects on potential natural ecosystems in Sweden. *Climatic Change* 78:381–406.

Koch, J. M. 2007. Restoring a Jarrah understorey vegetation after bauxite mining in Western Australia. *Restoration Ecology* 15:S26–S39.

Koch, J. M., and R. J. Hobbs. 2007. Synthesis: Is Alcoa successfully restoring a Jarrah forest ecosystem after bauxite mining in Western Australia? *Restoration Ecology* 15:S137–S144.

Kort, J., M. Collins, and D. Ditsch. 1998. A review of soil erosion potential associated with biomass crops. *Biomass and Bioenergy* 14:351–359.

Kremen, C., and A. Miles. 2012. Ecosystem services in biologically diversified versus conventional farming systems: benefits, externalities, and trade-offs. *Ecology and Society* 17:40.

Kröger, M. 2012. The expansion of industrial tree plantations and dispossession in Brazil. *Development and Change* 43:947-973.

Kröger, M., and J.-E. Nylund. 2012. The conflict over Veracel pulpwood plantations in Brazil – Application of Ethical Analysis. *Forest Policy and Economics* 14:74–82.

Kuemmerle, T., P. Olofsson, O. Chaskovskyy, M. Baumann, K. Ostapowicz, C. E. Woodcock, R. A. Houghton, P. Hostert, W. S. Keeton, and V. C. Radeloff. 2011. Post-Soviet farmland abandonment, forest recovery, and carbon sequestration in western Ukraine. *Global Change Biology* 17:1335–1349.

Kuiters, A. T., and P. A. Slim. 2003. Tree colonisation of abandoned arable land after 27 years of horse-grazing: the role of bramble as a facilitator of oak wood regeneration. *Forest Ecology and Management* 181:239–251.

Kuuluvainen, T., H. Tukia, and K. Aapala. 2005. Ecological restoration of forested ecosystems in Finland. Pages 285–298 *in* J. A. Stanturf and P. Madsen, editors. *Restoration of Boreal and Temperate Forests.* CRC Press, Boca Raton, USA.

Laestadius, L., S. Maginnis, S. Minnemeyer, P. Potapov, C. Saint-Laurent, and N. Sizer. 2011. Mapping opportunities for forest landscape restoration. *Unasylva* 62:47.

Laganiere, J., D. A. Angers, and D. Pare. 2010. Carbon accumulation in agricultural soils after afforestation: A meta-analysis. *Global Change Biology* 16:439–453.

Lagi, M., K. Z. Bertrand, and Y. Bar-Yam. 2011. *The Food Crises and Political Instability in North Africa and the Middle East.* New England Complex Systems Institute, Cambridge, USA.

Lamb, D. 2011. *Regreening the Bare Hills: Tropical Forest Restoration in the Asia-Pacific Region.* Springer, Dordrecht, Netherlands.

Lamb, D., and D. G. Gilmour. 2003. *Rehabilitation and Restoration of Degraded Forests.* International Union for the Conservation of Nature and World Wide Fund for Nature, Cambridge, UK.

Lamb, D., and P. Lawrence. 1993. Mixed species plantations using high value rainforest trees in Australia *in* H. Leith and M. Lohman, editors. *Restoration of Tropical Forest Ecosystems.* Kluwer Academic Publishers, Dordrecht, Netherlands.

Lambin, E. F., and P. Meyfroidt. 2010. Land use transitions: socio-ecological feedback versus socio-economic change. *Land Use Policy* 27:108–118.

Lambin, E. F., and P. Meyfroidt. 2011. Global land use change, economic globalization, and the looming land scarcity. *Proceedings of the National Academy of Sciences* 108:3465–3472.

Lambin, E. F., H. J. Geist, and E. Lepers. 2003. Dynamics of land-use and land-cover change in tropical regions. *Annual Review of Environment and Resources* 28:205–241.

Lambini, C. K., and T. T. Nguyen. 2013. A comparative analysis of the effects of institutional property rights on forest livelihoods and forest conditions: evidence from Ghana and Vietnam. *Forest Policy and Economics.*

Lara, A., and T. Veblen. 1993. Forest plantations in Chile: a successful model? Pages 118–139 *in* A. Mather, editor. *Afforestation Policies, Planning and Progress.* Belhaven Press, London and Florida, USA.

Lasanta-Martinez, T., S. M. Vicente-Serrano, and J. Cuadrat-Prats. 2005. Mountain Mediterranean landscape evolution caused by the abandonment of traditional primary activities: a study of the Spanish Central Pyrenees. *Applied Geography* 25:47–65.

Laurance, W. F., and R. O. Bierregaard, editors. 1997. *Tropical Forest Remnants: Ecology, Management, and Conservation of Fragmented Communities.* University of Chicago Press, Chicago and London.

Le, H. D., C. Smith, J. Herbohn, and S. Harrison. 2012. More than just trees: assessing reforestation success in tropical developing countries. *Journal of Rural Studies* 28:5–19.

Le, T. P., V. D. Nguyen, N. Q. Nguyen, L. V. Phan, E. Morrison, and S. Vermeulen. 2004. *Making the Most of Market Chains: Challenges for Small-scale Farmers and Traders in Upland Vietnam.* International Institute for Environment and Development, London.

Leach, M., and J. Fairhead. 2000. Challenging neo Malthusian deforestation analyses in West Africa's dynamic forest landscapes. *Population and Development Review* 26:17–43.

Lebel, L., and R. Daniel. 2009. The governance of ecosystem services from tropical upland watersheds. *Current Opinion in Environmental Sustainability* 1:61–68.

Leck, M., V. Parker, and R. Simpson. 1989. *Ecology of Soil Seed Banks.* Academic Press, San Diego, USA.

Lee, D. K. 2012. The forest sector's contribution to a 'low carbon green growth' vision in the Republic of Korea. *Unasylva* 63:9–16.

Lee, D. K., and S. J. Suh. 2005. Forest restoration and rehabilitation in Republic of Korea. Pages 383–396 *in* J. A. Stanturf and P. Madsen, editors. *Restoration of Boreal and Temperate Forests.* CRC Press, Boca Raton, USA.

Lee, D. K., H. S. Kang, and Y. D. Park. 2004. Natural restoration of deforested woodlots in South Korea. *Forest Ecology and Management* 201:23–32.

Lee, S. S. 1999. Forest health in plantation forests in Southeast Asia. *Australasian Plant Pathology* 28:283–291.

Lemenih, M., and F. Bongers. 2010. The role of plantation forests in fostering ecological restoration: experiences from East Africa. Pages 171–219 *in* F. Bongers and T. Tennigkeit, editors. *Degraded Forests in Eastern Africa: Management and Restoration.* Earthscan, London.

Leys, A. J., and J. K. Vanclay. 2011. Stakeholder engagement in social learning to resolve controversies over land-use change to plantation forestry. *Regional Environmental Change* 11:175–190.

Li, X., M. J. Yi, Y. Son, G. Jin, and S. S. Han. 2010. Forest biomass carbon accumulation in Korea from 1954 to 2007. *Scandinavian Journal of Forest Research* 25:554–563.

Lindenmayer, D. B., and J. Fischer. 2006. *Habitat Fragmentation and Landscape Change.* CSIRO Publishing, Collingwood, Australia.

Lindenmayer, D. B., and G. E. Likens. 2009. Adaptive monitoring: a new paradigm for long-term research and monitoring. *Trends in Ecology & Evolution* 24:482–486.

Linnell, J. D., J. E. Swenson, and R. Andersen. 2000. Conservation of biodiversity in Scandinavian boreal forests: large carnivores as flagships, umbrellas, indicators, or keystones? *Biodiversity & Conservation* 9:857–868.

Little, C., A. Lara, J. McPhee, and R. Urrutia. 2009. Revealing the impact of forest exotic plantations on water yield in large scale watersheds in South-Central Chile. *Journal of Hydrology* 374:162–170.

Liu, J., S. Li, Z. Ouyang, C. Tam, and X. Chen. 2008. Ecological and socioeconomic effects of China's policies for ecosystem services. *Proceedings of the National Academy of Sciences* 105:9477–9482.

Locke, H. 2010. Yellowstone to Yukon connectivity conservation initiative. Pages 161–181 *in* G. Worboys, W. Francis, and M. Lockwood, editors. *Connectivity Conservation Management: A Global Guide.* Earthscan, London.

Lof, M., J. Berquist, J. Brunet, M. Karlsson, and N. Welander. 2010. Conversion of Norway spruce stands to broadleaved woodland – regeneration systems, fencing and performance of planted seedlings. *Ecological Bulletins* 53:165–173.

Logan, W. E. M. 1967. Policy. *Unasylva* 21:86–87.

Long, R. P., P. H. Brose, and S. B. Horsley. 2012. Responses of northern red oak seedlings to lime and deer exclosure fencing in Pennsylvania. *Canadian Journal of Forest Research* 42:698–709.

Lovejoy, T. E., and R. Ashton. 2011. Stepping up the ambition for carbon management: a vision for carbon-rich, cross-continental biodiversity corridors. *Carbon Management* 2:101–103.

Lugo, A. E. 2009. The emerging era of novel tropical forests. *Biotropica* 41:589–591.

Lunt, I. D., M. Byrne, J. J. Hellmann, N. J. Mitchell, S. T. Garnett, M. W. Hayward, T. G. Martin, E. McDonald-Madden, S. E. Williams, and K. K. Zander. 2013. Using assisted colonisation to conserve biodiversity and restore ecosystem function under climate change. *Biological Conservation* 157:172–177.

Lynam, T., W. de Jong, D. Sheil, T. Kusumanto, and K. Evans. 2007. A review of tools for incorporating community knowledge, preferences, and values into decision making in natural resources management. *Ecology & Society* 12:1–15

McDonald, G. 1993. Reafforestation in Australia. Pages 163–180 *in* A. Mather, editor. *Afforestation Policies, Planning and Progress.* Bellhaven Press, London and Florida, USA.

MacDonald, D., J. R. Crabtree, G. Wiesinger, T. Dax, N. Stamou, P. Fleury, J. Gutierrez Lazpita, and A. Gibon. 2000. Agricultural abandonment in mountain areas of Europe: Environmental consequences and policy response. *Journal of Environmental Management* 59:47–69.

McElwee, P. 2009. Reforesting 'bare hills' in Vietnam: social and environmental consequences of the 5 million hectare reforestation program. *Ambio* 38:325–333.

McMahon, S. M., S. P. Harrison, W. S. Armbruster, P. J. Bartlein, C. M. Beale, M. E. Edwards, J. Kattge, G. Midgley, X. Morin, and I. C. Prentice. 2011. Improving assessment and modelling of climate change impacts on global terrestrial biodiversity. *Trends in Ecology & Evolution* 26:249–259.

McNamara, S., D. V. Tinh, P. D. Erskine, D. Lamb, D. Yates, and S. Brown. 2006. Rehabilitating degraded forest land in central Vietnam with mixed native species plantings. *Forest Ecology and Management* 233:358–365.

McNamara, S., P. Erskine, D. Lamb, L. Chantalangsy, and S. Boyle. 2012. Primary tree species diversity in secondary fallow forests of Laos. *Forest Ecology and Management* 281:93–99.

McNeely, J. A., and S. J. Scherr. 2003. *Ecoagriculture: Strategies to Feed the World and Save Wild Biodiversity*. Island Press, Washington and Covelo, USA and London.

McRae, B. H., S. A. Hall, P. Beier, and D. M. Theobald. 2012. Where to restore ecological connectivity? Detecting barriers and quantifying restoration benefits. *PLoS ONE* **7**:e52604.

McShane, T. O., P. D. Hirsch, T. C. Trung, A. N. Songorwa, A. Kinzig, B. Monteferri, D. Mutekanga, H. V. Thang, J. L. Dammert, and M. Pulgar-Vidal. 2011. Hard choices: Making trade-offs between biodiversity conservation and human well-being. *Biological Conservation* 144:966–972.

McVicar, T. R., L. Li, T. G. Van Niel, L. Zhang, R. Li, Q. Yang, X. Zhang, X. Mu, Z. Wen, and W. Liu. 2007. Developing a decision support tool for China's re-vegetation program: Simulating regional impacts of afforestation on average annual streamflow in the Loess Plateau. *Forest Ecology and Management* 251:65–81.

McVicar, T. R., T. G. Van Niel, L. Li, Z. Wen, Q. Yang, R. Li, and F. Jiao. 2010. Parsimoniously modelling perennial vegetation suitability and identifying priority areas to support China's re-vegetation program in the Loess Plateau: Matching model complexity to data availability. *Forest Ecology and Management* 259:1277–1290.

Maginnis, S., and W. Jackson. 2007. What is FLR and how does it differ from current approaches? Pages 5–20 *in* J. Rietbergen- McCracken, S. Maginnis, and A. Sarre, editors. *The Forest Landscape Restoration Handbook*. Earthscan, London.

Makarieva, A. M., V. G. Gorshkov, and B.-L. Li. 2006. Conservation of water cycle on land via restoration of natural closed-canopy forests: implications for regional landscape planning. *Ecological Research* 21:897–906.

Makarieva, A. M., and V.G. Gorshkov. 2007. Biotic pump of atmospheric moisture as driver of the hydrological cycle on land. *Hydrology and Earth System Sciences Discussions* 11: 1013-1033.

Malcolm, D. C. 1997. The silviculture of conifers in Great Britain. *Forestry* 70:293–307.

Malcolm, D. C., W. L. Mason, and G. C. Clarke. 2001. The transformation of conifer forests in Britain – regeneration, gap size and silvicultural systems. *Forest Ecology and Management* 151:7–23.

Manvell, A. 2001. Woodfuel in Niger: Crisis or adaptation in the last ten years? Paper for the SIDA–supported CIFOR research project Fuelwood: Crisis or Balance? 6–8 June 2001, Marstrand, Sweden.

Marín-Spiotta, E., and S. Sharma. 2013. Carbon storage in successional and plantation forest soils: a tropical analysis. *Global Ecology and Biogeography* 22:105–117.

Maron, M., R. J. Hobbs, A. Moilanen, J. W. Matthews, K. Christie, T. A. Gardner, D. A. Keith, D. B. Lindenmayer, and C. A. McAlpine. 2012. Faustian bargains? Restoration realities in the context of biodiversity offset policies. *Biological Conservation* 155:141–148.

Martin, P. A., A. C. Newton, and J. M. Bullock. 2013. Carbon pools recover more quickly than plant biodiversity in tropical secondary forests. *Proceedings of the Royal Society B: Biological Sciences* 280: Issue 1773, Article No.20132236.

Mason, W. L. 2007a. Changes in the management of British forests between 1945 and 2000 and possible future trends. *Ibis* 149:41–52.

Mason, W. L. 2007b. Silviculture of Scottish forests at a time of change. *Journal of Sustainable Forestry* 24:41–57.

Mather, A. 1993. Afforestation in Britain. Pages 13–33 *in* A. Mather, editor. *Afforestation: Policies, Planning and Progress.* Bellhaven press, London.

Mather, A. S. 2007. Recent Asian forest transitions in relation to forest-transition theory. *International Forestry Review* 9:491–502.

Mayers, J. 2000. Company-community forestry partnerships: a growing phenomenon. *Unasylva* 51:33–41.

Mayers, J., and S. Bass. 2004. *Policy that Works for Forests and People: Real Prospects for Governance and Livelihoods.* Earthscan, London.

Mayers, J., and S. Vermeulen. 2002. *Company-Community Forestry Partnerships; From Raw Deals to Mutual Gains?* International Institute for Environment and Development, London.

Meesters, A., A. Dolman, and L. Bruijnzeel. 2009. Comment on 'Biotic pump of atmospheric moisture as driver of the hydrological cycle on land' by A. M. Makarieva and V. C. Gorshkov. *Hydrology & Earth System Sciences* 13:1299–1305.

Meijaard, E., and D. Sheil. 2011. A modest proposal for wealthy countries to reforest their land for the common good. *Biotropica* 43:524–528.

Melo, F. P., S. R. Pinto, P. H. Brancalion, P. S. Castro, R. R. Rodrigues, J. Aronson, and M. Tabarelli. 2013. Priority setting for scaling-up tropical forest restoration projects: Early lessons from the Atlantic Forest Restoration Pact. *Environmental Science & Policy* 33:395–404.

Mendenhall, C. D., C. H. Sekercioglu, F. O. Brenes, P. R. Ehrlich, and G. C. Daily. 2011. Predictive model for sustaining biodiversity in tropical countryside. *Proceedings of the National Academy of Sciences* 108:16313–16316.

Mengistu, T., D. Teketay, H. Hulten, and Y. Yemshaw. 2005. The role of enclosures in the recovery of woody vegetation in degraded dryland hillsides of central and northern Ethiopia. *Journal of Arid Environments* 60:259–281.

Menzies, N. K. 1996. Forestry *in* J. Needham, editor. *Science and Civilisation in China.* Cambridge University Press, Cambridge, UK.

Mercer, D., and A. Underwood. 2002. Australian timber plantations: national vision, local response. *Land Use Policy* 19:107–122.

Mergen, F., H. G. Abbott, W. F. Mann, F. R. Moulds, A. H. Nordmeyer, J. D. Scott, and N. D. Vietmeyer. 1981. *Sowing Forests from the Air.* National Academy Press, Washington.

Merritt, D. J., and K. W. Dixon. 2011. An unorthodox approach to forest restoration: Response. *Science* 333:36–37.

Mesquita, A. O., and M. Passamani. 2012. Composition and abundance of small mammal communities in forest fragments and vegetation corridors in Southern Minas Gerais, Brazil. *Revista de Biología Tropical* 60:1335–1343.

Meyfroidt, P., and E. F. Lambin. 2011. Global forest transition: Prospects for an end to deforestation. *Annual Review of Environment and Resources* 36:343–371.

Michon, G. 2005. *Domesticating Forests: How Farmers Manage Forest Resources.* Institut de Recherchepour le Developpement, Center for International Forestry Research, World Agroforestry Center, Paris and Bogor, Indonesia.

Milder, J. C., A. K. Hart, P. Dobie, J. Minai, and C. Zaleski. 2014. Integrated landscape initiatives for African agriculture, development, and conservation: A region-wide assessment. *World Development* 54:68–80.

Millennium Ecosystem Assessment. 2003. *Ecosystems and Human Well-being: A Framework for Assessment.* Island Press, Washington.

Moinuddin, H., S. Pokhrel, X. Jiao, L. Linde, A. Fraser, M. A. Botengan, S. Katz, and N. Crishna. 2011. *Biodiversity conservation corridors initiative (BCI) report 2006–2011: Greater Mekong Subregion Core Environment Program.* Asian Development Bank, Manila.

Molina, A., G. Govers, J. Poesen, H. Van Hemelryck, B. De Bièvre, and Vanacker, V. 2008. Environmental factors controlling spatial variation in sediment yield in a central Andean mountain area. *Geomorphology* 98: 176-186.

Molina, A., G. Govers, F. Cisneros, and V. Vanacker. 2009. Vegetation and topographic controls on sediment deposition and storage on gully beds in a degraded mountain area. *Earth Surface Processes and Landforms* 34:755–767.

Morales, E. 2005. Early experience of total divestment: Chile. Pages 126–153 *in* M. Garforth and J. Mayers, editors. *Plantations, Privatization, Poverty and Power: Changing Ownership and Management of State Forests.* Earthscan, London.

Moreira, F., and D. Russo. 2007. Modelling the impact of agricultural abandonment and wildfires on vertebrate diversity in Mediterranean Europe. *Landscape Ecology* 22:1461–1476.

Mori, A. S., T. Furukawa, and T. Sasaki. 2013. Response diversity determines the resilience of ecosystems to environmental change. *Biological Reviews* 88:349–364.

Moxham, R. 2001. *The Great Hedge of India.* Robinson, London.

Muller, E., and J. Barborak. 2010. Mesoamerican biological corridor. Pages 183–191 *in* G. Worboys, W. Francis, and M. Lockwood, editors. *Connectivity Conservation Management: A Global Guide.* Earthscan, London.

Myers, J. A., and K. E. Harms. 2009. Seed arrival, ecological filters, and plant species richness: a meta analysis. *Ecology Letters* 12:1250–1260.

Nagendra, H., and J. Southworth. 2009. *Reforesting Landscapes.* Springer, Dordrecht, Netherlands.

Nair, K. S. S. 2007. *Tropical Forest Insect Pests: Ecology, Impact and Management.* Cambridge University Press, Cambridge, UK.

Nasi, R., P. Koponen, J. G. Poulsen, M. Buitenzorgy, and W. Rusmantoro. 2008. Impact of landscape and corridor design on primates in a large-scale industrial tropical plantation landscape. *Biodiversity and Conservation* 17:1105–1126.

Navarro, L. M., and H. M. Pereira. 2012. Rewilding abandoned landscapes in Europe. *Ecosystems* 15:900–912.

Nave, L., C. Swanston, U. Mishra, and K. Nadelhoffer. 2013. Afforestation effects on soil carbon storage in the United States: a synthesis. *Soil Science Society of America Journal* 77:1035–1047.

Nawir, A. A., C. Anyonge, D. Race, and S. Vermeulen. 2003. *Towards Equitable Partnerships between Corporate and Smallholder Partners: Relating Partnerships to Social, Economic and Environmental Indicators.* Workshop Synthesis. Food and Agriculture Organization of the United Nations, Rome.

Nearing, M., F. Pruski, and M. O'N'eal. 2004. Expected climate change impacts on soil erosion rates: a review. *Journal of Soil and Water Conservation* 59:43–50.

Nellerman, C., and E. Corcoran. 2010. *Dead Planet, Living Planet: Biodiversity and Ecosystem Restoration for Sustainable Development. A Rapid Response Assessment.* United Nations Environmental Program, Nairobi.

Nelson, E., G. Mendoza, J. Regetz, S. Polasky, H. Tallis, D. Cameron, K. M. A. Chan, G. C. Daily, J. Goldstein, P. M. Kareiva, E. Lonsdorf, R. Naidoo, T. H. Ricketts, and M. Shaw. 2009. Modeling multiple ecosystem services, biodiversity conservation, commodity production, and tradeoffs at landscape scales. *Frontiers in Ecology and the Environment* 7:4–11.

Newton, A., E. Cantarello, N. Tejedor, T. Kitzberger, C. Echeverría, G. Williams-Linera, D. Golicher, G. Bolados, L. Malizia, Manson R. , F. López-Barrera, N. Ramirez-Marcial, M. Martinez-Icó, G. Henriquez, and R. Hill. 2011. Landscape-scale dynamics and restoration of dryland forest ecosystems.*in* A. C. Newton and N. Tejedor, editors. *Principles and Practice of Forest Landscape Restoration: Case studies from the drylands of Latin America.* International Union for the Conservation of Nature, Gland, Switzerland.

Nguyen, H., J. Herbohn, J. Firn, and D. Lamb. 2012. Biodiversity–productivity relationships in small-scale mixed-species plantations using native species in Leyte province, Philippines. *Forest Ecology and Management* 274:81–90.

Nguyen, H., D. Lamb, J. Herbohn, and J. Firn. 2014. Designing mixed species tree plantations for the tropics: balancing ecological attributes of species with landholder preferences in the Philippines. *PLoS ONE* 9(5): e98600. doi:10.1371/journal.pone.0098600.

Nichols, J. D., M. Bristow, and J. K. Vanclay. 2006. Mixed-species plantations: prospects and challenges. *Forest Ecology and Management* 233:383–390.

Nilsson, S., and W. Schopfhauser. 1995. The carbon-sequestration potential of a global afforestation program. *Climatic Change* 30:267–293.

Noble, J. 1997. *The Delicate and Noxious Scrub.* Commonwealth Scientific and Industrial Research Organisation, Canberra.

Normile, D. 2010. Saving forests to save biodiversity. *Science* 329:1278–1280.

Nosetto, M., E. Jobbágy, T. Tóth, and R. Jackson. 2008. Regional patterns and controls of ecosystem salinization with grassland afforestation along a rainfall gradient. *Global Biogeochemical Cycles* 22. doi:10.1029/2007GB003000.

Obidzinski, K., and A. Dermawan. 2010. Smallholder timber plantation development in Indonesia: what is preventing progress? *International Forestry Review* 12:339–348.

Oldeman, L. R., R. T. A. Hakkeling, and W. G. Sombroek. 1990. *World Map of the Status of Human-Induced Soil Degradation.* International Soil Reference and Information Center, Wageningen, Netherlands.

Oliver, T., D. B. Roy, J. K. Hill, T. Brereton, and C. D. Thomas. 2010. Heterogeneous landscapes promote population stability. *Ecology Letters* 13:473–484.

Orsi, F., R. L. Church, and D. Geneletti. 2011. Restoring forest landscapes for biodiversity conservation and rural livelihoods: A spatial optimisation model. *Environmental Modelling & Software* 26:1622–1638.

OSS. 2008. *The Great Green Wall Initiative for the Sahara and the Sahel.* OSS, Tunis.

Ostberg, S., W. Lucht, S. Schaphoff, and D. Gerten. 2013. Critical impacts of global warming on land ecosystems. *Earth System Dynamics* 4:347–357.

Ostrom, E. 2010. Polycentric systems for coping with collective action and global environmental change. *Global Environmental Change* 20:550–557.

Ostrom, E., and M. Cox. 2010. Moving beyond panaceas: a multi-tiered diagnostic approach for social-ecological analysis. *Environmental Conservation* 37:451–463.

Oviedo, G. 2005. Land ownership and forest restoration. Pages 84–93 *in* S. Mansourian, D. Vallauri, and N. Dudley, editors. *Forest Restoration in Landscapes.* Springer, New York.

Pacoe, M. 2009. Australia's biggest scam comes crashing down. *Sydney Morning Herald,* 12 May 2009, Sydney.

Palmer, J. 1991. Jari: Lessons for land managers in the tropics. Pages 419–430 *in* A. Gomez-Pompa, T. Whitmore, and M. Hadley, editors. *Rainforest Regeneration and Management.* UNESCO and Parthenon Publishing, Carnforth, UK.

Palo, M., and E. Lehto. 2012. *Private or Socialistic Forestry: Forest Transition in Finland vs. Deforestation in the Tropics.* Springer, Dordrecht, Netherlands.

Pan, Y., R. A. Birdsey, J. Fang, R. Houghton, P. E. Kauppi, W. A. Kurz, O. L. Phillips, A. Shvidenko, S. L. Lewis, and J. G. Canadell. 2011. A large and persistent carbon sink in the world's forests. *Science* 333:988–993.

Pannell, D. J., D. J. McFarlane, and R. Ferdowsian. 2001. Rethinking the externality issue for dryland salinity in Western Australia. *Australian Journal of Agricultural and Resource Economics* 45:459–475.

Pannell, D. J., A. M. Roberts, G. Park, and J. Alexander. 2013. Designing a practical and rigorous framework for comprehensive evaluation and prioritisation of environmental projects. *Wildlife Research* 40:126–133.

Pant, M. M. 1979. Social forestry in India. *Unasylva* 31:19–24.

Pappila, M., and I. Pölönen. 2012. Reconsidering the role of public participation in the Finnish forest planning system. *Scandinavian Journal of Forest Research* 27:177–185.

Paquette, A., and C. Messier. 2010. The effect of biodiversity on tree productivity: from temperate to boreal forests. *Global Ecology and Biogeography* 20:170–180.

Parkes, D., G. Newell, and D. Cheal. 2003. Assessing the quality of native vegetation: the 'habitat hectares' approach. *Ecological Management & Restoration* 4:S29–S38.

Parkyn, S. M., R. J. Davies-Colley, A. B. Cooper, and M. J. Stroud. 2005. Predictions of stream nutrient and sediment yield changes following restoration of forested riparian buffers. *Ecological Engineering* 24:551–558.

Parris, H., S. Whitten, C. Wyborn, R. Hill, and D. Freudenberger. 2011. *An Overview of Key Socio-Economic Factors, Principles and Guidelines in Wildlife 'Corridor' Planning and Implementation.* CSIRO, Canberra.

Parrotta, J., J. M. Turnbull, and N. Jones. 1997. Catalysing native forest regeneration on degraded tropical lands. *Forest Ecology and Management* 99:1–7.

Paruelo, J. M. 2012. Ecosystem services and tree plantations in Uruguay: A reply to Vihervaara et al. (2012). *Forest Policy and Economics* 22:85–88.

Parungo, F., Z. Li, X. Li, D. Yang, and J. Harris. 1994. Gobi dust storms and The Great Green Wall. *Geophysical Research Letters* 21:999–1002.

Pasari, J. R., T. Levi, E. S. Zavaleta, and D. Tilman. 2013. Several scales of biodiversity affect ecosystem multifunctionality. *Proceedings of the National Academy of Sciences* 110:10219–10222.

Paterson, S., and B. A. Bryan. 2012. Food-carbon trade-offs between agriculture and reforestation land uses under alternate market-based policies. *Ecology and Society* 17:21.

Paul, K. I., P. J. Polglase, J. G. Nyakuengama, and K. P. Khanna. 2002. Change in soil carbon following afforestation. *Forest Ecology and Management.* 168:241–257.

Pawson, S., A. Brin, E. Brockerhoff, D. Lamb, T. Payn, A. Paquette, and J. Parrotta. 2013. Plantation forests, climate change and biodiversity. *Biodiversity and Conservation* 22:1203–1227.

Pearce, F. 2012. *The Land Grabbers: the New Fight over Who Owns the Earth.* Beacon Press, Boston, USA.

Peh, K. 2010. Invasive species in Southeast Asia: the knowledge so far. *Biodiversity and Conservation* 19:1083–1099.

Perfecto, I., and J. Vandermeer. 2010. The agroecological matrix as alternative to the land-sparing/agriculture intensification model. *Proceedings of the National Academy of Sciences* 107:5786–5791.

Perz, S. G., and D. L. Skole. 2003. Social determinants of secondary forests in the Brazilian Amazon. *Social Science Research* 32:25–60.

Petit, B., and F. Montagnini. 2006. Growth in pure and mixed plantations of tree species used in reforesting rural areas of the humid region of Costa Rica, Central America. *Forest Ecology and Management* 233:338–343.

Pew Research. 2007. Global unease with major world powers. *Pew Global Attitudes Project, June 27 2007.* Available at: www.pewglobal.org/2007/06/27/global-unease-with-major-world-powers [Accessed 16 May 2014].

Phalan, B., A. Balmford, R. E. Green, and J. P. W. Scharlemann. 2011. Minimising the harm to biodiversity of producing more food globally. *Food Policy* 36, Supplement 1:S62–S71.

Philpott, S., P. Bichier, R. A. Rice, and R. Greenberg. 2008. Biodiversity conservation, yield and alternative products in coffee agroforestry systems in Sumatra, Indonesia. *Biodiversity and Conservation* 17:1805–1820.

Pichancourt, J. B., J. Firn, I. Chadès, and T. G. Martin. 2014. Growing biodiverse carbon-rich forests. *Global Change Biology* 20:382–393.

Piotto, D., E. Viquez, F. Montagnini, and M. Kanninen. 2004. Pure and mixed forest plantations with native species of the dry tropics of Costa Rica: a comparison of growth and productivity. *Forest Ecology and Management* 190:359–372.

Poffenberger, M., and B. McGean. 1993. *Community Allies: Forest Co-Management in Thailand.* Center for Southeast Asian Studies, University of California, Berkeley, USA.

Polasky, S., E. Nelson, J. Camm, B. Csuti, P. Fackler, E. Lonsdorf, C. Montgomery, D. White, J. Arthur, B. Garber-Yonts, R. Haight, J. Kagan, A. Starfield, and C. Tobalske. 2008. Where to put things? Spatial land management to sustain biodiversity and economic returns. *Biological Conservation* 141: 1505–1524.

Poore, M. E. D., and C. Fries. 1985. *The ecological effect of eucalyptus.* FAO Forestry Paper 59. UN Food and Agriculure Orgsanisation, Rome.

Potter, L. 1987. Degradation, innovation and social welfare in the Riam Kiwa valley, Kalimantan, Indonesia. Pages 164–176 *in* P. Blaikie and H. Brookfield, editors. *Land Degradation and Society.* Methuen, London and New York.

Pretty, J. N., A. D. Noble, D. Bossio, J. Dixon, R. E. Hine, F. W. T. Penning de Vries, and J. I. L. Morison. 2006. Resource-conserving agriculture increases yields in developing countries. *Environmental Science & Technology* 40:1114–1119.

Pretzsch, H. 2005. Diversity and productivity in forests: evidence from long-term experimental plots. Pages 41–64 *in* M. Scherer-Lorenzen, C. Körner, and E.-D. Schulze, editors. *Forest Diversity and Function: Temperate and Boreal Systems.* Springer-Verlag, Berlin.

Pretzsch, H., and G. Schutze. 2009. Transgressive overyielding in mixed compared with pure stands of Norway spruce and European beech in Central Europe: evidence on stand level and explanation on individual tree level. *European Journal of Forest Research* 128:183–204.

Prévosto, B., D. R. Hill, and P. Coquillard. 2003. Individual-based modelling of *Pinus sylvestris* invasion after grazing abandonment in the French Massif Central. *Plant Ecology* 168:121–137.

Prévosto, B., E. Dambrine, C. Moares, and T. Curt. 2004. Effects of volcanic ash chemistry and former agricultural use on the soils and vegetation of naturally regenerated woodlands in the Massif Central, France. *Catena* 56:239–261.

Printiss, D. 2013. Longleaf pine savanna restoration in Florida, USA. Pages 53–59 *in* A. F. Clewell and J. Aronson, editors. *Ecological Restoration: Principles, Values and Structure of an Emerging Profession.* Island Press, Washington.

Program on Forests. 2011. *Investing in Trees and Landscape Restoration in Africa: Overview.* Program on Forests, World Bank, Washington.

Puettmann, K. J., K. D. Coates, and C. Messier. 2009. *Critique of Silviculture: Managing for Complexity.* Island Press, Washington.

Pullar, D., and D. Lamb. 2012. A tool for comparing alternative forest landscape restoration scenarios. Pages 3–20 *in* J. A. Stanturf, P. Madsen, and D. Lamb, editors. *A Goal-Oriented Approach to Forest Landscape Restoration.* Springer, Dordrecht, Netherlands.

Purvis, A., and A. Hector. 2000. Getting the measure of biodiversity. Nature 405:212–219.

Quijas, S., B. Schmid, and P. Balvanera. 2010. Plant diversity enhances provision of ecosystem services: a new synthesis. *Basic and Applied Ecology* 11:582–593.

Rackham, O. 1986. *The History of the Countryside.* J.M. Dent, London.

Ramankutty, N., A. T. Evan, C. Monfreda, and J. A. Foley. 2008. Farming the planet: 1. Geographic distribution of global agricultural lands in the year 2000. *Global Biogeochem.* Cycles 22:GB1003. doi:10.1029/2007GB002952

Ravindranath, N., and I. Murthy. 2010. Greening India Mission. *Current Science (Bangalore)* 99:444–449.

Reeson, A. F., L. C. Rodriguez, S. M. Whitten, K. Williams, K. Nolles, J. Windle, and J. Rolfe. 2011. Adapting auctions for the provision of ecosystem services at the landscape scale. *Ecological Economics* 70:1621–1627.

Reij, C., G. Tappan, and M. Smale. 2009. *Agroenvironmental Transformation in the Sahel: Another Kind of Green Revolution.* International Food Policy Research Institute, Washington.

Reitbergen-McCracken, J., S. Maginnis, and A. Sarre. 2007. *The Forest Landscape Restoration Handbook.* Earthscan, London.

Rey, F. 2003. Influence of vegetation distribution on sediment yield in forested marly gullies. *Catena* 50:549–562.

Richards, A. E., and S. Schmidt. 2010. Complementary resource use by tree species in a rain forest tree plantation. *Ecological Applications* 20:1237–1254.

Richards, A. E., D. I. Forrester, J. Bauhus, and M. Scherer-Lorenzen. 2010. The influence of mixed tree plantations on the nutrition of individual species: a review. *Tree Physiology* 30:1192–1208.

Richards, K. R., and C. Stokes. 2004. A review of forest carbon sequestration cost studies: a dozen years of research. *Climatic Change* 63:1–48.

Richardson, D. M., and M. Rejmanek. 2011. Trees and shrubs as invasive alien species: a global review. *Diversity and Distributions* 17:788–809.

Richter, D., and R. Houghton. 2011. Gross CO2 fluxes from land-use change: implications for reducing global emissions and increasing sinks. *Carbon Management* 2:41–47.

Ripple, W. J., and R. L. Beschta. 2012. Trophic cascades in Yellowstone: The first 15 years after wolf reintroduction. *Biological Conservation* 145:205–213.

Ritchie, E. G., B. Elmhagen, A. S. Glen, M. Letnic, G. Ludwig, and R. A. McDonald. 2012. Ecosystem restoration with teeth: what role for predators? *Trends in Ecology & Evolution* 27:265–271.

Ritchie, M. E., and H. Olff. 1999. Spatial scaling laws yield a synthetic theory of biodiversity. *Nature* 400:557–560.

Robbins, P. 2001. Tracking invasive land covers in India, or why our landscapes have never been modern. *Annals of the Association of American Geographers* 91:637–659.

Rodrigues, R. R., R. Lima, S. Gandolfi, and A. G. Nave. 2009. On the restoration of high diversity forests: 30 years of experience in the Brazilian Atlantic Forest. *Biological Conservation* 142:1242–1251.

Rodrigues, R. R., S. Gandolfi, A. G. Nave, J. Aronson, T. E. Barreto, C. Y. Vidal, and P. H. S. Brancalion. 2011. Large-scale ecological restoration of high-diversity tropical forests in SE Brazil. *Forest Ecology and Management* 261:1605–1613.

Rodwell, J., and G. Patterson. 1994. *Creating New Native Woodlands.* Forestry Commission, London.

Roe, D. 2008. The origins and evolution of the conservation-poverty debate: a review of key literature, events and policy processes. *Oryx* 42:491–503.

Roessiger, J., V. C. Griess, and T. Knoke. 2011. May risk aversion lead to near-natural forestry? A simulation study. *Forestry* 84:527–537.

Ross, H., M. Buchy, and W. Proctor. 2002. Laying down the ladder: a typology of public participation in Australian natural resource management. *Australasian Journal of Environmental Management* 9:205–217.

Ros-Tonen, M. A., T. F. Insaidoo, and E. Acheampong. 2013. Promising start, bleak outlook: The role of Ghana's modified taungya system as a social safeguard in timber legality processes. *Forest Policy and Economics* 32:57–67.

Rothe, A., and D. Binkley. 2001. Nutritional interactions in mixed species forests: a synthesis. *Canadian Journal of Forest Research* 31:1855–1870.

Ruddiman, W. F. 2007. The early anthropogenic hypothesis: Challenges and responses. *Reviews of Geophysics* 45(4). doi:10.1029/2006RG000207

Rudel, T. K., D. Bates, and R. Machinguiashi. 2002. A tropical forest transition? Agricultural change, out-migration, and secondary forests in the Ecuadorian Amazon. *Annals of the Association of American Geographers* 92:87–102.

Rudel, T. K., O. T. Coomes, E. Moran, F. Achard, A. Angelsen, J. C. Xu, and E. Lambin. 2005. Forest transitions: towards a global understanding of land use change. *Global Environmental Change-Human and Policy Dimensions* 15:23–31.

Rudel, T. K., L. Schneider, and M. Uriarte. 2010. Forest transitions: An introduction. *Land Use Policy* 27:95–97.

Samways, M. J., C. S. Bazelet, and J. S. Pryke. 2010. Provision of ecosystem services by large scale corridors and ecological networks. *Biodiversity and Conservation* 19:2949–2962.

Sandberg, A. 2007. Property rights and ecosystem properties. *Land Use Policy* 24:613–623.

Sanderman, J., and A. Chappell. 2013. Uncertainty in soil carbon accounting due to unrecognized soil erosion. *Global Change Biology* 19:264–272.

Sandker, M., B. M. Campbell, M. Ruiz-Pérez, J. A. Sayer, R. Cowling, H. Kassa, and A. T. Knight. 2010. The role of participatory modeling in landscape approaches to reconcile conservation and development. *Ecology & Society* 15:1–16.

Sang, P. M., D. Lamb, M. Bonner, and S. Schmidt. 2013. Carbon sequestration and soil fertility of tropical tree plantations and secondary forest established on degraded land. *Plant and Soil* 362:187–200.

Sannai, J. 2006. Assisted natural regeneration in China. Pages 29–33 in *Advancing Assisted Natural Regeneration (ANR) in Asia and the Pacific*. FAO Regional Office for Asia and the Pacific, Bangkok.

Sayer, J., T. Sunderland, J. Ghazoul, J.-L. Pfund, D. Sheil, E. Meijaard, M. Venter, A. K. Boedhihartono, M. Day, and C. Garcia. 2013. Ten principles for a landscape approach to reconciling agriculture, conservation, and other competing land uses. *Proceedings of the National Academy of Sciences* 110:8349–8356.

Schama, S. 1995. *Landscape and Memory*. Harper Collins, London.

Scherr, S. 2000. A downward spiral? Research evidence on the relationship between poverty and natural resource degradation. *Food Policy* 25:479–498.

Scherr, S. J., S. Shames, and R. Friedman. 2012. From climate-smart agriculture to climate-smart landscapes. *Agriculture and Food Security* 1:12.

Schirmer, J., and L. Bull. 2014. Assessing the likelihood of widespread landholder adoption of afforestation and reforestation projects. *Global Environmental Change* 24:306–320.

Schirmer, J., and P. Kanowski. 2005. A mixed economy commonwealth of states: Australia. Pages 101–125 *in* M. Garford and J. Mayers, editors. *Plantations, Privatization, Poverty and Power: Changing Ownership and Management of State Forests*. Earthscan, London.

Schmidhuber, J., and F. N. Tubiello. 2007. Global food security under climate change. *Proceedings of the National Academy of Sciences* 104:19703–19708.

Schmidt, L. 2008. *A Review of Direct Sowing Versus Planting in Tropical Afforestation and Land Rehabilitation*. Forest and Landscape Denmark, University of Copenhagen, Horsholm, Denmark.

Schneemann, B., and C. McElhinny. 2012. Shrubby today but not tomorrow? Structure, composition and regeneration dynamics of direct seeded revegetation. *Ecological Management & Restoration* 13:282–289.

Schutz, J. P., M. Gotz, W. Schmid, and D. Mandallaz. 2006. Vulnerability of spruce (*Picea abies*) and beech (*Fagus sylvatica*) forest stands to storms and consequences for silviculture. *European Journal of Forest Research* 125:291–302.

Scott, D. F., L. A. Bruijnzeel, and J. Mackensen. 2005. The hydrological and soil impacts of forestation in the tropics. Pages 622–651 *in* M. Bonell and L. A. Bruijnzeel, editors. *Forests, Water and People in the Humid Tropics: Past, Present and Future Hydrological Research for Integrated Land and Water Management*. Cambridge University Press, Cambridge, UK.

Scott, J. C. 1998. *Seeing Like A State: How Certain Schemes to Improve the Human Condition have Failed*. Yale University Press, New Haven, USA and London.

Scott, R., P. Pattanakaew, J. F. Maxwell, S. Elliott, and G. Gale. 2000. The effect of artificial perches and local vegetation on bird dispersing seed deposition into regenerating sites. Pages 326–337 *in* S. Elliott, J. Kerby, D. Blakesley, K. Hardwick,

K. Woods, V. Anusarnsunthorn, editors. *Forest Restoration for Wildlife Conservation.* University of Chiang Mai, Thailand.

Sedjo, R. A., and D. Botkin. 1997. Using forest plantations to spare natural forests. *Environment: Science and Policy for Sustainable Development* 39(10):14–30.

Sekercioglu, C. H., S. R. Loarie, F. O. Brenes, P. R. Ehrlich, and G. C. Daily. 2007. Persistence of forest birds in the Costa Rican agricultural countryside. *Conservation Biology* 21:482–494.

Sendzimir, J., C. P. Reij, and P. Magnuszewski. 2011. Rebuilding resilience in the Sahel: regreening in the Maradi and Zinder regions of Niger. *Ecology and Society* 16:1.

SER. 2004. *The SER International Primer on Ecological Restoration.* Society for Ecological Restoration International, Tucson, USA.

Sergio, F., T. Caro, D. Brown, B. Clucas, J. Hunter, J. Ketchum, K. McHugh, and F. Hiraldo. 2008. Top predators as conservation tools: ecological rationale, assumptions, and efficacy. *Annual Review of Ecology, Evolution, and Systematics* 39:1–19.

Sgro, C. M., A. J. Lowe, and A. A. Hoffmann. 2011. Building evolutionary resilience for conserving biodiversity under climate change. *Evolutionary Applications* 4:326–337.

Shackelford, N., R. J. Hobbs, J. M. Burgar, T. E. Erickson, J. B. Fontaine, E. Laliberté, C. E. Ramalho, M. P. Perring, and R. J. Standish. 2013. Primed for change: developing ecological restoration for the 21st century. *Restoration Ecology* 21:297–304.

Shao, Q., L. Huang, J. Liu, W. Kuang, and J. Li. 2011. Analysis of forest damage caused by the snow and ice chaos along a transect across southern China in spring 2008. *Journal of Geographical Sciences* 21:219–234.

Sheil, D., and D. Murdiyarso. 2009. How forests attract rain: an examination of a new hypothesis. *Bioscience* 59:341–347.

Shono, K., E. A. Cadaweng, and P. B. Durst. 2007. Application of assisted natural regeneration to restore degraded tropical forestlands. *Restoration Ecology* 15:620–626.

Shoo, L. P., P. Scarth, S. Schmidt, and K. A. Wilson. 2012. Reclaiming degraded rainforest: A spatial evaluation of gains and losses in subtropical eastern Australia to inform future investment in restoration. *Restoration Ecology* 21:481–489.

Shrestha, K. K., and P. McManus. 2008. The politics of community participation in natural resource management: lessons from community forestry in Nepal. *Australian Forestry* 71:135–146.

Sidle, R. C., A. D. Ziegler, J. N. Negishi, A. R. Nik, R. Siew, and F. Turkelboom. 2006. Erosion processes in steep terrain – truths, myths, and uncertainties related to forest management in Southeast Asia. *Forest Ecology and Management* 224:199–225.

Sikor, T. 2001. The allocation of forestry land in Vietnam: Did it cause the expansion of forests in the northwest? *Forest Policy and Economics* 2:1–11.

Sikor, T., and D. Muller. 2009. The limits of state-led land reform: An introduction. *World Development* 37:1307–1316.

Silveira, L., and J. Alonso. 2009. Runoff modifications due to the conversion of natural grasslands to forests in a large basin in Uruguay. *Hydrological Processes* 23:320–329.

Simberloff, D., J. A. Farr, J. Cox, and D. W. Mehlman. 1992. Movement corridors: conservation bargains or poor investments? *Conservation Biology* 6:493–504.

Simula, M. 2009. *Towards Defining Forest Degradation: Comparative Analysis of Existing Definitions.* Food and Agriculture Organization of the United Nations, Rome.

Sinclair, C. 2012. *Guide to Participatory Modelling.* Cooperative Research Center for Forestry, University of Tasmania, Hobart, Australia.

Sirami, C., L. Brotons, I. Burfield, J. Fonderflick, and J.-L. Martin. 2008. Is land abandonment having an impact on biodiversity? A meta-analytical approach to bird distribution changes in the north-western Mediterranean. *Biological Conservation* 141:450–459.

Siregar, U., I. Siregar, S. Budi, Y. Hero, D. Suharjito, and Hardjanto. 2012. Incorporating social and natural science in the restoration of an Indonesian conservation forest: a case study from Jambi. Pages 41–62 *in* J. A. Stanturf, P. Madsen, and D. Lamb, editors. *A Goal-Oriented Approach to Forest Landscape Restoration.* Springer Dordrecht, Netherlands.

Sitch, S., C. Huntingford, N. Gedney, P. E. Levy, M. Lomas, S. L. Piao, R. Betts, P. Ciais, P. Cox, P. Friedlingstein, C. D. Jones, I. C. Prentice, and F. I. Woodward. 2008. Evaluation of the terrestrial carbon cycle, future plant geography and climate-carbon cycle feedbacks using five Dynamic Global Vegetation Models (DGVMs). *Global Change Biology* 14:2015–2039.

Smith, D. M. 1986. *The Practice of Silviculture.* John Wiley, New York.

Smith, P., P. J. Gregory, D. van Vuuren, M. Obersteiner, P. Havlik, M. Rounsevell, J. Woods, E. Stehfest, and J. Bellarby. 2010. Competition for land. *Philosophical Transactions of the Royal Society B: Biological Sciences* 365:2941–2957.

Sommer, R. S., N. Benecke, L. Lõugas, O. Nelle, and U. Schmölcke. 2011. Holocene survival of the wild horse in Europe: a matter of open landscape? *Journal of Quaternary Science* 26:805–812.

Sonneveld, B. G. J. S., and D. L. Dent. 2009. How good is GLASOD? *Journal of Environmental Management* 90:274–283.

Southworth, J., and C. Tucker. 2001. The influence of accessibility, local institutions, and socioeconomic factors on forest cover change in the mountains of western Honduras. *Mountain Research and Development* 21:276–283.

Spears, J. 1983. Replenishing the world's forests. Tropical reforestation: an achievable goal. *Commonwealth Forestry Review* 62:201–217.

Speight, M. R., and F. R. Wiley. 2001. *Insect Pests in Tropical Forestry.* CABI Publishing, Wallingford, UK.

Sreedharan, C. K., and J. R. Matta. 2006. Strategies for sustaining joint forest management: lessons from Tamil Nadu Afforestation project, India. Pages 44–57 *in Strategies and Financial Mechanisms for Sustainable Use and Conservation of Forests: Experiences from Latin America and Asia.* UN Food and Agriculture Organization, Proceedings of an inter-regional workshop, Chiang Mai, Thailand 20–22 November 2006.

Staaland, H., Ø. Holand, C. Nellemann, and M. Smith. 1998. Time scale for forest regrowth: Abandoned grazing and agricultural areas in southern Norway. *Ambio* 27:456–460.

Stanton, R. 2001. Plantations for Australia: the 2020 vision – an assessment of the past, present and future role of Australia's national plantation strategy. Pages 301–311 *in Forests in a Changing Landscape. 16th Commonwealth Forestry Conference,* Fremantle, Western Australia.

Stanturf, J. A., E. S. Gardiner, J. P. Shepard, C. J. Schweitzer, C. J. Portwood, and L. C. Dorris Jr. 2009. Restoration of bottomland hardwood forests across a treatment intensity gradient. *Forest Ecology and Management* 257:1803–1814.

Steffen, W., Å. Persson, L. Deutsch, J. Zalasiewicz, M. Williams, K. Richardson, C. Crumley, P. Crutzen, C. Folke, and L. Gordon. 2011. The Anthropocene: From global change to planetary stewardship. *Ambio* 40:739–761.

Steinberg, P. F. 2009. Institutional resilience amid political change: The case of biodiversity conservation. *Global Environmental Politics* 9:61–81.

Stewart, H., D. Race, and A. Curtis. 2011. New forests in changing landscapes in south-east Australia. *International Forestry Review* 13:67–79.

Stewart, P. 1993. Afforestation in Algeria. Pages 92–104 *in* A. Mather, editor. *Afforestation: Policies, Planning and Progress.* Belhaven Press, London and Florida, USA.

Stickler, M. 2014. *Brief: Rights to trees and livelihoods in Niger.* World Resources Institute, Washington.

Stirzaker, R. J., R. A. Vertessy, and A. Sarre. 2002. *Trees, Water and Salt: An Australian Guide to Using Trees for Healthy Catchments and Productive Farms.* Rural Industry Research and Development Corporation, Canberra.

Stone, R. 2008. Ecologists report huge storm losses in China's forests. *Science* 319:1318–1319.

Stoneham, G., V. Chaudhri, A. Ha, and L. Strappazzon. 2003. Auctions for conservation contracts: an empirical examination of Victoria's Bush Tender trial. *The Australian Journal of Agricultural and Resource Economics* 47:477–500.

Suding, K. N. 2011. Toward an era of restoration in ecology: Successes, failures, and opportunities ahead. *Annual Review of Ecology, Evolution, and Systematics* 42:465–487.

Sudmeyer, R., and J. Simons. 2008. *Eucalyptus globulus* agroforestry on deep sands on the southeast coast of Western Australia: The promise and the reality. *Agriculture, Ecosystems & Environment* 127:73–84.

Swift, T. L., and S. J. Hannon. 2010. Critical thresholds associated with habitat loss: a review of the concepts, evidence, and applications. *Biological Reviews* 85:35–53.

Tak, K., Y. Chun, and P. M. Wood. 2007. The South Korean forest dilemma. *International Forestry Review* 9:548–557.

Tasser, E., J. Walde, U. Tappeiner, A. Teutsch, and W. Noggler. 2007. Land-use changes and natural reforestation in the Eastern Central Alps. *Agriculture, Ecosystems & Environment* 118:115–129.

Temperton, V., R. J. Hobbs, T. Nuttle, and S. Halle. 2004. *Assembly Rules and Restoration Ecology.* Island Press, Washington.

Thompson, K., J. P. Bakker, and R. M. Bekker. 1997. *The Soil Seed Banks of North West Europe: Methodology, Density and Longevity.* Cambridge University Press, Cambridge, UK.

Thrall, P. H., D. A. Millsom, A. C. Jeavons, M. Waayers, G. R. Harvey, D. J. Bagnall, and J. Brockwell. 2005. Seed inoculation with effective root-nodule bacteria enhances revegetation success. *Journal of Applied Ecology* 42:740–751.

Tilman, D., C. Balzer, J. Hill, and B. L. Befort. 2011. Global food demand and the sustainable intensification of agriculture. *Proceedings of the National Academy of Sciences* 108:20260–20264.

Toh, I., M. Gillespie, and D. Lamb. 1999. The role of isolated trees in facilitating tree seedling recruitment at a degraded sub-tropical rainforest site. *Restoration Ecology* 7:288–297.

Totman, C. 1989. *The Green Archipelago: Forestry in Pre-industrial Japan.* University of California Press, Berkley, USA.

Tougiani, A., C. Guero, and T. Rinaudo. 2009. Community mobilisation for improved livelihoods through tree crop management in Niger. *GeoJournal* 74:377–389.

Townsend, P., R. Harper, P. Brennan, C. Dean, S. Wu, K. Smettem, and S. Cook. 2012. Multiple environmental services as an opportunity for watershed restoration. *Forest Policy and Economics* 17:45–58.

Trac, C. J., S. Harrell, T. M. Hinckley, and A. C. Henck. 2007. Reforestation programs in southwest China: Reported success, observed failure, and the reasons why. *Journal of Mountain Science* 4:275–292.

Tscharntke, T., A. M. Klein, A. Kruess, I. Steffan-Dewenter, and C. Thies. 2005. Landscape perspectives on agricultural intensification and biodiversity – ecosystem service management. *Ecology Letters* 8:857–874.

Tscharntke, T., Y. Clough, T. C. Wanger, L. Jackson, I. Motzke, I. Perfecto, J. Vandermeer, and A. Whitbread. 2012. Global food security, biodiversity conservation and the future of agricultural intensification. *Biological Conservation* 151:53–59.

Tucker, N., and T. M. Murphy. 1997. The effects of ecological rehabilitation on vegetation recruitment: some observations from the wet tropics of North Queensland. *Forest Ecology and Management* 99:133–152.

Tunjai, P., and S. Elliott. 2012. Effects of seed traits on the success of direct seeding for restoring southern Thailand's lowland evergreen forest ecosystem. *New Forests* 43:319–333.

Turner, I. M., and R. T. Corlett. 1996. The conservation value of small, isolated fragments of lowland tropical rain forest. *Trends in Ecology & Evolution* 11:330–333.

Turner, I. M., K. S. Chua, J. S. Y. Ong, B. C. Soong, and H. T. W. Tan. 1996. A century of plant species loss from an isolated fragment of lowland tropical rain forest. *Conservation Biology* 10:1229–1244.

Uchida, E., J. C. Xu, and S. Rozelle. 2005. Grain for Green: cost effectiveness and the sustainability of China's conservation set-aside program. *Land Economics* 81:247–264.

UNCCD. 2012. *Zero Net Land Degradation: a Sustainable Development Goal for Rio+20 to Secure the Contribution of Our Planet's Land and Soil to Sustainable Development Including Food Security and Poverty Eradication.* United Nations Convention to Combat Desertification, Bonn, Germany.

UNEP. 2007. *Global Environmental Outlook – 4.* United Nations Environmental Program, Nairobi.

UNEP. 2011. *UNEP Billion Tree Campaign Reaches 12 Billion Milestone.* UNEP News Center, Nairobi.

UN Forum on Forests. 2006. *Report of the sixth session 27 May 2005 and 13 to 24 February 2006.* Economic and Social Council, Official Record 2006, Supplement 22 (E/2006/42; E/CN.18.2006/18).

UN General Assembly. 2012. *Sixty-sixth session, Agenda item 19. Resolution adopted on 27 July 2012. Annex: The Future We Want (para. 19).* Available at: www.uncsd2012.org/thefuturewewant.html [Accessed 16 May 2014].

United Nations. 2011. *World Urbanisation Prospects: The 2011 Revision.* United Nations Department of Economic and Social Affairs, Population Division, Population Estimates and Projections Section. Available at: http://esa.un.org/unup [Accessed 20 June 2014].

Valentin, C., F. Agus, R. Alamban, A. Boosaner, J. P. Bricquet, V. Chaplot, T. de Guzman, A. de Rouw, J. L. Janeau, D. Orange, K. Phachomphonh, P. Do Duy, P. Podwojewski, O. Ribolzi, N. Silvera, K. Subagyono, J. P. Thiabaux, T. Tran Duc, and T. Vadari. 2008. Runoff and sediment losses from 27 upland catchments in Southeast Asia: Impact of rapid land use changes and conservation practices. *Agriculture, Ecosystems & Environment* 128:225–238.

Valkonen, S., and L. Valsta. 2001. Productivity and economics of mixed two-storied spruce and birch stands in Southern Finland simulated with empirical models. *Forest Ecology and Management* 140:133–149.

Vallejo, R. 2005. Restoring Mediterranean Forests. Pages 313–319 *in* S. Mansourian, D. Vallauri, and N. Dudley, editors. *Forest Restoration in Landscapes.* Springer, New York.

Van Auken, O. 2000. Shrub invasions of North American semiarid grasslands. *Annual Review of Ecology and Systematics* 31:197–215.

Van Dijk, A., and R. J. Keenan. 2007. Planted forests and water in perspective. *Forest Ecology and Management* 251:1–9.

Van Oosten, C. 2013. Restoring landscapes—governing place: A learning approach to forest landscape restoration. *Journal of Sustainable Forestry* 32:659–676.

Van Uytvanck, J., D. Maes, D. Vandenhaute, and M. Hoffmann. 2008. Restoration of woodpasture on former agricultural land: The importance of safe sites and time gaps before grazing for tree seedlings. *Biological Conservation* 141:78–88.

Van Uytvanck, J., A. Van Noyen, T. Milotic, K. Decleer, and M. Hoffmann. 2010. Woodland regeneration on grazed former arable land: A question of tolerance, defence or protection? *Journal for Nature Conservation* 18:206–214.

Vanacker, V., F. von Blanckenburg, G. Govers, A. Molina, J. Poesen, J. Deckers, and P. Kubik. 2007. Restoring dense vegetation can slow mountain erosion to near natural benchmark levels. *Geology* 35:303–306.

Vandermeer, J., and I. Perfecto. 2006. The agricultural matrix and a future paradigm for conservation. *Conservation Biology* 21:274–277.

Vellend, M. 2004. Metapopulation dynamics following habitat loss and recovery: forest herbs in ancient and recent forests. Pages 163–177 in O. Hannay, K. Verheyen, B. Bossuyt and M. Hermy (editors) *Forest Biodiversity: Lessons from History for Conservation.* CAB International, Wallingford, UK.

Verstraeten, G., J. Poesen, K. Gillijns, and G. Govers. 2006. The use of riparian vegetated filter strips to reduce river sediment loads: an overestimated control measure? *Hydrological Processes* 20:4259–4267.

Vertessy, R. A., L. Zhang, and W. R. Dawes. 2003. Plantations, river flows and river salinity. *Australian Forestry* 66:55–61.

Vesk, P. A., and M. Westoby. 2004. Sprouting ability across diverse disturbances and vegetation types worldwide. *Journal of Ecology* 92:310–320.

Victor, D. G. 2006. A vision for world forests: Results from the Council on Foreign Relations Study. *Journal of Sustainable Forestry* 21:113–135.

Viera, D. L. M., and A. Scariot. 2006. Principles of natural regeneration of tropical dry forest for restoration. *Restoration Ecology* 14:11–20.

Vihervaara, P., A. Marjokorpi, T. Kumpula, M. Walls, and M. Kamppinen. 2012. Ecosystem services of fast-growing tree plantations: a case study on integrating social valuations with land-use changes in Uruguay. *Forest Policy and Economics* 14:58–68.

Vojta, J., and L. Drhovská. 2012. Are abandoned wooded pastures suitable refugia for forest species? *Journal of Vegetation Science* 23:880–891.

Walker, B., and D. Salt. 2006. *Resilience Thinking: Sustaining Ecosystems and People in a Changing World.* Island Press, Washington.

Wang, J., H. Ren, L. Yang, and D. Li. 2011. Factors influencing establishment by direct seeding of indigenous tree species in typical plantations and shrubland in South China. *New Forests* 42:19–33.

Wang, S., Z. Zhang, T. R. McVicar, J. Zhang, J. Zhu, and J. Guo. 2012. An event-based approach to understanding the hydrological impacts of different land uses in semi-arid catchments. *Journal of Hydrology* 416:50–59.

Wang, X. M., C. X. Zhang, E. Hasi, and Z. B. Dong. 2010. Has the Three Norths Forest Shelterbelt Program solved the desertification and dust storm problems in arid and semiarid China? *Journal of Arid Environments* 74:13–22.

Wang, X., J. Bennett, J. Xu, and H. Zhang. 2012. An auction scheme for land use change in Sichuan Province, China. *Journal of Environmental Planning and Management* 55: 1269–1288.

Watson, J. E., T. Iwamura, and N. Butt. 2013. Mapping vulnerability and conservation adaptation strategies under climate change. *Nature Climate Change 3*:989–994.

Wegner, L., and G. Zwart. 2011. Who will feed the world? The production challenge. *Food Chain* 1:187–205.

Weiher, E., and P. Keddy. 1999. *Ecological Assembly Rules: Perspectives, Advances, Retreats.* Cambridge University Press, Cambridge, UK.

Wessels, K. 2009. Letter to the editor: Comments on 'Proxy global assessment of land degradation' by Bai et al.(2008). *Soil Use and Management* 25:91–92.

West, G. B., and J. H. Brown. 2005. The origin of allometric scaling laws in biology from genomes to ecosystems: towards a quantitative unifying theory of biological structure and organization. *Journal of Experimental Biology* 208:1575–1592.

Weyerhaeuser, H., A. Wilkes, and F. Kahrl. 2005. Local impacts and responses to regional forest conservation and rehabilitation programs in China's northwest Yunnan province. *Agricultural Systems* 85:234–253.

White, A., and A. Martin. 2002. *Who Owns the World's Forests? Forest Tenure and Public Forests in Transition.* Forest Trends, Washington.

Whitten, S. M., D. Freudenberger, C. Wyborn, V. Doerr, E. Doerr, and A. Langston. 2011. *A Compendium of Existing and Planned Australian Wildlife Corridor Projects and Initiatives, and Case Study Analysis of Operational Experience.* CSIRO, Canberra.

Williams, M. 2006. *Deforesting the Earth: From Prehistory to Global Crisis.* University of Chicago Press, Chicago, USA.

Williams-Linera, G., C. Alvarez-Aquino, A. Suárez, C. Blundo, C. Smith-Ramírez, C. Echeverria, E. Cruz-Cruz, G. Bolados, J. Armesto, K. Heinemann, L. Malizia, P. Becerra, R. del Castillo, and R. Urrutia. 2011. Experimental analysis of dryland forest restoration techniques. Pages 131–181 *in* A. C. Newton and N. Tejedor, editors. *Principles and Practice of Forest Landscape Restoration: Case Studies from the*

*Drylands of Latin America.* International Union for the Conservation of Nature, Gland, Switzerland.

Willoughby, I., R. L. Jinks, G. Kerr, and P. G. Gosling. 2004. Factors affecting the success of direct seeding for lowland afforestation in the UK. *Forestry* 77:467–482.

Wilson, K. A., M. Lulow, J. Burger, Y.-C. Fang, C. Andersen, D. Olson, M. O'Connell, and M. F. McBride. 2011. Optimal restoration: accounting for space, time and uncertainty. *Journal of Applied Ecology* 48:715–725.

Wily, L. A. 2005. From meeting needs to honoring rights: The evolution of community forestry. Pages 388–395 *in* J. Sayer, editor. *Earthscan Reader in Forestry and Development.* Earthscan Publications, London.

Windle, J., and J. Rolfe. 2008. Exploring the efficiencies of using competitive tenders over fixed price grants to protect biodiversity in Australian rangelands. *Land Use Policy* 25:388–398.

Winjum, J. K., and P. E. Schroeder. 1997. Forest plantations of the world: their extent, ecological attributes, and carbon storage. *Agricultural and Forest Meteorology* 84:153–167.

Wollenberg, L. 2014. *The Landscape Fund: A New Way to Invest in Sustainable Agriculture.* CGIAR Research Program in Climate Change and Food Security. Available at: http://ccafs.cgiar.org/blog/landscape-fund-new-way-invest-sustainable-agriculture#.Uw_QqdZ-_IV [Accessed 29 May 2014].

Woodall, G. 2010. *Improving the Direct Sowing of Commercial Native Plants in Agricultural Lands of Southern Australia.* Rural Industries Research and Development Corporation, Canberra.

Woods, K., and S. Elliott. 2004. Direct seeding for forest restoration on abandoned agricultural land in northern Thailand. *Journal of Tropical Forest Science* 16:248–259.

Woods, R. 1990. Second rotation decline in *P. radiata* plantations in South Australia has been corrected. *Water, Air, & Soil Pollution* 54:607–619.

Worboys, G., W. Francis, and M. Lockwood, editors. 2010. *Connectivity Conservation Manual: A Global Guide.* Earthscan, London.

World Bank. 2001. *World Development Report 2000/2001: Attacking Poverty.* Oxford University Press, New York.

World Bank. 2013. *China: Promoting a More Resilient Forest Ecosystem in Hunan.* World Bank, Beijing.

WRM. 2011. *Chile: A Disastrous Forestry Model with New 'Branches' in Latin America.* World Rainforest Movement, Montevideo.

Wuethrich, B. 2007. Reconstructing Brazil's Atlantic rainforest. *Science* 315:1070–1072.

Wunder, S. 2005. Payments for ecological services: some nuts and bolts. . Center for International Forestry Research, Bogor, Indonesia.

Wunder, S. 2013. When payments for environmental services will work for conservation. *Conservation Letters* 6: 230–237.

Wunder, S., and S. Wertz-Kanounnikoff. 2009. Payments for ecosystem services: a new way of conserving biodiversity in forests. *Journal of Sustainable Forestry* 28:576–596.

Wunscher, T., S. Engel, and S. Wunder. 2008. Spatial targeting of payments for environmental services: A tool for boosting conservation benefits. *Ecological Economics* 65:822–833.

Xinhua, C., and Z. Jingchun. 1988. Sowing from the air: China's experience. *Tigerpaper* 15(1). FAO Forestry News, Bangkok.

Xu, J. 2011. China's new forests aren't as green as they seem. *Nature* 477:371.

Xu, Z., M. T. Bennett, R. Tao, and J. Xu. 2004. China's Sloping Land Conversion Programme four years on: current situation, pending issues. *International Forestry Review* 6:317–326.

Xu, Z. G., J. T. Xu, X. Z. Deng, J. K. Huang, E. Uchida, and S. Rozelle. 2006. Grain for green versus grain: Conflict between food security and conservation set-aside in China. *World Development* 34:130–148.

Yang, X., and L. Ci. 2008. Comment on 'why large-scale afforestation efforts in China have failed to solve the desertification problem'. *Environmental Science & Technology* 42:7722–7723.

Yin, R., J. Xu, Z. Li, and C. Liu. 2005. China's ecological rehabilitation: the unprecedented efforts and dramatic impacts of reforestation and slope protection in western China. *China Environment Series* 6:17–32.

Zahawi, R. A., K. D. Holl, R. J. Cole, and J. L. Reid. 2013. Testing applied nucleation as a strategy to facilitate tropical forest recovery. *Journal of Applied Ecology* 50:88–96.

Zaimeche, S. 1994. Change, the state and deforestation: the Algerian example. *Geographical Journal* 160:50–56.

Zhang, D., and E. A. Owiredu. 2007. Land tenure, market, and the establishment of forest plantations in Ghana. *Forest Policy and Economics* 9:602–610.

Zhang, D., and M. Polyakov. 2010. The geographical distribution of plantation forests and land resources potentially available for pine plantations in the US South. *Biomass and Bioenergy* 34:1643–1654.

Zhang, Y., K. Ma, M. Anand, and B. Fu. 2011. Multifractal pattern and process during a recent period of forest expansion in a temperate mountainous region of China. *Ecological Informatics* 6:384–390.

Zhang, Y., H. Y. H. Chen, and P. B. Reich. 2012. Forest productivity increases with evenness, species richness and trait variation: a global meta-analysis. *Journal of Ecology* 100: 742–749.

Zhou, G. Y., J. D. Morris, J. H. Yan, Z. Y. Yu, and S. L. Peng. 2002. Hydrological impacts of reafforestation with eucalypts and indigenous species: a case study in southern China. *Forest Ecology and Management.* 167:209–222.

Zhou, G., X. Wei, Y. Luo, M. Zhang, Y. Li, Y. Qiao, H. Liu, and C. Wang. 2010. Forest recovery and river discharge at the regional scale of Guangdong Province, China. *Water Resources Research* 46: doi:10.1029/2009WR008829.

Zhou, Z., Z. Gan, Z. Shangguan, and Z. Dong. 2009. China's Grain for Green Program has reduced soil erosion in the upper reaches of the Yangtze River and the middle reaches of the Yellow River. *International Journal of Sustainable Development & World Ecology* 16:234–239.

Ziegler, A. D., T. W. Giambelluca, L. T. Tran, T. T. Vana, M. A. Nullet, J. Fox, T. D. Vien, J. Pinthong, J. Maxwell, and S. Evett. 2004. Hydrological consequences of landscape fragmentation in mountainous northern Vietnam: evidence of accelerated overland flow generation. *Journal of Hydrology* 287:124–146.

Zuazo, V. H. D., and C. R. R. Pleguezuelo. 2009. Soil-erosion and runoff prevention by plant covers: a review. Pages 785–811 *in* E. Lichfouse, M. Navarette, P. Debaeke, S. Veronique, C. Alberola (editors), *Sustainable Agriculture.* Springer, Dordrecht, Netherlands.

# Index

A page reference in **bold** indicates a table and figures are denoted in *italics.*


abandoned land, farming: biodiversity in 98; natural forest regrowth on 27, *29*, 73; reforestation, historical 250; and reforestation planning **136**

*Acacias* 48, 90, 167

agricultural land: demand for vs. reforestation 18–22; dryland salinity, reforestation to combat 138–42; intensive cropping and natural forest regrowth 88; marginal agricultural land 12, 16, **136**, 187–8; and reforestation planning **136**; *see also* abandoned land, farming; farmers

agricultural systems 21–2

Algeria: area of planted production forests **39**; reforestation, case history 41–3

Arriagada, R.P. *et al* (2012) 222

Asia Pacific Economic Cooperation (APEC) **14**

Assisted Natural Regeneration 94, 179–80

Atlantic Forest Restoration Pact **15**, 16, 122, 130

Australia: aerial seeding, eucalypt forests 126–7; area of planted production forests **39**; direct seed sowing, eucalypts *128*; diverted successional development 124, *124*; dryland salinity 108, 141–2, 183; farmers' mistrust of plantations 214–15; *Flindersia australis* plantation, Australia *4*; Gondwana Link 156–7; large area natural forest regrowth **74**; national reforestation plans 15–16, **15**; reforestation, case history 44–6; second rotation problem 56

Ausubel, J.H. *et al* (2013) 19–21

Bai, Z.G. *et al.* 2008 8

Benayas, J.M.R. *et al* (2007) 98

Benitez, P.C. *et al* (2007) 10, 11, **11**

Biancalani, R. *et al.* 2011 8

BioCarbon Fund (World Bank) 245

bio-cultural restoration 85–6

biodiversity: on abandoned agricultural lands 98; buffer reforestation and 149–52, *150*; corridor/stepping stone reforestation 152–4; ecosystem functioning and 16–17; ecosystem services and 17–18; environmental offset programmes 195; large-bodied species, re-introduction of 174–5; local and regional diversity of species 17–18; loss, within homogenised landscapes

135, 136–7; loss and pests/ diseases 56–7; loss caused by deforestation 1–2, 13; minimum threshold areas 173–4; native vs. exotic species selection 103; within natural regrowth forests 98–9; within novel ecosystem structures 190; recovery trends of biota 175, *176*; Rewilding Project, Europe 98–9, 178; shared land approach and 133–4; species-area relationships *151*, 173; transnational corridor initiatives 154; upper limits of reforestation and 175–8
Blaikie, P. and Brookfield, H. 8
Bonn Challenge 13, **14**, 239
Brazil: area of planted production forests **39**; community involvement 241; national reforestation plans **15**, 16; natural forest regrowth 94; NGO co-ordinated large-scale reforestation 234; reforestation, case history 47–9; regulatory policies 225–6; restoring species-rich forests 122; taxation incentive scheme 222
broad-leaved species, soil carbon accretion 181
Buchy, M. and Hoverman, S. 228
buffer strips reforestation: biodiversity conservation 149–52, *150*; disturbance risk and 151–2, **152**; qualifying factors 151; species-area relationships *151*
Bull, J.W. *et al* (2006) 222

Campbell, J.E. *et al* (2008) 10, **11**
canopy cover, nurse trees 115–16
carbon capture: above-ground biomass (carbon sequestering) 179–80; below-ground soil organic carbon 180–2; carbon storage by forests 30–1, 98, 179–80, 249; climatic influences 181–2; creation of carbon storage markets 221; funding initiatives and 245–6; long rotation plantings 182; management implications 182–3; mixed-species planting 179, 181, 182; overview 178–9; as primary aim of reforestation 200; short rotation plantings 179, *180*; soil carbon accretion, coniferous species 181; timber plantations 179, *180*, 182
Cardinale, B.J. *et al* (2012) 17
Carle, J. and Holmgren, P. 61–2
Chazdon, R. *et al* (2009) 98
Chile: area of planted production forests **39**; plantations as protection for natural forests 58; reforestation, case history 49–52; water yields and reforestation 166
China: aerial seeding 127; atmospheric water vapour, observable effects 170; *Cunninghamia lanceolata* plantation, China *4*; dry season flows and reforestation 168; forest transition achieved 64; Grain for Green Project (Sloping Land Conversion Program) 16, 26, 29, 145–6, 172, 225; Loess Plateau, reforestation to prevent soil erosion 143–4, *144*, *145*; multi-species plantings, Hunan province 193; national reforestation plans **15**, 16; perceived profitability of tree growing 217; public views on environmental degradation 26; reverse auctions, reforestation tendering 224; silvicultural history 34; Three Norths Shelterbelt 16, 148–9; windbelts 172

Clapp, R. 50
Clewell, A.F. and Aronson, J. 119–20
climate change: carbon storage by forests 30–1; as challenge to Ecological Restoration 190; global vegetation patterns 250–2, **251**; impact on reforestation programmes 198–200, **199**, 249–50, **251**
Collier, P. 21
community involvement: Algeria, reforestation 42–3; Costa Rica, reforestation 85–6; devolution of power to 213–14, 241; Grain for Green Project (Sloping Land Conversion Program), China 146; land tenure 211, 212; local-level decision-making 228–9; Nepal, reforestation 77–8, 207–8, 241; Niger, reforestation 79–81, 207–8, 241; reforesting degraded farmland, Ghana 235–6; representativeness 230–1; role of institutions 228; South Korea, reforestation 39–40; typology of participation 229–30; village forest councils, India 232; *see also* stakeholders
Conacher, A. and Conacher, J. 8
coniferous species, soil carbon accretion 181
conservation objectives: balanced with land production 133–4; and natural forest regrowth 98–9; and poverty alleviation 242–3; reforested buffer strips 149–52, *150*
coppice management systems 70–1, 97, 107–8
corridor/stepping stone reforestation: composition of 153–4; critiques of 154; edge effects 155; Gondwana Link, Australia 156–7; guidelines for corridor design 155–7; overview 152–3; transnational corridor initiatives 154
Costa Rica: natural forest regrowth (Guanacaste Conservation Area) 84–6; payments for ecosystem services 26–7, 222; soil conductivity and reforestation 167–8
*Cunninghamia lanceolata* plantation, China *4*

Dauber, J. *et al* 2012 9, **11**
deforestation: and agricultural land demands 57; biodiversity loss and 1–2, 13; and GDP 64; hydrological flows and 138, 162, **164**, 194, **208**; socio-economic consequences 2; soil erosion and 142–7
degraded land: definitions of 8; effects of deforestation and 2; estimates of the amount of 7, 8; global assessments of 9; management techniques 168; as non-technical problem **208**; perceptions of 6–7; quantification of 9; reforestation priorities 160; reforesting, case study Ghana 235–6; restrictions, reforestation 9; socio-economic consequences 9
Deininger, K. (2011) 20
Del Lungo, A.J. *et al* (2006) 58–9
desertification and reforestation 147–9
desiccation theory 169
designing *see* planning
disturbances: buffer strips, risk from 151–2, **152**; effects on natural forest regrowth 71–2, 83, 93–4, 99; minimisation, carbon capture 182; resilience to 246–7, **247**; spruce-beech mixtures, resilience to 117

Dorrough, J. *et al* (2008) 91
dryland salinity: causes of 138; in grasslands 141; planting trees over shallow saline water tables 140; tree belts mixes with agricultural land 139–40, *140*; tree belts on hillslopes 139; in Western Australia 141–2; woodlots in rotation with agriculture 138–9

ecological processes: afforestation 5; deforestation 1–2; natural forest regrowth 88–90; need for large-scale reforestation 2, 6, 134; *see also* ecosystem functioning; ecosystem services
ecological restoration: corridor reforestation 155; defined 119; forests as permanent carbon stores 179–80; framework species method 120–1; maximum diversity method 121, 122; vs. novel ecosystems 190–2; options available, large-scale reforestation 191–2; overview 105, 119–20; reforested buffer strips 152; success/failure rates 195
economics: economic pathways to reforestation 66; economic resilience 246, **247**; economies of scale 248–9
ecosystem functioning: and biodiversity 16–17; diversity scale 17–18; ecological resilience 246–7, **247**; restoration of 12–13
ecosystem services: area-treated-to-area-benefited ratio 183; biodiversity and 17–18; climate change impacts 198–200, **199**; cultural services 13; difficulties in market creation **220**; functioning restoration 12–13; funding initiatives and 245–6; Grain for Green Project (Sloping Land Conversion Program) 146; incentive schemes for 220–2, **220**; increasing demand for 189–90; markets supplied by reforestation 26–7; multi-species plantings 111–12; novel ecosystems 119, 121–4, *124*, 190–2, 193; number of species selected *104*, 105, *106*; provided by plantations 58, 63–4, 109; provisioning services 13; regulating services 13; supporting services 13
edge effects 149–50, *150*, 155
El Salvador **74**
employment, forestry-related 54
enrichment plantings 96, *109*
environmental concerns: increased public awareness of 24–6; offset programmes 195–8
erosion controls: effectiveness of reforestation 171–2, **208**; soil erosion following deforestation 142–7; water erosion 170–1; wind erosion 147
Ethiopia **74**
eucalypts: aerial seeding, Australia 126–7; direct seed sowing, Australia *128*; plantations 33, 164
Europe: biodiversity after reforestation of abandoned agricultural lands 176–7; natural forest regrowth 98; Rewilding Project 98–9, 178
evapo-transpiration: counteracted, enhanced groundwater recharge 167; reduction of 194; and water resources 31, 138, 141, 144, 162, 163
exotic species: compared to native species 101–3, **102**; exotic weeds 89, 97; invasive 103; monoculture plantations 108

farmers: farmer managed natural regeneration 80–2; mistrust of reforestation 12, 209, 249; perceptions of degradation 7; *see also* agricultural land
Finland 233
fires: biodiversity destruction 98; control, Costa Rica 84; impacts on natural forest regrowth 73, 90; small forest patches, risk of *95*
Five Million Hectare Reforestation Program, Vietnam 239
*Flindersia australis* plantation, Australia *4*
floods: reforestation and 166, **208**; riparian plantings and 147
fog capture 169
food production: and agricultural productivity 21; vs. reforestation choice 187–8; rising and land demands 18–21
Forest Landscape Restoration model 204–5, 219
forest products: changing markets 59–60; forest scarcity pathway 65; fuelwood being collected *97*; increased market for 26, 33–4; rising markets for 12; silvicultural systems and 96–8; *see also* plantations; pulp/paper mills; timber production
forest transition: defined 64; pathways to 65–6, *65*
forests, defined 64
founder effect 120
framework species method 120–1
France 75–6
funding, sourcing 243–6; *see also* incentive schemes

Gabarone Declaration **14**, 15
Gamfeldt, L. *et al* (2013) 17
Germany 117
Ghana 201, 235–6
Gilmour, D. 77
Global Assessment of Soil Degradation (GLASOD) 9
Global Environmental Fund 245
global targets 239–40
globalisation reforestation pathway 66
governments: funding initiatives and 244–5; role in reforestation 240–1; *see also* state, the
Grain for Green Project (Sloping Land Conversion Program), China 16, 26, 29, 145–6, 172, 225
Grainger, A. 10, **11**
grasslands: dryland salinity in 141; as protection against soil erosion 143, 171
grazing, livestock 90
Great Green Wall for the Sahara and Sahel **14**, 148
Greenwood, W. and Buttle, J. 167
Guinea **74**

habitat generalists 99
habitat specialists 99
Hecht, S.B. and Saatchi, S.S. 94
Honduras **74**
Hong Kong *92*
Houghton, R. 10, **11**
hydrological flows: affects, scale of reforestation 165–6, *165*; consequences of deforestation/reforestation 138, 162, **164**, 194, **208**; dry season flows 166–9; dryland salinity and reforestation 138–42; floods 166; precipitation and reforestation 169–70; responses, different types of reforestation 163–4; soil conductivity and reforestation 167–8; soil erosion caused by 142–7

Ilstedt, U. *et al* (2007) 167
incentive schemes: direct (financial) incentives schemes

218–19; indirect incentive schemes 217–18; obtaining value for money 222–3; overview 216, **216**; payments for ecosystem services 220–2, **220**; reverse auctions 223–4
India: Great Indian Hedge 153; national reforestation plans 15–16, **15**; natural forest regrowth 86–8; non-profit incentives for tree growing 218; soil conductivity and *Acacia* plantations 167; village forest councils 232
indigenous management systems 76–7
Indonesia 7, 213, 244
Indufor 62–3
institutions, reforestation: attributes of **227**; case studies 231–6; defined 226; institutional resilience 247–8, **247**; inter-connected funding initiatives 245–6; key role 226–7; local-level decision-making 228–9; monitoring and feedback 231; polycentric system of governance 228–9; provincial/district level 227–8; representativeness 230–1; scale of 227; typology of participation 229–30
international agreements/ initiatives 13–15
International Union for the Conservation of Nature 13
IUCN (International Union for the Conservation of Nature) 24

Jackson, R.B. *et al* (2005) 162
Jandl, R. *et al* (2007) 181
Japan 34
Johnson, D.L. *et al.* (1997) 8

kamikaze forestry 138–9
Katz, E. 190
Kenya **74**
Koch, J.M. 127

Laestadius, L. *et al* (2011) 10, 11, **11**
Laganier, J. *et al* (2010) 181
Lambin, E.F. and Meyfroidt, P. 9
land grabbing 19, 20
land sharing 133
land sparing 133
land tenure: community tenure 211, 212; consent for reforestation 31; establishing, within reforestation policies 211–13; impact on natural forest regrowth 94; impact on reforestation 12, 23–4; land allocation programmes, Vietnam 214; and property rights 12, 23–4; *see also* agricultural land
landholders: changes in areas of plantations owned by different landowners **59**; economic incentives, market conditions 244–5; economic interests vs. conservation aims 242–3; involvement, policy development 241; knowledge about tree growing 217; perceived profitability of tree growing 217–18; *see also* agricultural land; smallholders; stakeholders
landscapes: buffer strips reforestation 149–52; composition of (the landscape mosaic) 134–6; corridor/ stepping stone reforestation 152–7; creation of homogenous landscapes 31, 102, 135, 136–7; definitions 135; economies of scale and 248–9; elements affecting reforestation 134–6, **136**; identifying priority reforestation areas 157–9; multi-functional, creating 185–6;

planning arrangements 159–60; preferred locations for reforestation **158**; reforestation and combating climate change 31–2; reforestation of riparian strips 147; slopes and erosion-reducing reforestation 142–7
Laos 82–4
large-scale reforestation: definitions 5–6; questions likely to arise 186, **186**
legal obligations, landowners to reforest 29

Madagascar **74**
Malaysia 116
markets supplied by reforestation 243–6
Martin, P.A. *et al* (2013) 98
maximum diversity method 121, 122
meat consumption 19
Mergen, F. *et al* (1981) 126
mining companies: aerial seeding 126–7; ecological restoration and 190; land rehabilitation and performance bonds 197; mine site rehabilitation 29, 224, 225, 245; planting choices and successional development 123, 124
mixed-species planting *see* multi-species plantings
monocultures: benefits/limitations of monocultures 17–18, 34; economic interests vs. conservation aims 242–3; exotic species 108; long rotations, planting methodologies 108–11; pests and diseases 56–7; plantations 53, 105; short rotation, planting methodology 106–8; transfer into multi-species plantations 116; understory diversity 110
multi-species plantings: advantages of **112**; carbon storage capacities 179, 181, 182; different types for production purposes 112–13, **113**, *114*; ecological functioning 111–12; even-aged species mixtures 116–18; existing monocultures into multi-species plantations 116; Hunan province, China 193; overview 111; reforested buffer strips 152; temporary nurse trees 115–16; tree cover with an under-planted crop 115; to withstand climate change 200

native species, compared to exotic species 101–3, **102**
natural forest regrowth: on abandoned agricultural land 27, *29*, 73; acceleration of successional development 94–6; composition of *71*; conservation objectives, achievement of 98–9; Costa Rica, case study 84–6; definitions 3; disadvantages of 73; ecological constraints 88–90; effects of repeated disturbances 71–2, 83, 93–4, 99; European landscapes 98; examples of over large landscape areas **74**; France, case study 75–6; further research areas 99; India, case study 86–8; within intervening strips, timber plantations 137; within the landscape mosaic 135; Laos, case study 82–4; local species extinction 70; natural regrowth pathways 71–2, *72*; Nepal, case study 76–8, *79*; Niger, case study 79–82; Norway, case study 75; overview 68; vs. planting/seeding strategies 188; reasons for failure of 89; regrowth potential 91–3; from remnant forests 70; seeding

of 69–70; seedlings used 91–2; socio-economic constraints 90; species used 92–3; Thailand, case study 82–4; vegetative regrowth (old stumps) 70–1
natural forests: deforestation of and plantation productivity 57–8; ecosystem services 57; for supply of commodity timbers 60
Navarro, L.M. *et al* (2012) 176, **176**
Nepal: community involvement 77–8, 207–8, 241; natural forest regrowth 76–8, *79*, 93; soil conductivity and reforestation 168
Newton, A. *et al* (2011) 90
NGOs (non-governmental organizations): in developing countries, environmental focus 26; funding initiatives 245; on reforestation 13; role in reforestation 228, 230, 240–1
Niger: community involvement 79–81, 207–8, 241; natural forest regrowth 79–82, 90, 93
Nilsson, S. and Schopfhauser, W. 10, **11**
nitrogen 56
Norway 75
*Nothofagus* 166
nurse trees 115–16
offset programmes: difficulties in monitoring 197; Environmental Restoration as 195; quantitative assessments of 196; successional development 196–7

Oldeman, L.R. *et al* 1990 9
One Billion Tree Program 239
opportunity costs 141, 158
Ostrom, E. 228
Ostrom, E. and Cox, M. 226

Palo, M. and Lehto, E. 66
Pannell, D.J. *et al* (2001) 183
Parris, H. *et al* (2011) 154
perch trees 95
Peru **74**
Perz, S.G. and Skole, D.L. 94
phase farming with trees 138–9
Philippines 15–16, **15**, 116
pinus 148, 166, 181
planning: computer models 157–8, 202–3; cost-benefit analysis 157–8; environmental data 158, 203; Forest Landscape Restoration model 204–5, 219; new multi-functional landscapes 200–1; participatory models 203–4; priority locations, reforestation 158, **158**; top-down/bottom-up planning 202–3
plantations: area of planted production forests in selected countries **39**; areas allocated to production and protection functions **64**; areas of planted forest, expansion of **35**; carbon sequestering 179, 182; catalytic effect of, Vietnam 111–12; changes in areas of plantations/ natural forests **35**; commercial advantages of 33–4; common cause of plantation failures **107**; creation of regional biodiversity 137; declining productivity 55–6, 189–90; definitions 3; ecosystem services 63–4; future production trends 61–3, **61**, *62*; global figures (historical) 34–5; history of 34; industrial (historical) 2; inter-connected institutional arrangements and 248; location of 136–7; long rotation monocultures 108–11; management techniques 53, 63; multi-species plantings for production purposes 112–13, **113**, *114*; natural regrowth,

intervening strips 137; overview 4, 101; pests and diseases 56; as protection for natural forests 57–8; short rotation monocultures 106–8; silvicultural methods 35; smallholders and 137; *see also* monocultures; reforestation, twentieth century
planting methodologies: even-aged species mixtures 116–19; long rotation monocultures 108–11; multi-species plantings 111–14, 116; short rotation monocultures 106–8; temporary nurse trees 115–16; three main approaches 104–5, *104*, *106*; tree cover with an under-planted crop 115
policies: common misperceptions about reforestation **208**; direct (financial) incentives 218–19; ecosystem services, payments for 220–2; empowerment of communities 213–14, 241; enabling policies 210–15, **211**; to establish land tenure/property rights 211–13; financial and non-financial incentives **216**; framework development 207–8; incentives 216–20, **216**; indirect incentives 217–18; inter-connected institutional arrangements 209, 245–6, 247–8; policy tools 207; popularisation of tree planting 214–15; regulatory policies 224–6; removing disincentives 210–11
polycultures *see* multi-species plantings
population changes 18–19, 27, *29*
populus 148
poverty, impact on reforestation 22–3, 242–3
precipitation and reforestation 169–70
property rights *see* land tenure
*Prosopis juliflora* 87–8, 89–90
public perceptions: global survey, environmental attitudes 24, **25**; increased awareness environmental concerns 24–6
Puerto Rico **74**, 166
Puettmann, K.J. *et al* (2009) 131
pulp/paper mills: Brazil 47–8; plantation locations and 53, 59, 60; planting methodologies 33–4, 102–3, 107; rotation length and carbon sequestration 179, *180*

recolonisation of forests *see* natural forest regrowth
reforestation: change in forest areas in relation to GDP 24–5, *25*, 64; choosing a method 130–1; common misperceptions **208**; definition 5; factors favouring reforestation 24–31, 238–9; global assessments of availability of 10–11, **11**; historical 2; local assessments of availability of 12; opportunity costs 12, 22–3, 33; potentially limiting factors 18–24, 238; risks associated with **18**, 31–2
reforestation, twentieth century: Algeria, case history 41–3; Australia, case history 44–6; Brazil, case history 47–9; Chile, case history 49–52; distribution of benefits from 54; land used 52–3; objectives of 52, 54; plantations 53–4; policy factors 54; South Korea, case history 39–41; species used 53; United Kingdom, case history 35–9; *see also* scale, of reforestation
rehabilitation (forest restoration) 105, 121
remote sensing techniques 9
restoration, definitions 5

Rewilding Project, Europe 98–9, 178
Rio+20 13, **14**
riverine areas, reforestation of 147
rivers 1–2
rubber (*Hevea brasiliensis*) 48
Rwanda **15**

sawlog production 53, 59, 115
scale, of reforestation: ecological restoration and 191–2; economies of scale and 248–9; hydrological flows and 165–6, *165*
Schmidt, L. 129
second rotation problem 55–6
seeding, direct: aerial distribution 125–7; comparative advantages/disadvantages with seedlings **126**; direct seed sowing, eucalypts (Australia) *128*; dispersal and natural forest regrowth 69–70, 84–5; ground applications 127–9; methods for improving outcomes 129–30; overview 125; variable success rates 125
seedlings: comparative advantages/disadvantages with direct seeding **126**; deliberate planting of 96; density of, natural regrowth 91–2, 93; naturally regenerated seedlings, Hong Kong *92*; shrub cover and 94; species used 92–3; *see also* planting methodologies
seeds: dispersal and natural forest regrowth 95; genetic diversity 103–4; wildlife dispersal of 69, 70
semi-natural planted forests 3
set-aside lands 9
shelter breaks 147
shifting cultivation 7, 82–3
short rotation plantings 179, *180*
slopes and erosion-reducing reforestation 170–1
smallholders: demand for agricultural land 21–2; involvement, policy development 207–8; reforestation activity 58–9; reforestation pathways 65
Smith, P. *et al* (2010) 19–20
socio-economic factors: constraints on natural forest regrowth 90; corridor/stepping stone reforestation 154; of deforestation 2; degradation 9; natural forest regrowth 93; social resilience 246–7, **247**
soil seed banks 69, 89, 97
South Korea: area of planted production forests **39**; forest transition achieved 64; national reforestation plans **15**, 16; reforestation, case history 39–41
Spain **74**
spatial targets 239–40
spruce-beech mixtures 117
Staaland, H. *et al* (1998) 75
stakeholders: in Forest Landscape Restoration 204; involvement, policy development 209, 240–1; multi-stakeholder institutions 230–1; in participatory planning models 203–4; planning and 158, 201–2; *see also* smallholders
state, the: national reforestation plans 15; as plantation owners 53–4; role of in plantation growth 57–9; state forest pathway 65; *see also* governments
steep slopes: Loess Plateau, China 143–4, *144*, *145*; reforestation and (soil erosion prevention) 143; soil erosion and 142, 170–2
streamflow, following reforestation 162–3, 166–7
stumps, (vegetative regrowth): constraints on regrowth from 88–9; Laos and Thailand, case study 83; natural forest regrowth

70–1; Niger *81*; Niger, case study *81*
successional development: directly seeded forests 129; within environmental offset programmes 196–7; facilitation of 94; within novel ecosystems 123, 124, *124*
Sweden *29*, 73, *109*

Tanzania **74**
taungya system 115, 235, 236
teak (*Tectona grandis*) 34, 143, 167–8, 171
Thailand 82–4, 93–4
timber production: carbon capture and 179, *180*, 182; commodity timbers 60; vs. creation multi-purpose forests 189–90; illegal logging policies 210–11; intervening strips, natural forest regrowth 137; reforestation area (economies of scale) 161–2; sawlog production 53, 59, 115; *see also* plantations
topography considerations 143–4, *144*, *145*

Ukraine **74**
understories: diversity, within monocultures 110; native species development, Vietnam 110–11, *110*; vegetation and soil erosion prevention 142–3
United Kingdom 35–9, **39**
United Nations Convention on Biological Diversity (CBD) 13, **14**
United Nations Convention to Combat Desertification (UNCCD) 8, 13, **14**
United Nations Environment Programme (UNEP) 8, 26
United Nations Environmental Program (UNEP) 13, **14**
United Nations Forum on Forests (UNFF) 13, **14**
United Nations Framework Convention on Climate Change (UNFCCC) 13, **14**
United States of America (USA) 60, **74**
Uruguay 201–2, 215

vegetation and soil erosion prevention 142–3, 172
vegetative regrowth *see* stumps, (vegetative regrowth)
Vietnam: forest transition achieved 64; land allocation programmes 214; national reforestation plans **15**; nurse trees 116; a privately owned tree nursery *215*; soil conductivity and reforestation 168; Tam Dao Mountains, understory native species development 110–11, *110*
Vihervaara, P. *et al* (2012) 201–2

watershed protection: through reforestation 12, 26–7; *see also* dryland salinity
Watson, J.E. *et al* (2013) 199
weeds: contamination by exotic weeds 89, 97; shrub cover and control of 94
wind erosion 147, 172
windbreaks 147

Zhang, Y. *et al* (2011) 248
Zhou, G. *et al* (2002) 164
Zhou, G. *et al* (2009) 172